SILVERWHITE

LENNART MERI

Silverwhite

The Journey to the Fallen Sun

*Translated and adapted from the Estonian
by Adam Cullen*

HURST & COMPANY, LONDON

First published in the United Kingdom in 2025 by
C. Hurst & Co. (Publishers) Ltd.,
New Wing, Somerset House, Strand,
London, WC2R 1LA
© Lennart Meri, 2025

All rights reserved.
Printed in Scotland by Bell and Bain Ltd, Glasgow
The right of Lennart Meri to be identified as the author of this publication is
asserted by him in accordance with the Copyright, Designs and Patents Act, 1988.

A Cataloguing-in-Publication data record for this book
is available from the British Library.

ISBN: 9781805262947

www.hurstpublishers.com

Cartography: Professor Raivo Aunap, University of Tartu

CONTENTS

Foreword		vii
Acknowledgements		xi
I.	Ships Take Flight	1
II.	Beneath Two Suns	49
III.	Whispers Reborn as Words	137
IV.	The Circle Is Complete	165
V.	Red Sails	191
VI.	The Silverwhite Way	225
VII.	'Their Greatest Might Is in the Ships' *(Verse 366 of Livonia's oldest rhyming chronicle)*	281
VIII.	Sixteen Years with al-Idrisi	327
IX.	Tallinn in the Haze of Legends	351
X.	Son's Land and Mother's Land	393
XI.	The First Captains	403
XII.	Where Does Poetry End, Where Does History Begin?	437
Afterword to Silverwhite		441
Notes		443
Bibliography		445
Index		459

FOREWORD

Lennart Meri wrote this book in the mid-1970s, when history and geography were conspiring against his country's survival. Occupied by the Soviet Union in 1940, Estonia, along with Latvia and Lithuania, was wiped off the map. If you could find them at all on the world atlas, the Baltic states appeared only as "Soviet Socialist Republics" – small provinces of the Kremlin's empire. Outsiders rarely travelled there. Linguistic, cultural and demographic russification was in full swing, meaning that Estonians seemed fated to become a minority in their own country, backward and irrelevant, mere footnotes to the grand story of proletarian internationalism. That Estonia would only fifteen years later regain its independence – let alone that Meri would become its foreign minister and then president – seemed as unlikely as the re-emergence of Atlantis.

It was an act of defiance, therefore, to root Estonia in time and space: a place and people whose existence is evidenced for millennia – and thus long predates the Russian colonial masters. Meri does not state this aim explicitly in his book; he does not need to. He was writing first and foremost for Estonians, accustomed to read between the lines of whatever the Soviet censors allowed to be published. His tools are scholarly: archival research, linguistic analysis, archaeological investigations. His tone is quizzical, not polemical. The reader is invited to assess the evidence; to entertain conjecture, even whimsy; to appreciate the almost wilfully provocative conclusions. The result is a manifesto of the imagination, for Estonia's past, present and future existence.

Silverwhite's central character is a meteorite, which struck the Estonian island of Saaremaa with devastating force, probably around 1500 BCE, leaving a spectacular crater which today is one of Estonia's foremost tourist attractions. Meri's central thesis is that Saaremaa may be the legendary Thule, described by the Greek geographer Pytheas 1,000 years later. The *Oxford English Dictionary* claims that the origins of the word are a mystery. Not so, says Meri firmly, for "*Tuli*, 'fire', is an Estonian word with roots that reach back to the proto-Uralic language." In a nod to the Estonians'

FOREWORD

ethnic cousins, he continues, "It is also *tuli* in Finnish, *tul* in Mordvinian, *tul* in Mari, and *tyl* in Udmurt. In Komi, *tyl-kort* is a fire-striker."

Sceptics might regard this and other evidence – textual, folkloric, physical and archaeological – as fragmentary at best. Meri's enthusiasm and penmanship fit it all together into an attractive, even convincing, mosaic. Modern scholarship may draw different conclusions. But that is to miss the point. As Meri writes. "History begins where poetry ends. Sometimes, they brush one another."

Like Meri himself, *Silverwhite* is dazzlingly erudite, but not rigorously academic. It is packed with facts, yet punctuated with aperçus and personal touches. Its scope is sweeping. The reader is drawn along the ancient Baltic-Volga waterway, from the far-flung heights of classical Arab civilisation to the farthest corners of Viking expeditions, with a rich cast of witnesses from medieval chroniclers to the geographers of antiquity.

But the focus is tight. Again, just like Meri. He was the most cosmopolitan Estonian I have ever met, more fluent in his five foreign languages than most people can ever hope to be their native tongue. Dauntingly well-read, he seemed able to quote from memory on any subject, in any context (not always appropriately for those hamstrung by social mores, but always memorably). Yet this was no skittish xenophilia; his polymathic interests were rooted in the scenes and stories of his native land.

Meri was Estonia's paramount personality in the years following the restoration of independence. But he bestrode a wider stage, as the best-known politician in the Baltic states, and indeed across all the former captive nations. Only the Czech playwright-president Václav Havel could compete with him in moral stature, cultural depth and political clout.

Meri's hallmark achievement – under-appreciated at the time – was a speech to an international dinner audience in Hamburg in February 1994, warning of the dangers of Russian imperialism. It was stunningly unfashionable (the Russian delegation, led by an obscure official called Vladimir Putin, stormed out of the room in protest). In retrospect, it is unbearably prescient.[*] This edition – the first English translation – is published in 2025, a time when storm clouds are darkening over all of Europe, and Estonians' insights could scarcely be more valuable. May this book give them wings.

To say that Meri was irrepressible is true, yet misleading. For he had been repressed. Meri was deported to Siberia aged twelve for the "crime"

[*] Address by H.E. Lennart Meri, President of the Republic of Estonia, at a Matthiae-Supper in Hamburg on February 25, 1994 https://is.gd/lennartmeri

FOREWORD

of his family background. He survived on stolen potatoes. I once visited him in Kadriorg, Estonia's presidential palace, to find him reading the records of the KGB's interrogation of his father Georg-Peeter Meri, a pre-war diplomat and Estonia's foremost translator of Shakespeare. I found the flimsy, faded transcripts unbearably poignant. They told of a dignified, civilised man defending his honour, life and country against a barrage of ignorant, repetitive questions, usually ending around four in the morning.

Official protocol held no terrors for someone with such horrors in their past. No respecter of status, he could deploy a bluntness, verging on rudeness, when necessary. But this was a political instrument, not a character trait. Deeper features of his personality – restless, mischievous, ingenious – shine from the pages of Adam Cullen's meticulous and sympathetic translation. It is an honour to commend *Silverwhite* on its belated appearance to English-language readers.

Edward Lucas

London,
March 2025

ACKNOWLEDGEMENTS

The translator wishes to thank Mart Meri, Peep Ehasalu, Prof. Thomas DuBois, and Jaan Undusk for their invaluable advice, as well as the Cultural Endowment of Estonia and the Baltic Centre for Writers and Translators for an ideal winter haven.

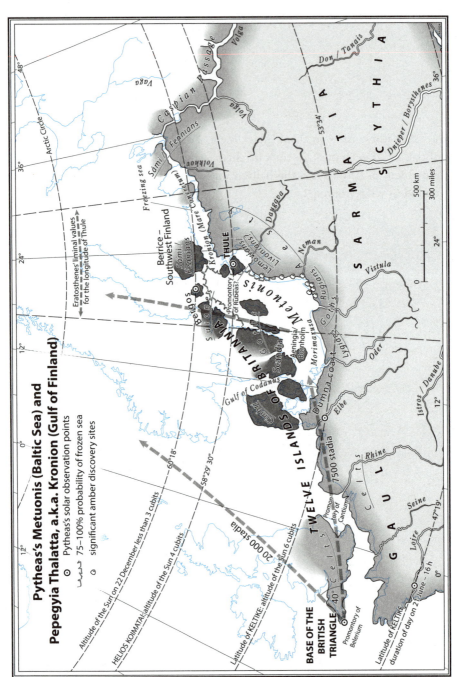

Map design © Professor Raivo Aunap, reproduced with sincere thanks.

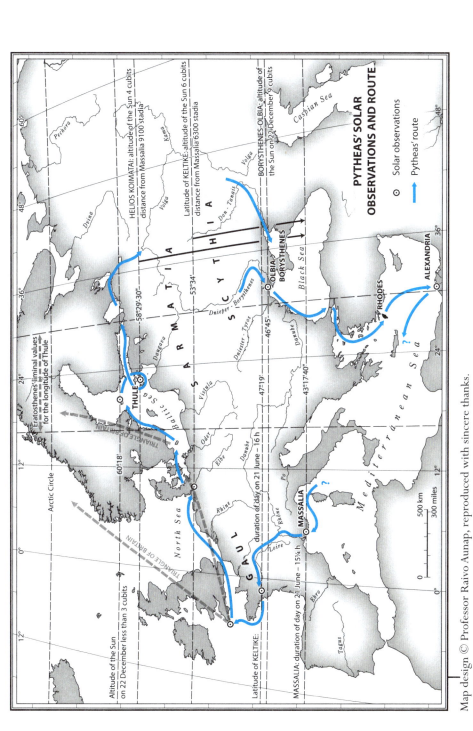

Map design © Professor Raivo Aunap, reproduced with sincere thanks.

I

SHIPS TAKE FLIGHT

WE COME TO THE SEA

It wasn't his turn to sit there, to be fair. And all things considered, standing watch over the sea and waiting for the dawn is a job for simple men. The fields are drying, hammers ring in the forge till late hours – blacksmiths are essential in peace and bloodshed alike. Who can coerce a smith?

Yet, curiosity tends to wander and night sets thoughts adrift. Enough blades have been hammered for the men and necklaces for the women, spearheads boiled and ship nails cut. When the buttercups burst into bloom, restlessness takes hold. At least he can have his fill of staring at the sea from atop a limestone cliff. Not the harbour with its bleached hulls nor the jetty strewn with seaweed, from which night gusts carry the reek of tar, but rather open sea itself beyond the distant spits; the rambling lands of youth that give to one man, take from another, and kill and bury a third.

The coastal watchman stirs the embers, pushes a log to the edge to keep it from erupting into flame, and gazes back out at the strait, resting on his elbow.

It kills and it buries and, yet, it is like one's own field and family. Naissaar[1] on our doorstep and the Daughter Islands[2] betrothed to neighbours. Varangu Island here and another, sailors swear, in Sápmi,[3] where souls brandish swords to make the heavens flicker and glow. One sea united and equally briny, every nook accessible by ship. If you can make it there, then so can others. And that is why he reclines here.

The smith glances up at the sky. Great Odamus[4] has spun to the west and is suspended above the headland. A dawn breeze will be rising soon. Then, a cautious *crack!* makes him prick his ears.

SILVERWHITE

What do you know? – old red-tail's come to stand watch as well. Hunger drew him to the hunt. A pair of yellow eyes smoulders in the darkness, the feeble glow of the embers glinting back as if they were beads. The clever old fox knows that none cares to take his hide in spring. So many were trapped and skinned, sailed across the sea and sold for silverwhite. The forests were emptied of his brethren, but they still teemed in dark corners near the fortress. Elders forbid hunting fox in those places. The men laugh, sure, but most abide.

The smith leans towards his birch-bark box to share a morsel with the bird-eater, but his hand freezes in mid-air.

A glint, nearly imperceptible, sparkles across the bay on Naissaar. There, on Sealhead Promontory. He rises to his knees, eyes fixed on the dark slash protruding from the island, hand groping for the tar-soaked pine logs, and as he does, the distant spark blossoms as if nourished by the wind, swelling into a flame before exploding into a roaring bonfire. There's no time to waste. Dry juniper crackles into flame, tarred logs hurl sparks into the paling sky, dawn recedes, and the bridge between bonfires scintillates into the bay's bottomless blue.

Beacons alight on other points. Suurepea, Kakumaja. A wail rises on Long Leg Road, leading up to the stronghold. Revala[5] awakes.

Gradually, as if pushing the morning breeze before them, the first sails come into sight from behind the islands.

* * *

It all began forty or fifty generations earlier and quite differently.

He grips a short-poled spear. It has been hewn with care. His features are strange, angular. He's hairy. And a bit on edge. The sounds of the forest have faded, falling somewhere behind him, leaving him to face, alone, the silence that crouches ahead. He slows, pauses, listens. Every one of his senses is trained for change and to swiftly repel any yet-unknown danger. Either he will best it, or he won't. Occasionally, he studies the sky amid the treetops. What an unusual sky, entirely unlike the one behind him – empty! Gradually, a distant roar fills the void. He's never heard anything like it. Then, the cries of birds. He's never heard such high-pitched screeches before. The wind rises and the man's flaring nostrils pick up a stimulating scent, one he has never smelled before. What drives him onward when everything else is pulling him back? As taut as a bow, he pushes aside the last hazelnut branches. He was expecting something, but this?

SHIPS TAKE FLIGHT

Before him glints the endless bosom of the sea, simultaneously hostile and kind, repelling and inviting, more mysterious than the forest, more home-like than the starry sky, the end of one world and the beginning of another, greater one.

Perhaps you did not notice your hand freeze in mid-air and your foot, raised to take another step, sink slowly back to the moss. At that moment, a change took place within you; one that endures to this day. 'I envy the forces of nature – waterfalls, volcanoes, and wind – for the impact they have on people,' said an artist whose name I can't recall. Of all the forces of nature, the sea is the most powerful, but also the most intimate to humankind. Five millennia ago, Estonians arrived at its shore, brushed aside the green curtain of forest, and froze in place. Here we now stand, and each of us carries signs of the sea within us.

A WANDERING FIELD

My bones have come from the east and my flesh from the west, but it is as difficult to detail as a split personality after the hemispheres of the brain are divided.

Central Sweden was still devoid of human settlement when Estonians came to the shore of the Baltic Sea to fish and set traps. We spoke our own language and have retained close to a thousand words that can still be understood by peoples along the Volga and Pechora rivers and beyond the Ural Mountains. A thousand words isn't much in modern-day vocabulary, but during the Stone Age? As you can see, it sufficed for us. Furthermore, language and words are two different things. To me, language is like a sturdy log house with its walls, roof, doors, windows, nooks, and hiding places. Words are the occupants of that house. They reproduce, give birth, grow, go about their daily lives, walk in and out, and receive strangers, some of which remain and take up residence in that house as well. Occasionally, a word may perish at a young age or venture out into the world and never return. The house lives longer, changes less, and all its occupants must adapt to the dwelling. The house of Estonian, for example, had no tolerance for dual consonants at the beginning of words and permitted only the use of three vowels to craft successive syllables, stipulating that even those follow certain laws of music. Stress resides in the first syllable of words and the language's history in its pronunciation of the letter *i*, both here and beyond the snowy mountains. Estonian words

SILVERWHITE

may be pale or dark, short or tall, but when they meet beneath the roof of its house, then they all behave the same way.

Language is a powerful reality and linguistic kinship awakens strong feelings. During the Soviet occupation, Estonian artists, composers, and writers would wander to the east, as if searching for something they'd left behind and must regain for tomorrow. Afterward, we admire their paintings, listen to their tunes, and say to ourselves: ah-ha, *that* is us.

But how am I to explain it?

In south Estonia and west Siberia, I've come across tranquil bog lakes that reflect the trees and buzzing of insects. Facing the water's stillness, the eye is immediately drawn to a person on the opposite shore. (Though does a lake even have an opposite shore?) I see him picking his way along, sometimes vanishing behind trees and reeds, but still sense his slow progress that will ultimately bring us together. Yet when the moment arrives, he is gone. I am totally alone except for a dragonfly flitting its wings above its reflection. The man I just witnessed on the outside has now dissolved within me and lives on in my memory. No doubt he also spotted me and is now amazed by where I dissolved away to. To him, I have disappeared; to me, he has disappeared; and in reality, it is only a lie or at least partially correct, as neither of us has disappeared and neither of us is still there; we've both simply changed and transformed and are living jointly in a single body, marvelling together, listening to the lake's buzzing silence together, turning our backs on the lake and walking on to the next, where all might repeat and also might not. I try to commit to memory seeing him walking on the opposite side of the lake; seeing myself from the outside. Sometimes, we must observe ourselves, our language's past, and our history from the outside, though I truly do not know how to describe it.

Estonians' language, memory, and self-awareness have come from the east, our facial features from the west. It is easy to say and difficult to accept. A person lives in their native language, and language lives in a person. One does not choose one's language, but language may choose a person. Something like that also took place on the shore of the Baltic Sea when new settlers began to seep up from the south; ones archaeologists associate with the Battleaxe culture. They were of a different appearance and race, light-skinned and Europid, noticeably unlike our stubby, angular people. They were more skilled at making tools. Look, I say 'them'. For some reason, I place them outside myself. Why? They took our women and we theirs, we interbred and, contrary to popular belief, our outward appearance changed beyond recognition in a rather short time, perhaps

SHIPS TAKE FLIGHT

over the course of just a dozen generations. We were few, they were many. Only about a quarter of the genetic heritage we'd brought from the east was retained. Still, we preserved our language and through it our roots, which in one way or another are intertwined with languages spoken along the Volga and throughout the Urals. I view the tribe of the Battleaxe culture as an alien from outer space, as all I can do is think in language, not genes. Even so, by blood, they are closer to Estonians than the Mari people, not to mention the Khanty or the Mansi along the Obi River. I could also have put it this way: we came from the south, discovered a small tribe of Comb Ceramic people on the shore of the Baltic Sea, absorbed them, and adopted their language, which we use to this day. Such a statement would correspond to the facts but contradict feelings: as if I'd been cast out of my own life and were observing what is happening at home from beyond a distant threshold. Language and culture are significantly more important than race to Estonians. Perhaps I shouldn't make such a casual generalisation. I am speaking of feelings. Strictly speaking, roots are a meaningless poetical metaphor. But strictly speaking, poetic metaphor is also a crucial element of the Estonian cultural landscape.

I've conjured an image for myself. Grain is ripening in the warm August sunlight, wheat growing alongside rye on a boundless field. The wind rises, making the stalks bob as the sun dips lower in the sky. A golden wave slides across the field. It comes from the right and slowly rolls beyond the horizon to the left. The plants are different but the wave single and indivisible. Each stalk has its own root system, though it's not important right now if those hundreds of thousands all intertwine beneath the earth or not. Let us observe the field: every plant is living its own individual life and bending in the evening breeze. If we were to stop and stand in the middle of the field, the grain stretching high above our heads. we would only perceive the stalks' indistinct swaying. Yet, since we stand at a distance of several millennia, we see the wave's approach and exit; we see massive movement happening simultaneously with slow ripening; we see the unity of movement and stillness. Taken separately, every head of grain moves independently, and it would be irrational to measure, compare, and describe the movement. Taken as a whole, however, the heads are joined into a field of rye and wheat; into an undulating expanse, the orderly movement of which can be seen, observed, and measured. I use this image to attempt to unravel the mechanism of cultural influences. It brushes us like wind and continues forward in a wave without any need for us to leave the place where we stand. I'm not sure if I am able to make myself any clearer.

5

SILVERWHITE

If we, however, were to observe the field closely enough for centuries, then we would also see it moving as a whole. The soil loses its vigour, the rain washes away salts, and photosynthesis, which turns solar energy into edible carbohydrates, is no longer capable of feeding the land's cultivator. Humans plough a new plot and the field travels. Ancient hunters and fishermen likewise travelled when local energy supplies were exhausted. We tend to imagine migrations in an exceedingly modern way: intentional movement from one location to another, packed suitcase in hand. It's more suitable to compare ancient migrations to a field gradually reaching towards fresh soils – a process of which the field itself is as unconscious as were our forest-dwelling ancestors. Needless dramatics can also be tamed by dividing the kilometres over millennia. Continents, seas, and rivers are constantly before us on maps, and we consider the temporal factor only infrequently and perfunctorily. Two and a half thousand kilometres can easily fit into our experiences; six thousand years cannot. A distance of two and a half thousand kilometres lies between the Urals and the shore of the Baltic Sea, and if it is true that the former is Estonians' ancestral home, then we spoke a proto Finno-Ugric language for about a thousand years six millennia ago. Did we migrate to the Baltic Sea, or did only our language travel? Did the two alternate? Much is still unclear with the exception of one conclusion: if you boil the movement down to a steady progression, then its average speed was five centimetres per hour, which makes four hundred metres per year. The speed is hard to imagine when surrounded by straight modern-day highways, but such deliberation is more common to our forested region than the lightning-quick advances of peoples of the steppes or the Vandals descending from Scandinavia to the ruins of Carthage. Migrations play a part in cultural contact, but it is smaller than the day-to-day cultural radiation that I previously attempted to compare to the undulations of a field. Western culture is unique in its heterogeneity, and that is the engine driving the rapid development of science and technology today. Through our language, we have become immersed in European cultural dualism and added something important. Perhaps the stillness of the forests? A restless suspicion that the woods turn transparent during the full moon?

FROM BOAT TO SHIP

There are few peoples in the world who have been tied to one place and loyal to it for so long. We know by name our springs, mighty trees, bends in the rivers, and moraines that we insist on calling hills, and each has a story to tell about them. The land speaks to us in our own language, though it naturally isn't the voice of the forests or the waters that we hear, but the whispering of dozens of generations; the distant echo of their joys and accomplishments, troubles and struggles, truths and conflicting arguments. Sooner or later, such an ancient people develops a nostalgic desire to re-tie broken connections and attempt to restore dialogue where younger peoples can only demand monologues. In doing so, Estonians have collected an astoundingly great library of folklore; perhaps the greatest in the world per capita. Nevertheless, there was a time when the springs and the bare cliff faces still lacked the names we gave them; when Estonia was still bring created.

In the beginning, we lacked our own word for the sea.

How could the word have formed when our entire earlier existence was spent beneath the green canopy of the Eurasian forests? Yes, the heavens occasionally gleamed over broad rivers and some lakes were vast enough to conceal their far shore. But the forest was endless, and endlessness did not intimidate us. Finality was what was daunting. On this shore, we crashed into an invisible boundary – even the water ahead tasted bitter. Had we reached the furthest and most dreadful end of the world? One that even swallowed the sun each night?

* * *

The arrival of Baltic Finns in the Baltic region gave rise to a completely new phenomenon, which we call 'maritime culture' and the 'maritime cult'. It advanced under the strong influence of our new maritime neighbours. At the same time, forest and agricultural culture organically preserved its continuity with ancient, original elements. Estonians' character is a blend of Indo-European dynamism and Uralic stationariness, which led to one researcher calling us 'always a bit restless' and even 'brilliantly rippling' in our fantasies.

* * *

We did not number many, perhaps just a couple thousand souls. Even so, we were not the first to step foot upon this shore.

Who came before us? Perhaps the Basques? I do not know; I cannot remember.

SILVERWHITE

One of their settlements was in Kunda. Another, even older site was discovered on the bank of the Pärnu River near Sindi. Let us call them the Kunda people. They quietly live on within modern Estonians. We borrowed a new word from them for a new phenomenon: *meri* – 'sea'. The word is ten thousand years old, perhaps older. We can recognise it in many other languages with which Estonian has little or absolutely nothing in common. In Slavic languages, it is *more*; in French, *mer*; in Latin, *mare*.

We fished and hunted when we reached these shores, and we mainly travelled in boats that we'd learned to make for navigating large rivers and lakes.

It is a long way to go from boat to ship.

How this gap was crossed is unknown. We can only imagine it via indirect sources. The oldest vessel used by Finno-Ugric fishing tribes was the dugout canoe, *haabjas*. It was still in use on Kasari River at the beginning of the 20th century and could be found in Vepsian villages even later. While filming my documentary *The Waterfowl People* in August 1970, I was able to watch and record Khanty boatbuilder Aleksei Lazyamov making one in the watershed of the Obi and Yenissei rivers. It was, as the Estonian word betrays, made from the aspen tree (*haab*), which is soft and pliable. Lazyamov picked an aspen that was 35 centimetres thick for his boat. It took nearly two hours to fell such a tree in the Stone Age. This isn't mere speculation: archaeologists have tried using a stone axe. First, the craftsman shaped the log into a boat, sharpening the bow and the aft as well, for the most part. Next, he hollowed it, alternating between fire and axe. Yet, a 35-centimetre-wide boat would hardly be wide enough to fit a pair of human hips. The ancient fisherman thus invented a clever process still used in carpentry to this day. Lazyamov soaked his boat in water for several days, then dragged it to a bonfire on the shore. He heated the boat by the light of the fire, carefully rotating the vessel until dense steam rose from the surface. The heated wood turned flexible and submissive. Bending the sides wide, he wedged four or five strong poles horizontally between them to keep them from closing together again as the boat cooled. This was repeated several times, gradually replacing the poles with longer ones and finally achieving a much more spacious boat. Paddling a pea-pod-shaped dugout such as that (with a single oar) requires exceptional balance. When I attempted to paddle one across the pristine and fast-flowing Nüüna River for the first time many years ago, I found myself at the bottom of the river mere moments later and can still remember the hydrophytes waving among the stones, greenish and

SHIPS TAKE FLIGHT

secretive, as well as the silvery reflection of the surface above me. The Khanty joke that a dugout paddler must do everything they can to hold their tongue in the middle of their mouth.

Finno-Ugric tribes' millennium-long journey to the west was likely made with the aid of such dugouts along offshoots of larger rivers. The subject of this travelogue is ships and waterways, so let us merely note that the word 'journey' may evoke false associations. It was a slow seeping through the forests and along the river labyrinths, over which branches interlaced into a dark, arching ceiling. Rivers were the only road, but also their main source of nourishment. The ancient Greek geographer Strabo (63 BCE–24 CE) noted: 'The rivers are in such favourable position to one another that you can sail from the shore of one sea to another, portaging goods for only a short time, […] but for the most part, the journey is along the rivers, downstream one and upstream another.' This description, which holds true for Central Europe, can only conditionally apply to the forests of North Europe, because our rivers primarily run longitudinally from north to south or vice versa. Tributaries do brush close to one another near a river's headwaters, but watersheds had to be crossed by land. The dugout of the Khanty is brilliantly adapted to such travel: a towline is simply attached to the bow. If the need arises to cross from one river basin to another, then the dugout is pulled like a sled. The hunter/fisher's provisions and scant possessions fit neatly into the vessel. Crossing a watershed is a slow and difficult process: we had to forge through the primeval forest, climb across fallen trees, penetrate dense thickets, climb, and descend. A watershed comprises many ridges. Over aeons, the most favourable crossing-points were determined and given the names 'pushers' and 'pullers'. They play an important part in North European history, eventually leading to the founding, and even naming, of some settlements. One recognisable example is preserved in the Russian town Vyshny Volochok – *volok* means 'portage'. It is situated in the Valdai Hills, which was once one link in a long chain of waterways connecting the Volga and the Baltic Sea.

As simple and practical as the dugout may have been, it was unfit for seafaring.

However, another type of vessel emerges from indirect sources and the ethnography of Siberian peoples. Its construction was much more complex, but also lighter. First, the frame was built. I have seen such boats in villages on Cape Billings and in Inchoun, on the shore of the Chukchi Sea. Driftwood was used as material, and since the value of

SILVERWHITE

wood is so great in that region, the craftsmanship was especially clean and painstaking. Poles were mortised and usually sewn together with leather straps. Humans began using drills in the early Stone Age, as a result of which the Estonian and Khanty words for the tool coincide: the Khanty *pur* and Estonian *pura* derive from the same proto-language. Since researchers have determined the approximate time that both tribes separated, the word must date back to an even more distant past. This mortised-and-sewn boat frame is extremely resilient, as the joints are allowed slight flexibility. One might think that elasticity is merely a consequence of primitive craftsmanship, but rather the contrary. Aleuts, who built their boats on the same principle, joined three keels together (or, rather, the load-bearing part of the frame that later became our ancient keel). This was not due to a scarcity of wood, but a need for greater flexibility. Even in tranquil weather, there can be rather heavy breakers on an open ocean shoreline. The calm and orderly wave, nearly imperceptible in open water, rises into a sudden wall as it reaches the shore and crashes upon itself. Stiff-hulled boats cannot cross the barrier. They are trapped on shore, which is quite an odd sight in sunny weather.

Seal skins were sewn together as a hull. The size of the vessel depended on its purpose. Seal hunting was done in a one- or two-person boat. It was covered and equipped with inflated balloons; a watertight skirt sealed around the paddler's body. Such a vessel is also known as a kayak. It could even flip and, although Northern peoples (including Estonians) historically did not know how to swim, its occupants would not be in danger of drowning. All it took was a strong pull to right the vessel: an inflated bladder was tied to the paddle, giving the stroke a point of support.

These light boats were also used to hunt larger sea animals such as walruses, with a living weight of 800 kilograms, and whales, the weight of which could reach up to 27 tonnes. For this, the hunters used open-topped boats that were significantly larger but of similar construction. The first Chukchi barge ever recorded, in 1660, could carry up to thirty people. This is quite an impressively sized vessel and, as we will see later, comparable in capacity to ancient Baltic ships built in the first millennium CE.

Perhaps I should emphasise once again that it took nearly two hours of hard labour to fell an aspen with a stone axe. A boat was extremely precious. Launching one was a festive event surrounded by a garland of magical rituals. Breaking a bottle of champagne on the iron side of a mammoth ocean-going vessel is a distant echo of sacrificial Early Stone

SHIPS TAKE FLIGHT

Age acts. A boat was the result of tedious common efforts; a boat would nourish. Boats played a central part in ceremonious moments in fishers' later lives as well. The Khanty would carve a special wedding paddle, which likely took no less time than crafting an entire new boat. It would be naive to see only the joy of ornamenting in the process. In every society, even the simplest, labour is greatly valued. A brightly painted wooden paddle with lacework carving and wooden bands would have to 'pay off' somehow in terms of the superstitions of the time. I realise it may be a banal association, but art has a regulating role in society akin to that of vitamins in physiology. Without vitamins, there is no life; without art, there is no human. Art is an inevitability. Art is the gene of culture, which reproduces itself and thereby nourishes culture – the only environment in which the human soul can breathe. I digress. All I wish to say is that the most ancient elements of Estonian ship worship must be sought from the 'Finno-Ugric' era, in which we hadn't yet reached the shore of the Baltic Sea or were only just arriving.

My exploration of framed boats strayed eastward only because that is where I have touched them with my own hand. Their area of distribution, to use scientific terms, extends to the British Isles and even further. Celts landed on Iceland before the Scandinavians in similar skin-hulled boats that tolerated waves and storms far better than the first fully wooden seafaring vessels. Depending on their diffusion, they are known by different names: *umiakk* in northeast Asia, 'coracle' in Wales and England, *curragh* in Ireland. Julius Caesar saw them during the Gallic Wars, and in some places around the British Isles, skin-covered vessels were still in use until the 20th century. They have been depicted in Scandinavian cliff drawings. The simple and crucial inventions of human culture appear to be amazingly universal. Perhaps they indeed are, stemming from a similarity of production methods in a common climate, in this case maritime. Only later did navigation technology become more fine-tuned and adapt to the character of a particular sea: wave frequency, storms, tides (flat-bottomed vessels), and, finally, with the invention of the sail, the prevailing winds.

The following lines were written 2,400 years ago:

Their boats which ply the river and go to Babylon are all of skins, and round. They make these in Armenia, higher up the stream than Assyria. First they cut frames of willow, then they stretch hides over these for a covering, making as it were a hold; they neither broaden the stern nor narrow the prow, but the boat is round, like a shield. They then fill it with reeds and send it floating down the river with a cargo; and it is for the most part palm wood casks of wine that they carry down. Two men standing upright steer the boat, each with a paddle, one drawing it to him, the other thrusting it from him. These boats are of all sizes, some small, some very large; the largest of them are of as much as five thousand talents burden. There is a live ass in each boat, or more than one in the larger. So when they have floated down to Babylon and disposed of their cargo, they sell the framework of the boat and all the reeds; the hides are set on the backs of asses, which are then driven back to Armenia, for it is not by any means possible to go upstream by water, because of the swiftness of the current; it is for this reason that they make their boats of hides and not of wood. When they have driven their asses back into Armenia, they make more boats in the same way. Such then are their boats. (Herodotus 1920)

Apparently, Baltic Finnic peoples once used skin-covered boats as well. Archaeologists have not managed to find them, though such vessels are generally extremely rare finds. This isn't merely due to the material, but also the coastline, which does not coincide with that of today.

I should amend my earlier claim that the aspen dugout turned out to be unseaworthy. Wooden planks could be used to increase the height of the railings, tied using leather straps or other available materials. In the village of Amnya, 79-year-old Khanty boatbuilder Nikita Yernykhov used boiled cedar roots for the stitching. Additional timbers were installed and also sewn fast to strengthen the uppermost planks. Finally, the craftsman hammered wedges between the timbers and the planks to make the structure rigid. The wedges were rubbed with fish oil to make them slide easier.

Thus, the sewn boat was born – a distant ancestor of the ancient Baltic ship. The most essential knots used to construct the sewn boat and the ancient ship were similar or, rather, identical. Now, it was no longer necessary to hollow out the boat's spine, and this led to the development of the keel.

Dugouts and sewn boats played a part in bringing Baltic cultures closer, and they played it very well. The early Bronze Age (1000–600 BCE) has been called the first Viking Age on the Baltic. This term is anything but exact, especially now that the actual Viking Age has presented itself to researchers in an entirely different light, losing its earlier aggressive reputation and romantic sheen. It only aims to emphasise the intensity of seafaring: during that era, the Baltic Sea essentially became an inland lake. It's possible that Estonians were unable to appreciate with respect

SHIPS TAKE FLIGHT

the great benefits of our little sea. The continuous shoreline is never tragically far away and offers shelter from storms behind bountiful islands and peninsulas. Its waves are not insurmountable, its winds and gales stay within the bounds of human capacities. The Baltic Sea is nature's sandbox for future steersmen, more manageable than the Atlantic or the Mediterranean at the dawn of seafaring. It has been called the Nordic Mediterranean, and this comparison possesses a fair degree of truth from the perspective of our travelogue.

There is no shortage of evidence of dense seafaring during the early Bronze Age. This proof speaks worlds, quite literally.

The first artificial satellites were rocketed into space during my youth, and overnight the words 'satellite' and 'Sputnik' made their home in every language. If some sudden catastrophe were to destroy our culture, then future cultural historians would be able to draw important conclusions about mid-20th-century physics, chemistry, electronics, and communications devices using just those two words, and consequently also inferences regarding biological development, perception, and even poetic metaphors. This all assumes they are capable of determining the words' definition; of finding the content embedded in 'satellite'. Words are tiny independent worlds inhabited by the history of ancient generations. In our daily lives, we usually pay no attention to what a fascinating treasure rolls off our tongues and is casually cast to the winds. Linguists hold the key to secrets hidden within our language. Language transforms in sync with the development of lifestyle, some letters are replaced by others, words are worn shorter and may disappear from use entirely together with the concept they symbolised, or may acquire a more modern meaning and ultimately be overhauled completely. So as not to drift too far from seafaring, let us use the concept of 'northwest' as an example. In Estonian, the direction is named *loe* (*loode* in genitive form). Inhabitants of the western Estonian islands also used *loode-vesikaar*, literally 'northwest water-direction'. With this vernacular, we discover that the concept originally signified high water, which in Estonia's geographical location meant the sea rising with winds from the northwest. Such a leap would remain hypothetical if we didn't turn to the Ludic language for help. The Luds' historical area of settlement was a narrow strip between Lake Ladoga and Lake Onega/Äänisjärvi. In Ludic, the word appears as *lōdeh*, genitive *lōdhen*, but means 'west'! There is no other possibility: high water in that part of the Gulf of Finland and Neva Bay only comes with a western wind, not north-western. In Finland, on the other hand, the word *luode*

SILVERWHITE

means 'northwest' and 'low tide', but was earlier 'northwest' and 'flood'. And in the end, the word's history takes us right back to seafaring, having been borrowed from the opposite shore. For Scandinavians, *flod* meant 'high water' and never acquired a directional meaning. The ancient word remains quite recognisable in several modern languages: English 'flood', German *Flut*, and even French *flot* – wave, flow, influx, float. The distant Nordic word only entered the French language through the Franks in the 12th century. In linguistics, a word's migration, evolution, or changing direction is shown with the greater-than sign. Let us also follow this practice in our travelogue: *flod* 'flood' > *loe* 'northwest'.

THE SEA LISTENS, THE ANIMAL COMPREHENDS

What is a taboo word?

They are used by every fisher and hunter to this day, albeit more jokingly than with any seriousness. The custom dates to ancient society when hunters and fishers believed that mentioning an animal's name could limit the quarry and incur grievous misfortune. Taboo words were still common in Estonian coastal tradition a mere century ago. The restriction has a direct, though unexpected, connection to the ancient ship. From archaic texts:

> With some skippers, you weren't allowed to say the word 'ruffe' when they were caught in the net – 'hubbub in the net' is what you had to say. (Muhu)
>
> A long, long time ago, seal hunters and village men alike used a special name for every animal, wind, ship-mark, and everything having anything to do at all with their area of activity when they were going or there. (Reigi)
>
> Hunters called a seal 'fish', its blood 'alder', and its meat 'blunt' [...]. You weren't to point at a seal with your bare hand. You had to be wearing a mitten or be holding your other hand, meaning show its direction with your wrist; otherwise, your wrist would let you know the moment you'd violated the tradition. (Jämaja)
>
> No one was to call animals by their Estonian names on a fishing shore. They all had their own names for that such as dog – *taily*, [...] cow – *karova*, ox – boy, goat or ram – *pukk*, horse – club foot, pig – *vinya* [...]. Back then, if anyone stepped foot on that shore, be he from here or elsewhere, and called an animal by its proper Estonian name

SHIPS TAKE FLIGHT

by accident or on purpose, then the punishment, without exception, was getting them drenched from head to toe. (Hanila)

The last text is particularly interesting because, as we will see later, the custom of greeting Neptune also came to take place when crossing the equator.

But back to the Bronze Age.

A loanword in and of itself doesn't mean that the borrower lacked a corresponding notion. Transfers were influenced by fashion, though this only tells us that one suffix was favoured over another for mysterious reasons, that similar words' associative effect was affected, and, last but not least, that ideological systems also influenced borrowing. The last text is a prime example of this. Hanila Estonians used the German loanword *pukk* and the 'club foot' euphemism for a ram and a horse, respectively, which does not point to the absence of those animals in their barns or language. In my documentary film *The Waterfowl People*, Komi-Zyrian hunter Prokop Bogdanov even called sables 'soft-paws' at home in his hunting lodge, saying that he wouldn't use 'the other word' so long as the season was at hand. The hunting season lasted five months and the replacement word was part of his speech more often than the real one. Taboos and restrictions were much stricter in the Bronze Age, of course, with violations punishable by death. Out of taboo, ancient Scandinavians adopted a Baltic-Finnic word that vividly outlines our area of cultural contact. The word is *kala* ('fish'), one of the oldest in the Baltic-Finnic languages. Its form can even be recognised in the Kamassian language in the distant Altai Mountains: *kola* (Collinder). The word is believed to be at least seven or even ten thousand years old. Still, what matters is not the word's age, but the mechanism of taboo borrowing. Ancient Scandinavians were especially tenacious in their use of replacement words when hoping to catch a big, plump fish or even a sea creature. Over time, the Baltic-Finnish *kala* – which Nordic peoples pronounced *hwala* – was only used for only larger sea mammals, before settling on the largest of catches: *kala* > *whala* > whale (Pokorny 1969; Ariste 1972). The Baltic-Finnish *kala* boomeranged back to Estonian as *vaal* ('whale').

These are the deep tracks of seafaring. From now and till the end of time, we must regard the Baltic Sea as a unifying factor, not dividing; as a great lake with combinations of cultural contact criss-crossing every shore and enriched by everything but mere chance. The relative importance of chance is greater on the sea than on a river, and ceases to exist entirely

SILVERWHITE

when summarised into a statistic: the sum of coinciding coincidences is regularity. We must see the ancient hunter and fisher behind our linguistic examples. He came to the shore of the sea. It took time for him to adapt to the foreign environment, and even more time to come to truly know and govern it. Yet they had time aplenty, much more than we do, and one day the density of cultural contact became so great that the Baltic-Finnic *poiss* ('boy') – which sounded a little different at the time, more like *poi* – found a home in Nordic languages. The English 'boy' also derives from this word and era (Paul Ariste to the author, 1.3.1973). The Finno-Ugric brooch known as a *sõlg*, which became traditional among our peoples when we all still resided near the Volga, took root as a name and a type of jewellery in Scandinavia as well – telling evidence of the high quality of Baltic-Finnic craftsmanship, but also of our knowledge of the circle and basic geometry. It was a time of tremendous changes. The Scandinavian languages have been called un-Indo-European, revealing the long-term effect of nearby Baltic-Finnic languages. Equally un-Finno-Ugric are the Baltic-Finnic languages, which speak to the deep influence of Scandinavian languages in turn. Several extinct Nordic languages indeed live on in Baltic-Finnic languages, albeit disguised as loanwords – a source available only to linguists. It was a period of ideological renewal in local intellectual life: a painful break from the old, a farewell to the forest, and a turn towards the sea and the field. During that period, Estonian adopted Nordic loanwords such as *põld* ('field'), *muld* ('soil'), *ader* ('plough'), *rukis* ('rye'), *leib* ('bread'), *taigen* ('dough'), *tuba* ('room'), *ahi* ('oven'), *katel* ('kettle'), *vokk* ('loom'), *säng* ('bed'), *hame* and *särk* ('shirt'). Unusual shifts occurred in words that reflected societal relationships. There came a need to borrow Nordic words such as *mõõk* ('sword'), *kaup* ('goods'), *vara* ('wealth'), *kuld* ('gold'), *raha* ('money'), and *kuningas* ('king'), the last naturally used for a leader or elder, which survived in Estonians' memory until the 19th century and was the cause of odd semantic complications with the imperial regime during peasant uprisings.

Given these circumstances, it is simpler to understand why Estonian was forced to borrow the words *rand* ('beach'), *kallas* ('shore'), *kalju* ('cliff'), *kari* (in the sense of 'stony shallows'), *loodevesi* ('tidewater'), and others like them: we had entered a new natural environment. Nautical terminology is the most convincing evidence of this. Estonian is rich in such terms and, more importantly, it reflects the most crucial shipbuilding ties regarding the ancient Baltic ship.

SHIPS TAKE FLIGHT

First, a heavy centre plank was set in place – the ship's 'mother', which only later grew the small keel characteristic of Baltic vessels. The stem was hewn from a massive trunk. Baltic vessels' dimensions, finish, and religious importance developed in tandem with shipbuilding abilities. Next, the gunwales were put in place – *parras*, a very old Estonian word. The first side planks were joined to the centre one, their ends pushed into notches carved into the stems at the bow and stern. Each plank overlapped the one below it like roof tiles. Planks were sewn together in the earliest era of ancient shipbuilding; ship nails were only utilised later. Both sides rose until they reached the necessary height. Only then were the ribs put into place and tied (later nailed) to the planks.

The ancient ship's ribs had no load-bearing role. They didn't form a skeleton with the keel, but simply reinforced the sides. Although the technique may seem primitive, it isn't. Ships built in that fashion were flexible – a particularly important feature for larger vessels. However, entrusting a load-bearing role to the structure's outer shell also set an upper limit on the vessel's development, both in time and size.

Ancient Baltic ships were oar-powered. Oars were fitted into rowlocks. The vessel was steered with a rudder, or *saps*. It is an important word that appears to be a unique link between the older Finno-Ugric boat type and the ancient ship, but it is certainly – as we will see later – a connection between the ancient ship and the modern naval tradition.

* * *

> What's being made for us there
> in the shadow of Viru's alder,
> behind Harju's oaks?
> A ship's being made for us there:
> five were the files filing,
> a hundred the saws sawing,
> a thousand the knives drawing.
> The ship was finished
> and launched onto the waters,
> the youth were set a-rowing,
> the elders set a-watching.
> The youth rowed, their strokes nimble,
> the elders watched, their heads bobbing.
> They rowed their way to Riga,
> out past Tallinn's market.

SILVERWHITE

> The sea's black ox bellowed,
> the wave's spotted cow mooed,
> the cocks of Riga-town sang:
> Row the ship to Finland,
> from Finland to Saaremaa,
> from Saaremaa to Germany,
> from Germany to the Russian path,
> from the Russian path to Viipuri,
> from Viipuri to Holland,
> from Holland to your own land!
>
> (Haljala)

* * *

But where, exactly?

Many archaeological finds on the banks of the Volga that date to the earliest days of the Bronze Age bear astonishing similarities to those along the Baltic Sea. Researchers agree on two conclusions:

1. The relics spread both from west to east and vice versa.
2. The number of finds increases steadily and peaks in the first millennium of the Common Era.

It would be naive to think we were remembering our ancient path of Finno-Ugric migration. There was no path. However, there was the logic of landscape, and nothing about that has changed. There were giant rivers that flowed the wrong way, for the most part: from our cosy taiga into the unbearable tundra or the even more hostile steppes. Woven through them was a maze of tributaries that generally crossed latitudinally, not exiting the accustomed environment. These provided the most favourable opportunities for movement. An ancient migration westward had followed these tributaries, seemingly in consecutive waves that left kindred peoples behind in the riverbends – tribes that spoke nearly identical languages just yesterday.

Objects travelled along these rivers and people came right in their wake. Let's sidestep the notion of trade, as defining it would take up the rest of this book's pages. Objects travelled exponentially further than their owners, and there's nothing strange about it. It wasn't a consumer society. The lifetime of an object was significantly longer than that

SHIPS TAKE FLIGHT

of a human. A bronze axe was a great treasure, and when William the Conqueror invaded England, warriors were still waving stone weapons in the Battle of Hastings (among others). Periodising history into the Stone, Bronze, and Iron Ages is like portioning an animal on a butcher's block. Objects were passed from hand to hand, needs grew out of habit, and gradually people's journeys also grew longer, the landscape becoming more and more familiar in ever-larger sections. When homogeneous ethnic settlement and linguistic proximity disappeared at the end of the Bronze Age, the waterway was already worn in deep enough to continue operating automatically. The exchange of objects is one form of contact between producer and consumer, the waterway its footprint on the landscape. People's actual needs are a prerequisite. However, contact can also significantly accelerate the intensification of need into habit. Passes through the Alps were already 'sacred' in the Bronze Age: it was more beneficial for locals to defend trade routes than rob the traders. We know from ethnographic sources that both coasts of the African continent traded with each other. Goods journeyed to meet somewhere in the heart of the landmass before finally reaching the right hands.

We may imagine the early days of the waterway connecting the lands of the Volga to the Baltic Sea in a similar way. When the Finno-Ugric tribes dispersed, former ties were stretched over an enormous area to the point of breaking – or, rather, without breaking, gradually transforming into traditions and surviving with new content and value. From 1969 to 1972, archaeological digs were performed in the 1st- and 2nd-century CE stone *tarand* burial sites in Kõmsi, near Lihula (Lõugas 1972: 163–72). One handsome bronze brooch with a leaf-shaped head was forged in the headwaters of the Volga and Oka rivers. Similar jewellery was discovered near Stockholm slightly later. However, more intriguing from the standpoint of the Baltic–Volga waterway are Kõmsi's burial ceramics. The Corded Ware ornamentation was common among the northwest tribes of the Volga and tenaciously endured from the Bronze Age till the time of written historical records. In Estonia, the technique has been found only in coastal areas; it is entirely absent from Latvia and Finland. That's not what matters most: bronze jewellery could be passed from one merchant to another along a long waterway, but clay pots could not. It would be absurd. Thus, we encounter an unknown fate – the Unknown itself. A craftsman of the Volga journeyed to Estonia and created a clay pot there in Kõmsi with the technique and patterns of their distant homeland (Vello Lõugas to the author, 20.3.1973). Was it a woman? Pottery was

SILVERWHITE

traditionally women's work. When? During the 1st or 2nd century. Why? How? Whence? What became of that person? Archaeologists have composed a summary based on this and similar finds: 'Outstandingly active cultural and trade ties between areas of the Central Volga and the Urals and the Baltics existed during the first millennia CE' (Smirnov 1970: 176–80).

That was the climax. But the beginning?

Archaeological finds give us a relatively precise picture of the most ancient Estonian ship's age and dimensions: length 7.5 metresm, width 2.5 metres, age 2,700 years.

The discoveries in question were made by Vello Lõugas at a ship burial site on Saaremaa. Its mighty bow would have ploughed through the waves and the low gunwales indicate modest seaworthiness – sufficient for the Baltic Sea and large rivers, but no further. Ceramics shards point to local tradition, and even the grave's stone filling differs from that found on Gotland while closely resembling Baltic burial traditions in the late Bronze Age (Lõugas 1970b).

* * *

There was a huge crowd of people and horses when a ship was launched – sometimes up to 200 people and 100 horses. There was usually also a singer who would sing. When the song was nearly over, people would ready themselves and their horses to pull, and everyone'd tug at once during the last lyrics. (Kihelkonna)

* * *

More significant than the vessel's dimensions are the ideological conclusions: someone was buried in the ship. By the year 800 BCE, ships had acquired a central role in the lives of coastal peoples. One needed one's vessel even after death to reach the afterlife and remain loyal to seafaring there, also. Customs reflect true attitudes. It was no longer the beginning and wasn't the climax, but dynamic interaction was taking place on the Baltic Sea. Bronze is an alloy of copper and tin. The metals are not found together. Sweden's copper-rich coast acquired a particular attraction, but so did the southern coast of the Baltic, from which crucial waterways branched out – including to the southwest, into a mosquito-infested marshland, on one hill of which Rome's first thatched roofs were rising.

About a century remained until the catastrophe.

SHIPS TAKE FLIGHT

WHEN THE CASPIAN SEA EMPTIED INTO THE GULF OF FINLAND

Have any written records of our Bronze-era waterways been preserved?

When I was a student at the University of Tartu, the rail connection with Tallinn was serviced by a black locomotive that spewed black smoke. I'd go to the station at around midnight, buy a ticket, and curl up to sleep on the passenger car's upper baggage shelf. The next morning, I'd wake up in the capital. It was a big undertaking and a long journey. Now, there are nearly two dozen departures per day, the trip is cheaper and much quicker, and the distance has disappeared. Let us account for this possibility of disappearing distances.

There were two different methods practised by geographers in the Classical Era. Some sat at home, interviewed visiting merchants from beyond the seas or naval leaders returning triumphantly home, and recorded what they heard. By keeping this up for 20 years or so and casting aside contradicting claims, one could arrive at outstanding results that provided a rather objective, albeit indefinite, description of the greater world.

Of course, the opposite of theoreticians is practicians who, for one reason or another, actually travel and are able to describe distant shores from first-hand experience. The outcomes of such stories are no better or worse than the conclusions drawn by theoreticians, just different. Speed of travel was slow and the journeys themselves treacherous. They saw places more vividly and described them with greater objectivity, but it was merely a tiny shard broken off the greater world. Practicians are more credible in terms of details, theoreticians in drawing general conclusions.

We must also account for deliberate distortions. Merchants weren't interested in betraying secrets of their trade. To ward off competitors, they could invent blood-curdling dangers, one-eyed monsters, mountain chains that reached the heavens, or blood-thirsty giants that were allegedly found in little-known lands. Folkloric geographical clichés persisted nearly unchanged for millennia. As Western civilisation became more familiar with the rest of the world, they withdrew further and further to the peripheries, blending with local folklore. One might think this has no bearing on our discussion of seafaring, but far from it: the ancient merchant followed on the heels of the cynocephalus, a dog-headed creature, and the cynocephalus's retreat to Nordic regions tellingly coincides with the formation of trade routes for fur hunters.

SILVERWHITE

At first glance, factoring in all these conditions results in disappointment: records of the Baltic–Volga waterway could only be expressed in the vaguest of geographic fancies. Even the plural is a problem: only its most general and critical feature must be sought. And what could that be?

When classical geographers assert that the Caspian Sea was connected to the Baltic Sea, it may be seen – within the bounds of reasonable probability – as a literary expression of an ancient waterway, no more and no less.

One more limiting factor: according to the classical understanding, land was surrounded by Oceanos on every side. When the Baltic Sea first takes the stage, it appears as the ocean's twin sister, separated from the world sea by the sparse chain of little 'Scandinavia' islands. It isn't a separate sea, strait, or bay, but a coastal ocean to be equated with the northwest or even north-northwest part of Oceanos.

Do classical geographers have anything to say?

1. Anaximander (610–546 BCE): To him, the Caspian Sea does not yet exist.
2. Hecataeus of Miletus (550–476 BCE): The Caspian Sea joins Oceanos at the eastern edge of the world (at Vladivostok, by our understanding).
3. Herodotus (484–425 BCE): The Caspian is 'by itself, not joined to the other sea'.
4. Eratosthenes (276–195/194 BCE): The Caspian is a former inland sea; a giant bay in the world sea that drains into Oceanos near north Europe. How close? About 16,000 stadia from Britannia's eastern coast, thus in the vicinity of 'Ladoga' (or 'the White Sea').
5. Pomponius Mela (in about 44 CE): 'The Caspian Sea first breaks into the land like a river, with a strait as small as it is long'. To the west of the Caspian inlet is the Baltic Sea (Codanus Bay), the coast of which, he claims, is settled in all but the places rendered uninhabitable by the cold.
6. Jordanes (in about 551 CE): The Caspian Sea still flows into Oceanos in the northeast corner of Europe, where it is called the Vagus River.
7. Adam of Bremen (in about 1075 CE): The area between the Caspian and the Sea of Azov is equated with the eastern coastal areas of the Gulf of Finland, and the peoples of the steppes with the ancient Estonian county of Virumaa (Latin *Vironia*).

This far exceeds our hopes!

SHIPS TAKE FLIGHT

Let us remind ourselves: under certain conditions, distances tend to disappear without actually ceasing to exist. Ships were one such condition. The outline of the Caspian Sea is warped along the axis of maritime navigation. Greek merchants saw ships on the Volga that were not local river barges. Therefore, the Caspian had to be connected to the world sea. But in what direction?

Even Hecataeus's false notion that the Caspian drained into Oceanos to the east is significant: archaeologists have confirmed that in the earliest days of the Baltic–Volga trade route, it split off down tributaries of the Kama River and continued to the Irtysh, from which it crossed over the Kazakh steppes westward to the great cultural hubs of Central Asia (Tretyakov 1966). In other words, even Hecataeus's fantastical inaccuracy had foundations in reality: the lower course of the Volga was not in use at that time and the trade route touched lands beyond the Caspian Sea from the east. Later, there was a breakthrough that temporally and, as we shall soon see, causally coincides with the conquests of Alexander the Great, and the Caspian's connection to the world sea suddenly became more exact: together with the maritime route, the Caspian aligned longitudinally from south to north and retained that shape for several centuries. There's a logic to the deformation, of course. The more that use of the Baltic–Volga waterway intensified, the closer the Caspian was nudged to the Baltic. The loss of distance climaxed with the Arabs' conviction that the Baltic stretched to connect with the Kama and the Volga, all the way to the location of modern-day Kazan. The overall picture is lacking in detail, but as a result it presents a grand and convincing generalisation.

Though *is* it so lacking in detail?

Hecataeus gave the following report of an island

> upon which [...] amber is thrown up by the waves in the spring season. As to the remaining parts of these shores, they are only known from reports of doubtful authority. With reference to the Septentrional or Northern Ocean [which includes the Baltic Sea]; Hecataeus calls it, after we have passed the mouth of the river Parapanisus, where it washes the Scythian shores, the Amalchian sea, the word 'Amalchian' signifying in the language of these races, frozen. (Pliny the Elder)

There's an important part missing from the last sentence, though it's not worth complaining about. We are given the excerpt by Pliny the Elder (23/24–79 CE) in his *Naturalis Historia* (*Natural History*). It would seem pedantic to criticise him for the loss of one part of a sentence, given that entire nations have disappeared over the five hundred years between them. Instead, we should admire and amiably envy the consistency of cultural processes in the classical world, but we really have no time for that, either.

Furthermore, Pliny is not to blame for the excerpt being imperfect. He himself complains about it. We can assume that Pliny did not weave it into his text at random. He knew Hecataeus better than we do and quoted him where their information overlapped. This grants us the right to cast aside Greek suffixes and thus place the word 'peninsula' next to the amputated *Amalch*: the Yamal Peninsula. In Nenets, *ja mal* means 'the end of the world'. It's hard to find a more striking name for the giant peninsula on the western shore of the lower reaches of the Ob River. It was truly the trembling end of the world: bare, windy, empty, and hungry. Still, the Nenets did not occupy that region in Hecataeus's day. Much about the Samoyedic people's migration has yet to be explained. Together with archaeologists, linguists have concluded that Europe's northern shore was inhabited at that time by proto-Sámi tribes who spoke a language close to Samoyedic tongues. Nevertheless, while linguistically credible, connecting Amalch and Yamal raises doubts in geographical terms. To Hecataeus, the Yamal Peninsula was probably terra incognita. The name of the Yamal Peninsula has been tied to the Baltic-Finnic people's word for 'god': Estonian *jumal*; Finnish and Karelian *jumala*, Vepsian *jumal/gumal*; Sámi *jummel*; and Mari *jume/jeme*, all of which originally stood for the heavens. Nenets' beliefs add another point of support: the name for the spirit of rivers' headwaters was *Jav'mal*. The latter is related to a sacred site of a nomadic Yamal Peninsula tribe: *Jaumalhe* (Homich 1966: 198). The pre-Nenets, i.e. proto-Sámi settlements, elements of whose customs and beliefs were adopted by the Nenets, settled much broader swathes of the taiga. The river headwaters' Javmal and his sacred site Jaumalhe were thus closer to Hecataeus than his Amalch. It's also worth noting that it was during Hecataeus's time that the modern-day Nenets began their push to the north, which is confirmed by elegant bronze items originating from the steppes that have been discovered in archaeological sites along the lower Ob River.

The pre-Nenets peoples established a strong maritime culture, hunting walrus and whale and practising pottery. Much later, when seafarer Thorir Hund pillaged a sacred site on the shore of the Arctic Ocean near the Výnva/Northern Dvina River in 1026, he beheaded a totem statue. It wore a golden crown, which is unlikely, and was called Jomal, which is likely: *god Biarma er heitir Jómali*, according to the text. Although it came

SHIPS TAKE FLIGHT

later, Skaldic poetry cannot be suspected of mystification. Willem Barents came across the Nordic idol tradition on his famous expedition and christened the southern tip of Vaygach Island *Afgoden Hoek*, or Idol Point (Müller 1837: 115). In 1826 the missionary Ivan Veniaminov destroyed 14 stone and 256 wooden totems. The Nenets regarded the island as holy ground, which is reflected in their name for it: *Hebidjaja*.

Gradually, we begin to realise why Pliny included the incomplete and erroneous quote in his text. He believed that in spite of those details, it contained a degree of actual information. This possibility, as we will see, cannot be ruled out. As soon as we reach early medieval records, we can state with conviction that trade in precious walrus tusks and baleen, the ivory of the North, was in full swing on the shore of the Arctic Sea. For now, we still know little about that prosperous maritime culture. Polar hunters made contact with the distant steppes in Hecataeus's time. We'll never come to know the finer details. Or will we? Mute trade developed in the rivers' headwaters under the protection of *Jaumalhe > Jamalh > Amalch*. Merchants weren't interested in the local customs but in place names. If we restore the ancient meaning of *jumal*, 'god', then we get Sky Sea, *Taevameri*, in Hecataeus's quote. There is even a logic to it that is too poetic to be true, as a Sky Sea is precisely what it was: a desolate sea beneath a desolate sky, the upside-down reflection of the Nordic expanse, harsh, cold, and frozen. Sea Sky 'signifying in the language of these races, frozen'. Is that now the oldest Estonian word?

GREECE'S GATE TO THE CELTIC WORLD

Skraha meant 'fur' in the Gothic language. In Baltic-Finnic languages, it has been preserved as *raha* in both form and meaning, now signifying 'money'. Furs were Nordic gold and shaped economic ties that remained unchanged for centuries.

Let us now add amber, which was called *helm* in that era (now *merevaik*, 'sea sap'). Demand for the strange organic mineral was remarkably more volatile. The amounts that have been found in Cretan and Mycenaean archaeological sites gave rise to a fantastical hypothesis in the early 20th century that the Baltic Sea figured into the Trojan economic orbit. Later, it turned out that amber, contrary to popular belief, could also be found in England, Spain, Italy, Sicily, Syria, and Romania, giving rise to an unnecessary wave of scepticism in the wake of the furore. Ultimately, scientists performed chemical analyses and determined that only amber from the Baltic region has a succinic acid proportion of 3–8 per cent (Clark 1953: 251–2), which is of the same quality coveted by the fair women of Crete and Mycenae whose age would now be at least 3,500. No conclusions with regard to seafaring can be drawn from this. Amber was transported along Central European waterways to reach pre-Homeric Greece. In return, distant echoes of Cretan and Mycenaean architecture found their way to the territory of Estonia in the form of cists, stone ossuaries for the deceased, which spread through the region in about 500 BCE and which were built until the limits of recorded Estonian history.

Fashion is fashion.

Amber's popularity exploded with even greater force in the days of the Roman Empire. Pliny the Elder moaned about unchecked consumer society as follows: 'So highly valued is this as an object of luxury, that a very diminutive human effigy, made of amber, has been known to sell at a higher price than living men even, in stout and vigorous health. This single ground for censure, however, is far from being sufficient.'

Let us join his displeasure and turn to the most famous seafarer of the classical world, to whom we owe the mystery of Thule. Pytheas of Massalia was more than a theoretician and remarkably more than a practician. His *ultima Thule* had been the oldest enigma of Western cultural history. Yes, I do believe the pluperfect is justified in that statement.

Massalia was the modern-day Marseilles, the Greeks' city-colony at the mouth of the Rhône, the foothold of tin trade with Cornwall. British tin nourished Europe's Bronze Age and the Rhône valley was more a corridor for Europe than it was for France. At the place now called Côte d'Or, routes branched off into the Loire, Seine, and Meuse rivers, the last of which is called the Maas at its mouth. Even today, hot African winds blow deep into the interior along the Rhône valley, following the Garonne to the very shores of Cornwall. Olive trees imported by the Greeks still bear their fruit and the bony spines of the mountains on the shore of the Mediterranean glow through the earth's taut, translucent skin. Alas, the marshes have disappeared from the mouth of the Rhône and its lower reaches. They dominated when the colony was founded (6th century BCE): it was a Dutch landscape and the oldest settlement clung to the eastern hillsides of the limestone upland. In addition to British tin, Celtic grain and mysterious amber arrived in Massalia. Moving in the opposite direction were products of Greek craftmanship and, most of all, salt, which even made its way to the peoples of the Alps along winding paths. Trade rafts sailed up to 90 kilometres per day downstream and up

SHIPS TAKE FLIGHT

to 20 upstream in fair conditions. Even slower packhorses were employed during floods or in watersheds.

Three generations before Pytheas, Massalia was ruled by Carthage – the Phoenicians' city-state founded on the opposite shore of the Mediterranean. The Carthaginian Himilco embarked from Massalia to study the 'remote parts of Europe'; in other words, shores across the Strait of Gibraltar. It was a sea voyage. What does that mean? The shore played a role equal to the sea's, the ship hugging the coastline. Such a maritime route differed very little from a river route. Unbroken shoreline on the horizon was the best aid to an ancient helmsman and the most reliable compass both in the time before and after the instrument came into use. The reason for this does not lie in insufficient knowledge of the art of navigation, but the character of a sea versus an ocean: they are narrow and speckled with islands. Every sea shapes its navigation, the navigation its routes, the route its harbour locations, and a harbour location the possibilities for a city to perhaps, though not necessarily, develop around it. On seas, including the Baltic, coastal routes persisted for an extremely long time because they were in harmony with the nature of the sea. On the Mediterranean, they were still the primary means of navigation in the 16th century. This is important to keep in mind, otherwise we would over-modernise the voyages of both Himilco and Pytheas. The bronze plaque bearing Himilco's travelogue has been lost, but the most significant aspects of its content have been preserved in a later retelling. Himilco passed Cape St Vincent, the furthest southwest point of Portugal, which for many centuries marked the western ecumenical border. The Greeks called any environment suitable for human habitation 'ecumenical', though if we closely study the context of the word, we find that the Greeks saw themselves as humans above all, at least during the given period. Himilco reached the shore of Brittany (Bretagne) in present-day France and, for a while, secured the Carthaginians' monopoly in the tin trade. Tin came from Cornwall, a peninsula extending into the ocean far away, which is to say from Britain, but what does 'Britain' mean? Is it a new continent on the far side of the strait? Or is it a small archipelago that also includes the 'Scandinavian islands'?

Trends change, but humans' basic needs do not. The Rhône trade route functioned without variation during Pytheas's days and with only very slight changes a few dozen centuries later: the first English ship arrived in Marseilles in the Late Middle Ages and found there a pilot's journey ahead on the Mediterranean – a sea that was still relatively unknown to

SILVERWHITE

the English in 1590. Nevertheless, its hold was filled with cargo: the same that had travelled those waters centuries earlier (Braudel 1981: 628).

The Carthaginians' rule had shrunk, though not been broken, by the time of Pytheas. They still held Gibraltar, and Greeks' access to the Atlantic was a gamble. We also must not contemporise Gibraltar: it was not the Second or even the First World War fortress capable of locking the Mediterranean. In fair winds, ships could slip through the narrow stretch unhindered because the Carthaginian guard was unable to exit the harbour. Massalia's prosperity depended on the tin trade, but trade is not built upon chance. Tin came and exports left, rarely by sea, mainly down the Rhône valley, but even there, merchants would sooner or later run into the shore, the ocean, seafaring, and the threat of competing Carthage. Was seafaring inescapable? Is British tin located on an island? Where does amber come from? Such were the pressing questions in Massalia: overly self-evident to be worded and consequential enough to draw lines of force to Pytheas's activities. How far was Britain? What about the amber shore?

We must first determine what Pytheas might already know in 325 BCE, and what he could not.

TEMPLE OF THE MUSES

This question isn't so simple to answer, because Pytheas lived during the time of a great upheaval: one world collapsed, another was born. Was he the last link in an ancient cultural chain? The first in a new one? Or did he join the old and the new perceptions of the world together?

Alexander, a pupil of Aristotle who came into this world in Macedonia in 356 BCE, left it at the age of 33 in Babylon and went straight into history under the name Alexander the Great. In just a dozen years, the Greeks' tiny world on the Peloponnese and islands of the Mediterranean expanded to dominate the known world. Half of the Nile and all of the Indus suddenly ran through the empire of Alexander the Great, and historians were unable to guess how much of the world was still left, if any at all. Never-before-seen mountains, deserts, flora, and fauna entered Greeks' lives. Strange peoples, unknown languages, exotic deities, and new knowledge and sciences came into their cultural orbit. Even after the empire's political collapse, trade routes ensured the continued exchange and assimilation of disparate Mediterranean and Eastern cultures, unique in that all parties were givers, all were receivers, and none were losers.

SHIPS TAKE FLIGHT

Out of the seventy cities and harbours that Alexander the Great founded, Pytheas should be associated with Alexandria at the mouth of the Nile.

In the city, a structure was built that was meant to, and did, become a temple of the muses: the Mouseion. It was the predecessor of modern-day academies and included zoological and linguistic departments, a botanical garden, and an observatory. It is even better known as the predecessor of national libraries. When a vessel docked in Alexandria and had any unfamiliar manuscript on board, it was custom, and law, to make a copy in the harbour to be brought to the Mouseion. The institution grew even faster and greater than Alexander's empire, the Library incorporating not only earth, but other heavenly bodies.

Apollonius of Rhodes, who served as head of the Library, may have shared both time and space with Pytheas. He was more of a poet than a scholar and rather old-fashioned as the former. Written in Homer's epic style, his *Argonautica* describes a voyage to retrieve the Golden Fleece from a faraway land with amber on its shores. Poorly received in Apollonius's hometown, insulted and offended, he moved to Rhodes. The proud Greek probably wouldn't have been pleased that his posthumous fame came from cosmopolitan Rome. Eratosthenes began directing the Library in 245 BCE. He was older than Apollonius, primarily a geographer, and involuntarily a poet. He may also have encountered Pytheas in time and space. All we can prove is that he knew Pytheas's book, admired it, and used it profusely in his *Geographika*. From this threshold, the paths branch off: Apollonius set the foundation for a literary succession that preserved Pytheas's form and style but shed his name. Eratosthenes established a scholarly succession that preserved Pytheas's name; along with the name, the controversies; and along with the controversies, Pytheas's arguments and one quote. Before its destruction, his *About the Ocean* was also used by Tacitus (1st century CE) and perhaps the greatest classical geographer of all, Klaudios Ptolemaios, Ptolemy (2nd century CE), whose life's work was written right there in Alexandria. The Mouseion's Library was irrevocably turned to ash in the 3rd century CE to root out all heterodoxy. Was Pytheas's manuscript destroyed in the fire? It cannot be proved but is also unlikely. At least eight classical authors used his work in one way or another at various times, which means several copies existed. Arab geography seems to support this inference: there, Pytheas's *distant Thule* appears, albeit with altered pronunciation, which is natural, but also in a different connection, which seems to reference the use of an original document even in the 10th century CE.

SILVERWHITE

A COSMIC CHAIN

Pytheas is the source of classical geography's first and, in some respects, sole reports from the distant North, published in his *About the Ocean* around the year 325 BCE. His use of the singular in the title is significant. *Ōkeanos*, the 'world sea', was becoming a focal point of classical geography. Evidence gathered from experiences pointed towards land playing a definitive role around the world and enclosing all seas into inland bodies of water. The Sea of Azov, the Black Sea, the Sea of Marmara, and even the Mediterranean Sea, which the Greeks called Our Sea with tender self-evidence, are bordered by coastline in four directions and only relatively dissimilar to large lakes. Even when the shore temporarily disappeared beyond the horizon, the seafarer never lost the sense of occupying a space enclosed by four homely walls. True, doubt remained and the more experiences that came, the more uncomfortably it gnawed at one's mind. This doubt had its own name. Shepherds move from one pasture to another and the herd wades through a river. A herd of goats will seek a shallow crossing; a herd of ox will ford deeper sections. In ancient Greek, an ox ford was called a *bosporos*. The Sea of Azov, enclosed as it is, still connects to the Black Sea through the Cimmerian Bosporus (now the Kerch Strait); the Black Sea connects to the Sea of Marmara through the Thracian Bosporus (now simply the Bosporus Strait) and from there to the Pillars of Hercules and perhaps even further. Past the Pillars seethed the ocean, but its horizon was mute and provided no answer to the question: Does land encircle the seas? Or does an indivisible world sea encircle land as a 'cosmic chain', to use a metaphor that probably comes from Pytheas?

Having posited the theoretical question, Pytheas set out to find an answer through practical observation. It was a novel approach in Greek culture, and it turned out to be successful, providing Pytheas with a consequentially advantageous head start. His field trip to the Spanish coast allowed him to correctly demonstrate the dependence of tides on phases of the Moon. However, natural observations and journeys of discovery must be described in time and space. Here, Pytheas was truly a trailblazer. He used three measures to determine distance, of which only the sea day was borrowed from earlier authors. Unfortunately, the distance covered during Pytheas's sea days is ambiguous. Herodotus's sea day a hundred years earlier appears to have been merely 55 kilometres; by a few centuries later, the Greeks had increased it to 185 kilometres. Pytheas's sea day lies somewhere between the two limits and likely fluctuated according to weather conditions and how the type of vessel adapted to the sea. The

SHIPS TAKE FLIGHT

uncertainty of sea days as a system of measurement apparently forced Pytheas to seek new ways to determine distance. Again, he placed his faith in practical observation. When travelling from south to north, the position of stars, length of day, and solar zenith changed remarkably. A second unit of measurement that he adopted was the length of the summer solstice to quarter-hour precision. The third was the zenith of the midday sun on the winter solstice. Although this seems analogous to the second method, and certainly is in a sense, it's worth keeping them separate.

Could Pytheas have done the opposite, measuring the solar zenith in the summer and the length of days in winter? No, not really. The summer sun rises much higher in the heavens than in winter, which means the probability of error would grow to that degree and more. And vice versa: to measure the length of the day on the winter solstice, one would need to observe the sun on the horizon in the morning and evening of 22 December in areas where cloud cover can often hamper one's efforts. For example: in Tartu, there were only 17.1 hours of sunshine during the entire month of December 1976 (*Edasi*, 5.1.1977). In these conditions, determining the beginning and end of a day would be far less exact than measuring the height of the sun by its midday glow. The use of different methods in summer and winter speaks to Pytheas's precision. This was also emphasised by Hipparchus (190–120 BCE), one of the greatest classical astronomers. The method itself is important as well, of course, being as simple and clever as inventing the wheel: humans began to master distances that had earlier fled to the world of poetic metaphors. Today's sextants are derived from Pytheas's idea.

CELTIC TIN

We should voice due respect for the authority of Fridtjof Nansen and Richard Hennig, but also take out a large sheet of clean white paper.

According to Hennig's hypothesis, 'Pytheas's primary task was seemingly to determine whether Britain should be treated as an island or part of a continent'. This notion cannot be confirmed by fact. Yet even more fantastical is his conviction that 'Pytheas travelled slowly and with characteristic meticulousness around the entire island of Britain, which took a full forty days' (Hennig 1961: I, 184). This claim, which was drawn from arbitrary constructions, has, by today, fossilised into irrefutable dogma. The constructions are as follows.

SILVERWHITE

Greek seafarers calculated that a vessel could sail 1,000 nautical stadia (about 185 kilometres) in a day. Pytheas 'determined the island's circumference to be over 40,000 stadia'. From this, Hennig infers an actual research expedition, Pytheas's total freedom of movement, his reaching the northern tip of Scotland in the vicinity of Orkney, and the voyage continuing from there in the direction of Iceland or the Norwegian shore. At various times and with various outcomes, the latter three locations have laid claim to the title of Thule. The proof is refined. Since 40,000 stadia (7,400 kilometres) is more than double the circumference of Britain, Polybius and Strabo are subject to criticism. For, as Hennig argues, Pytheas initially only recorded 40 days of travel, and since he organised his expedition (350–320 BCE!) in extremely dangerous and unknown waters, they must have sailed at half the ordinary speed, which gives us the exact circumference of Britain. Polybius and Strabo were said to have only calculated the days into stadia later.

Let us start from the beginning. Pytheas informs his readers about sites where tin and amber were discovered. The two goods of great importance to the Greeks form a force field, along whose lines Pytheas might have moved. We may honour him as the greatest traveller of the classical world, but we would be blithely contemporising his trip as a modern-day research expedition. Polybius labels him a poor man. It would be even wilder to imagine Pytheas as a wealthy man carrying a purse and haggling in clannish England for cheap transport to Iceland. No, in the best case he might have been the travelling companion of those unfamiliar with trade, chained to the framework of traditional routes and the restrictions of common law. Added to the social limitations were those of a technical nautical nature. In Pytheas's time, sailing was only done with the wind, and this even on the Mediterranean. Sea voyages, just like fieldwork, were either permitted or forbidden on certain calendar days: Hesiod already taught that the time was right for trimming grapevines when Arcturus shone through the night. Fifty days after the summer solstice (11 August) was the prime time to embark on a sea voyage, he claimed. The phenological distribution of labour was based on rational experience. Ships usually embarked on just one voyage per year. This can be compared to fixed fair dates, which depended on how agricultural processes were spread out over the year and thus on the local environment first and foremost. The adoption of the Levant sail allowed vessels to travel in side winds, but this was done only on rare occasions even in the Middle Ages and sea voyages generally weren't attempted between October and April (Braudel 1981: 251).

SHIPS TAKE FLIGHT

Secondly, if a traveller is approaching an unfamiliar shore by sea, not by land, and the vessel is their only vehicle, then all coasts become islands. The territory of Estonia was still an 'island' when Adam of Bremen wrote his chronicle (1073–6) and Scandinavia remained one until the 15th century. When reading Pytheas, we therefore should not draw definite conclusions about a continent being in one place or an island in another.

In many texts by Strabo, Diodorus Siculus, Pliny the Elder, and Cosmas Indicopleustes, we are forced to sift glittering grains of truth from bad-tempered remarks. Waves casting amber onto the shore? The sea frozen over in the North? The distant northern sun never setting? 'For not only has the man who tells about Thule, Pytheas, been found, upon scrutiny, to be an arch-falsifier,' seethed envious Strabo, an outstanding geographer upon whose biography that malicious assessment remains a timeless stain. Likewise, Polybius used the term *idiotes anthropos*.

Many have attempted to pinpoint the location of mysterious Thule since Pytheas's death. The field of fantasy is almost limitless: Greenland to the west, the mouth of the Vistula to the east, Iceland to the north, and to the south ... Yes, the southern limit is the vaguest. Fridtjof Nansen naturally placed Thule in Norway and Winston Churchill, just as resolutely, claimed it lay in north England (Churchill 1957: 9).

Even the origins and meaning of the word Thule have remained a mystery. It is a 'proper name of unknown origin', as the *Oxford English Dictionary* laconically states. Etymology is the field of study that tracks the source and original definition of words. In the event of a toponym, toponymy is responsible. The two are inevitable when researching ancient seafaring, though the less, the better. Pytheas's voyage to Thule took place during an expansion of the Celts and several other tribes, and the ethnic map of Europe had drastically changed by the time of Strabo. The Greeks wrote Thule in two different ways: *Thoule* and *Thule*, using the letters theta, omicron, upsilon, lambda, and eta.

Tin and amber: two goods, two topics.

Pytheas left Massalia/Marseilles and firstly familiarised himself with the Cornwall tin trade. The sources are as follows (Mette 1952; Aalto and Pekkanen 1975, 1980):

(1) Britain is triangular in shape, very much as is Sicily, but its sides are not equal. This island stretches obliquely along the coast of Europe, and the point where it is least distant from the mainland, we are told, is the promontory which men call Cantium [Kent], and this is about one hundred stadia [17.8 km] from the land, at the place where the

SILVERWHITE

sea has its outlet, whereas the second promontory, known as Belerium [Land's End, the southern headland of the Cornwall peninsula], is said to be a voyage of four days from the mainland [...]. Of the sides of Britain the shortest, which extends along Europe, is seven thousand five hundred stadia [1,340 km], the second, from the Strait to the (northern) tip, is fifteen thousand stadia [2,679 km], and the last is twenty thousand stadia [3,572 km], so that the entire circuit of the island amounts to forty-two thousand five hundred stadia [7,590 km]. [...] The inhabitants of Britain [Celts] who dwell about the promontory known as Belerium are especially hospitable to strangers and have adopted a civilized manner of life because of their intercourse with merchants of other peoples. They it is who work the tin, treating the bed which bears it in an ingenious manner. This bed, being like rock, contains earthy seams and in them the workers quarry the ore, which they then melt down and cleanse of its impurities. Then they work the tin into pieces the size of knuckle-bones and convey it to an island which lies off Britain and is called Ictis; for at the time of ebb-tide the space between this island and the mainland becomes dry and they can take the tin in large quantities over to the island on their wagons. [...] On the island of Ictis the merchants purchase the tin of the natives and carry it from there across the Strait to Galatia or Gaul [France]; and finally, making their way on foot through Gaul for some thirty days, they bring their wares on horseback to the mouth of the river Rhône. (Diodorus Siculus)

(2) It is situated to the north-west, and, with a large tract of intervening sea, lies opposite to Germany, Gaul, and Spain, by far the greater part of Europe. Its former name was Albion; but at a later period, all the islands, of which we shall just now briefly make mention, were included under the name of 'Britanniae'. This island is distant from Gesoriacum [Boulogne], on the coast of the nation of the Morini, at the spot where the passage across is the shortest, fifty [Roman] miles [74 km]. Pytheas and Isidorus say that its circumference is 4,875 [Roman] miles [7,209 km]. It is barely thirty years since any extensive knowledge of it was gained by the successes of Roman arms, and even as yet they have not penetrated beyond the vicinity of the Caledonian forest. (Pliny the Elder)

(3) Pytheas of Massalia attributes both phenomena [the ebb and flow of tides] to the growing and shrinking of the moon. (Aetios)

(4) Pytheas of Massalia informs us that above Britain the tide rises 80 cubits [35.5 m]. (Pliny the Elder)

SHIPS TAKE FLIGHT

(5) Polybius, in his account of the geography of Europe, says he passes over the ancient geographers but examines the men who criticise them, namely, [...] Pytheas, by whom many have been misled; for after asserting that he travelled over the whole of Britain that was accessible Pytheas reported that the coastline of the island was more than forty thousand stadia, and added his story about Thule and about those regions in which there was no longer either land properly socalled, or sea, or air, but a kind of substance concreted from all these elements. (Strabo)

Texts 1 and 2 allow us to determine the probable direction and duration of Pytheas's journey: from the Rhône valley to the Loire River, from the mouth of the Loire by sea to the southwest tip of Cornwall – a total of thirty-five or more days by land and by sea. Pytheas approached England from the southwest. Text no. 1 leaves no doubt that it is this point, the promontory of Cornwall, aka Belerion, that is connected to the continent in the vicinity of the mouth of the Loire via an ancient trade route. His approach from the southwest, as we will see later, is of crucial importance.

The Cornwall peninsula is bordered by the English Channel to the south and the Bristol Channel to the north, the width of which is 170 kilometres at the mouth and cannot be perceived as an inlet. Pytheas informs his reader of every significant aspect of the Cornwall tin trade (text no. 1), just as Pliny alludes to it 300 years later: Britain is praised in Greek writing (text no. 2). However, a significant change has taken place over three centuries. Britain now includes territories apart from those mentioned by Pytheas: 'all the islands [...] were included under the name of "Britanniae"' (text no. 2). By Roman times, Britain was constricted to present-day England. Meticulous as always, Pliny doesn't refrain from noting that despite the Romans' thirty-year conquest, their knowledge of England was still quite lacking, and he also uses Pytheas's circumference of Britain for the island. We can characterise this error of formal logic as such: England is Britain, but Britain is not England.

Tin-abundant Cornwall was apparently the only part of England that the scientist of Massalia came to know. Two aspects point to this fact. Firstly, 'he travelled over the whole of Britain that was accessible' (text no. 5); secondly, he studied tides in the Bristol Channel, which is 'above' Britain (*supra Britanniam*; text no. 4).

How are we to understand the 'accessible' part of Britain? Naturally, the opposite of an accessible part is an inaccessible part. Does this mean bogs, cliffs, mountains, chasms? What is referred to is likely a social

SILVERWHITE

obstacle; not a natural one. Aliens were outlaws in places beyond those granted a market truce and forbidden to move freely. Cornwall, Europe's primary tin supplier during the Bronze Age and an ancient place of safe haven, was a striking exception to common law. There, the law of salvage did not apply. Inhabitants were 'especially hospitable to strangers and [had] adopted a civilized manner of life because of their intercourse with merchants of other peoples' (text no. 1). Pytheas was solely able to explore the narrow zone protected by a safe harbour, which, in the very best case, meant the 'accessible' part of Cornwall.

THE BRISTOL TIDES

Texts no. 3 and 4 appear to support my last hypothesis. Sea tides were one of Pytheas's main objects of research. He observed them on the coast of Spain, likely well before his voyage to Thule. However, the tides in the Bristol Channel, i.e. the northern coast of Cornwall, are some of the highest in the world and the very highest in Europe. Pytheas could not have overlooked a natural phenomenon of that sort. His phrasing gives us crucial footing for this conclusion: *supra Britanniam*, according to Pliny. Similar phrasing was applied to the Rhine, which constituted the Roman Empire's northeast border. Thus, it signified an obstacle or a boundary, sometimes a river, othertimes a mountain range, and here a long and narrow peninsula. The mighty tides 'above' sparked his interest. A researcher remains a researcher in every situation. Naturally, Pytheas did not see a (170-kilometre-wide) bay, but a shore that disappeared from sight to the northeast: being a rapidly narrowing inlet, as we now know, the Bristol Channel cuts almost 300 kilometres inland, which is the reason for its dramatic tides.

Pytheas measured the tides 'above' at 80 Roman cubits: 35.5 metres. No tide in the world is that high. Was it an error of measurement? Pliny conveys the given figure to us from the writings of Pytheas's pupil Posidonius. However, clarity can be found in the interpretation of tides. Tides comprise ebbs and flows. Thus, when ebbing water drops to the zero-mark on a tide staff in the Bristol Channel, its height at the next highest tide is 12.5 metres: the highest tides in Europe, which are irregular and depend on the positions of the sun, moon, and earth. Pliny avoids using the words 'ebb' and 'flow', preferring instead to use an expression that can be interpreted as 'tide'. What matters most is, of course, whether

SHIPS TAKE FLIGHT

Pytheas measured the height of tides or of ebbs and flows. Text no. 3 allows us to infer that Pytheas based his measurements on the lunar cycle, contrasting the flow with the ebb. If that is true, then 80 cubits would have measured the height of the flow (or ebb), which would make the height of the tides a full 160 cubits, 160 Greek *pēchys*, or 71 metres. This is absurd because Pliny correctly recorded the Greek unit of measurement. Yet, in addition to the cubit, Pytheas also uses the *palaistē* (palm, 74 millimetres) and *daktylos* (finger, 18.5 millimetres) without, as we later determine, specifying his units of measurement on every occasion. We also know that Pytheas differentiated between natural phenomena on which he received indirect data and those that he personally observed and measured, which means the height of tides 'above Britain' was determined first-hand (text no. 4). Knowing Pytheas's precision, on the one hand, and Pliny's imprecision in separating Greek units of measurement from the Roman system, on the other, we can ask ourselves what lies behind the Roman cubits: *pēchys* (44.4 centimetres), as Pliny believed? Or Pytheas's *palaistē*? A flow of 80 *palaistē* would be 5.92 metres, making the height of tides 'above Britain' 11.8 metres. The height of syzygy tides in the Bristol Channel is 12.5 metres after every fourteen days or more, and the height of the highest tides on the coastline to the east of Lynton matches Pytheas's measurements precisely. If we do not wish to overburden fortunate circumstance, then we could, and should, also deduce that Pytheas must have spent two or more weeks in Cornwall.

Tides only concern us inasmuch as they help to determine Pytheas's ensuing path.

Let us summarise the preceding conclusions as follows: Pytheas arrived in Cornwall from the southwest; Pytheas travelled on foot across the 'accessible' part of Britain and arrived at the Bristol Channel, where he measured the tide; Cornwall is indeed the 'accessible' part of Britain; England is the part of Britain that was once called Albion; Pytheas's 'Britain' consequently comprises several islands, lands, or continental areas apart from England; and Cornwall is undoubtedly the part of 'Britain' whose location, in relation to the European continent, was known most precisely by Pytheas and the tin merchants of Massalia.

We should also add that amber reached England by way of the Baltic Sea. There, jewellery was crafted in styles borrowed from Mycenaean Crete. This points to the possibility of contact, and that alone.

SILVERWHITE

THE BRITISH TRIANGLE

The first issue was tin. Pytheas grasped this. The second was amber, and Thule is a secondary outcome of it. Pytheas's own writing confirms it: first 'Britain', and only then 'add[ing] his story about Thule' (text no. 5). Let us first attempt to determine the amber region and Thule's location in relation to Britain. Our sources include Strabo, Diodorus Siculus, Pliny the Elder, Geminus, and Cosmas Indicopleustes.

According to Strabo, Thule lies north of 'Britain'. How are we to understand this? Literally! Stockholm lies north of London. London lies north of Marseilles. Ireland lies north of Britain (i.e. Cornwall). The words 'north of' should not be interpreted as two areas sharing the same longitude, but simply that they are positioned on different degrees of latitude. Latitude was indeed Pytheas's creation. The frequency with which his northern course is mentioned speaks foremost to the exceptional nature of the direction. Ancient seafaring, as we established earlier, hugged the coastline. On the rare occasions when vessels turned their bows to the open sea, they travelled from west to east or vice versa: all ancient navigation was longitudinal. Adhering to the east–west axis simplified pre-compass navigation. Pytheas was the first to measure the sun's zenith and the length of day on his journey from south to north; in other words, he was the first to utilise the curvature of the earth for both practical and theoretical purposes.

Regardless of what direction Pytheas took on his continuing journey to Thule, Cornwall was his point of departure.

As Thule lies to the north of 'Britain', we must first determine what he means by the moniker.

Strabo believed that Pytheas incorrectly described the entire region beyond the Rhine, including Scythia. Consequently, Pytheas's voyage from Cornwall to the amber shore must have been in a generally northeast direction. Alas, sources cast no further light on this. After Gaul was conquered and joined to Rome, Pytheas's records lost their unique value, were no longer referred to, and are lost to us today. Nevertheless, his descriptions of 'Britain' as a grouping of islands has been preserved. From this, we can infer that his journey proceeded by sea along the coast of the European continent. North of the mouth of the Somme River, the English cliffs came into sight once again, rising into the 115-metre-high White Cliffs of Dover before disappearing into the horizon to the northeast. Measuring the tides along the Bristol Channel, Pytheas could

38

also have determined that the northern shore of Cornwall ran in a generally northeast direction. Based on those two observations, as well as several later ones, Pytheas devised an ingenious working hypothesis. Firstly, that Britain, 'like Sicily', was triangular. All coasts to the port side of Pytheas's course fit into the bounds of his triangle. Secondly, that Britain was an island 'like Sicily'. The latter is not extrapolated from the former. For reasons that we will touch upon henceforth, Pytheas dubbed every triangularly bounded shoreline 'Britain'. The belief, even in his eyes, that this conclusion was merely a working hypothesis is apparent from the idealised triangular ratio 1 : 2 : 2⅔, giving the angles of 20, 40, and 120. Sicily had nothing in common with that.

The only impeccably definable point of the triangle is Cornwall, a peninsula that by Pytheas's day was most accurately pinpointed in relation to the European continent, the Loire–Rhône trade route, and Massalia. Consequently, the Cornwall peninsula corresponds to the southwest 40-degree corner. Tracing the shortest side (7,500 stadia) northeast 'along Europe' (text no. 1), the point lands near the Skåne peninsula in south Sweden near the present-day small town of Ystad (c. 55° 24' 36" N). The second side of the triangle juts out at 120 degrees along the Skåne coast. It doesn't point true north (if it's even reasonable to demand such precision from such great generalisations) but strays to the east. Even so, it coincides nicely with the longitudinal axis of the Baltic Sea, which was most likely Pytheas's intention.

This 'Britain' achieves its goal as a working hypothesis. An unknown area is determined in relation to a known area. In this case, the amber region on the unknown Baltic Sea is determined in relation to Cornwall, which the Greeks knew much better. There was no better process for calculating distance and location on the sea in Pytheas's day, especially in the southwest-to-northeast direction.

While travelling to Thule, Pytheas passed Scandiae, Dumna, Bergos, and Berrice, the largest of all, 'from which the crossing to Thule starts' (Pliny the Elder). If any toponym seems familiar, then let us not get ahead of ourselves. Interpretations based upon phonetic similarities generally turn out to be false. Instead, let's settle for the very least and see what Pytheas has to say about Thule. And remember: just as Columbus did not set sail to discover America, neither did Pytheas venture out to discover Thule. The former was searching for a shorter route to India's riches and the latter more precise information on the riches of the amber coast and the Baltic Sea.

SILVERWHITE

What do we know about Thule?

Thule is six days' sail north of 'Britain' and is near the frozen sea (Strabo).

Thule is one day's sail from the frozen ocean and is the most remote of all lands known to the Greeks (Pliny the Elder). It is not located near Britain or Ireland (Strabo).

Thule is in the northern part of the temperate zone where summer nights last two to three hours from sunset to sunrise (Geminus of Rhodes).

Thule is not in the Far North: 'there are grain and honey' from which 'the people get their beverage' (mead), and the grain is threshed 'in large storehouses' (Strabo).

Thule lies 1,648.5 kilometres to the north of the Borysthenes (Strabo). Borysthenes was the Greeks' Dnieper and also their city-colony at the mouth of the river. Eratosthenes, who composed his map just half or three-quarters of a century later on the basis of Pytheas's discoveries, positioned Thule and the mouth of the Dnieper on the same meridian, which would lead us to the area of Lake Ladoga. True, his meridian was not exact and tends to stray to the northwest of Ladoga.

What else?

Thule had a mixed economy. Farming was engaged in and livestock raised, but foraging also played a sizeable role. Researchers now have a consensus on the type of grain cultivated: millet. In former Livonia, millet has been found in archaeological digs at the site of Holme Castle on Mārtiņsala Island in the Daugava River, where it was cultivated before the Common Era.

To Estonians, Estonia lies not in the north or the south, but precisely where it is. To the south is the southern shore of the Baltic Sea, and the ever-blue coast of the Mediterranean disappears into a mythical fog. A Mediterranean Greek explorer must have perceived the same when he entered the latitude of white nights. To him, it was the furthest border that any envoy of the civilised world had ever crossed. Pytheas's information conveys astonishment; a deep sense of majestic Nordic natural phenomena; the discovery of a sea that began to slowly freeze and be rendered motionless as winter approached. Greeks' customary sea voyages from the Mediterranean to the Black Sea did not change the position of stars in the sky, the course of the sun, or the length of days. Pytheas's expedition from south to north shattered his static worldview. The environment was harsh and unfamiliar, and even more alien were the Nordic heavens. White nights? Astounding. Let us call it the dawn–dusk phenomenon.

SHIPS TAKE FLIGHT

The goal was amber, Thule was its unintended result.

Amber is found on the island of Basilia, which is 'directly opposite the part of Scythia which lies above Galatia' (Diodorus Siculus).

THE BALTIC GETS ANOTHER NAME

Let us employ the aid of another Mediterranean geographer, Pomponius Mela, who was born and died in Spain early in the 1st century CE. Unlike the famous Strabo, Mela was not a practician, but a compilatory geographer. And unlike the avid itinerant, he did not take a jealous approach to Pytheas. As a result, we owe to him several supplementary and more objective pieces of information regarding north Europe. Luckily, his thick earth-sciences work *Chorographia* is preserved in its entirety and offers a crucial basis for pinpointing Thule's location.

Mela wrote in Latin but leaned primarily upon Greek geography, sharing the conviction that the Caspian Sea was connected to the northwest part of the world ocean near the Baltic: 'and from the Scythian [Arctic] Ocean [Asia] lets in the Caspian' (Pomponius Mela 1998: II). We must, however, allow Mela to define his concept of Asia, as it does not coincide with ours: 'hugging the edge of the very Maeotis [Azov] all the way to the Tanais [Don], it becomes the riverbank where the Tanais is located'.

At the beginning of the common era, the Don River and the Don meridian formed Europe's boundary. Thus, the Finnic Volga tribes were located in Asia: 'The Scyths look north, too, and they possess the littoral of the Scythian Ocean all the way to the Caspian Gulf, except where they are forestalled by the cold.'

Although Pytheas's first focus was tin and his second was amber, it did not necessarily mean he must have visited the Baltic Sea. Amber trade routes split to the west and south near the mouth of the Elbe. Pytheas could merely have visited a market on the item's path. Let's see what Mela has to say about it.

He names the following continental European coastal areas from west to east: 'After Gaul the Germans reach as far as the Sarmatae, and they to Asia.' Gaul is Rome's northern neighbour, generally coinciding with present-day France. We shouldn't be taken aback by the latter half of the sentence, as we know the boundary of Asia was closer to the Baltic Sea than it is today. Luckily, Mela also describes the northeastern and eastern limits of Sarmatia: 'Sarmatia, wider toward the interior than toward the sea, is separated by the Vistula River from the places that follow.' This

SILVERWHITE

gives us a better idea of Sarmatia's location: its narrow strip on the Baltic shore lies to the east of Germania and widens deeper inland along the Vistula. Now comes an important point: 'the Scythian peoples – almost all designated under one name as the Belcae [*et in unum Belcae adpellati*] – inhabit the Asian frontier [i.e. the 'Don meridian', approximately the Vologda–Archangel line] except where winter remains continuous and the cold remains unbearable'.

We have circled back around to Pytheas's data and there is no longer any doubt: contrary to what geographical historians believe, the famous traveller of Massalia/Marseilles did reach the Baltic Sea.

The link is Belcae/Bergos. A place name? Tribe name? Toponym? Ethnonym? For the moment, that isn't important. What matters most is the two men's data coinciding. Pytheas's isn't localised, but Pomponius Mela places it north of the Vistula and to the northeast. If 'Belcae/Bergos' is located near the Baltic Sea, then we should also place Scandiae, Dumna, Berrice (Pliny the Elder), and Basilia (Diodorus Siculus) in the same region. A sea route is, of course, the thread that connects them. We cannot trace its definite path in the period 350–325 BCE, but we do know that ships clung to the shoreline even in the days of Adam of Bremen. From here, we will discuss it in greater detail. It would be foreign to more skilled seafaring in Pytheas's day. The sail was yet unknown on the Baltic. Therefore, we must seek the locations of toponyms in coastal areas. We'll do so in the next chapter, and only then because Pomponius Mela has something quite important to say about the Baltic Sea itself.

On the other side of the Albis [Elbe], the huge Codanus Bay is filled with big and small islands.

The seven Haemodae extend opposite Germany in what we have called Codanus Bay; of the islands there, Scandinavia, which the Teutoni still hold, stands out as much for its size as for its fertility.

Thule is located near the coast of the Belcae [a tribe?: *Thyle Belcarum litori adposita est*], who are celebrated in Greek poetry and in our own. On it – because there the sun rises far from where it will set – nights are necessarily brief, but all winter long they are as dark as anywhere, and in summer, bright. All summer the sun moves higher in the sky at this time, and although it is not actually seen at night, the sun nevertheless illuminates adjacent places when its radiance is close by.

Mela's impeccable description, which is based upon Pytheas's lost work, leaves no room for doubt: Thule can only be located on the Baltic

SHIPS TAKE FLIGHT

Sea. Unexpectedly, Pliny the Elder supports him and weaves yet another name: 'at a distance of three days' sail from the shores of Scythia, there is an island of immense size called Baltia'.

When reading classical authors, we must ask ourselves each and every time: what do they mean by one or other toponym? Form is more tenacious than content: the word can remain, but the substance may change. What did Pliny mean by 'Scythia'? This is how he defines it: 'The name "Scythian" has extended, in every direction, even to the Sarmatae and the Germans.'

The picture is starting to get clearer. Classical authors of the 1st century used 'Sarmatians' to refer to, among other peoples, the developing West Slavs, eastern neighbours to the Germans. The area was settled by both West Slavs and Balts, and largely overlaps with Lusatian culture, to use an archaeological term. Therefore, the 'island' (in quote marks because, as I've said, all coastal lands tended to metamorphose into islands during that period) of Baltia should be a three-day sea journey from the southern coast of the Baltic Sea.

To dispel any doubt, here is Pliny's full description:

> at a distance of three days' sail from the shores of Scythia, there is an island of immense size called Baltia, which by Pytheas is called Basilia. [...] scattered in the German Sea, are [islands] known as the Glaesariae, but which the Greeks [!] have more recently called the Electrides, from the circumstance of their producing *electrum* [*ēlektrōn*] or amber. The most remote of all that we find mentioned is Thule [...]. There are writers also who make mention of some other islands, Scandiae, Dumna, Bergos, and Berrice, the largest of all, from which the crossing to Thule starts. At one day's sail from Thule is the frozen ocean, which by some is called the Cronian Sea.

The various names given to the 'island of immense size' include Basilia, Balisia, Balcia, Baltia, Abaltia, and Abalcia. A toponym? An ethnonym that later, in neighbours' colloquial speech, expanded to cover the Balts' location and coast? It isn't of critical importance. Let us first and foremost recognise that Pytheas is the godfather of the Baltic Sea. Contrary to widespread and deep-rooted belief in geographic history, he reached the Baltic Sea and, in one form or another, recorded a name that should nowadays be over 2,300 years old. Based only on known information, the Baltic toponym wasn't adopted for the sea until Adam of Bremen wrote

SILVERWHITE

his chronicle (1073–6 CE): *Mare* or *Sinus Balticum*. Since then, Pytheas's *Basilia, Balisia > Balcia, Baltia > Baltic Sea* has come into international use.

Consequently, we may, within the bounds of reason, trust the Greek's other toponyms and position all other 'islands' that are directly or indirectly associated with amber on the Baltic Sea.

MENTONOMON AND THULE

With clever analysis, researchers have determined that at least eight classical geographers read Pytheas's original work and acquired their knowledge of the Baltic Sea and North Europe directly from him. All others used retellings of the text. A number of the place names Pytheas recorded appear only in his work and nowhere else. Others disappeared for long periods before resurfacing centuries later. This points to the credibility of Pytheas's information, but also to the difficulties in their interpretation. Pytheas alone mentions Berrice, from which persons embark for Thule; the 1,000-kilometre-long bay of Mentonomon; the amber-rich island of Abalus; Scythian Bannonianna and a tribe named the Guttones. Alternative forms include Berrice, Berricen, Uergon, and Uerigon; Baunonia, Raunonia, Raunoniam, and Raunomiaa (with the long *a*). To this day, the language of their origin is unclear.

Pytheas departed the tin-shore of Cornwall and portaged across Jutland to open water and the amber coasts of the Baltic Sea (Bruns and Weczerka 1962: maps 1 and 2; Hennig 1961: 1, 199; Dreijer 1960: 73). Was it the sea? A lake? The water was salty. We do not know why or from what original language he named the Baltic Sea 'Mentonomon'. Its five different spellings are preserved in Pytheas's texts alone. He saw the Baltic as a bay of the ocean with a length of over 1,000 kilometres, which was estimated closer to the end of the journey than the beginning. The Guttones tribe, which Pytheas places along the Baltic Sea, is identical to later authors' Suiones (Swedes) and Ptolemy's Sulonians. Their settlements are just south of the Laps, meaning the southern coast of the Gulf of Finland, which in the following millennium we come to know as Revala, Viru, and Votia. From this, we can infer that in Pytheas's interpretation, the Baltic Sea continued not into the Gulf of Bothnia, but the Gulf of Finland. As unbelievable as it may seem, the Gulf of Bothnia was yet unknown by the beginning of the second millennium CE and did not appear on older maps. The distance from the Skåne peninsula to the mouth of the Neva in

SHIPS TAKE FLIGHT

the Gulf of Finland is 1,100 kilometres, which noticeably corresponds to the 6,000 stadia (1,071 kilometres) length of Mentonomon.

The decisive factor in determining the location and distance of the Baltic Sea in relation to Cornwall is the southeast corner of Pytheas's constructed 'Britain', which rests on the Skåne coast. At first, if you remember, Pytheas travelled on a northeast bearing 'along Europe' from Cornwall, with a general direction of about 58 degrees north. After reaching the Scandia/Skåne coastline, he turned north and continued in a direction of approximately 9 degrees north. With these angles, Pytheas placed the coasts to the port of the vessel within the bounds of the hypothetical British triangle. The new course taken from Skåne ran along the eastern coast of Scandinavia.

As you may recall, *helm* was the ancient Estonian word for amber; one we adopted from the Kunda people who were here before us. The ancient Baltic ship is more or less familiar, and we can imagine our Greek on board one. Denmark, more precisely the Jutland peninsula, is the lock to the Baltic's back door. Circumventing it would have added a full 900 kilometres to the journey, and stormy Skagen at its tip would furthermore have been too treacherous for a vessel of that type. Another route was used to access the Baltic from the North Sea in the Bronze Age: horizontally across the base of the peninsula by traversing the Eider, Treene, Rheide, and Sly Firth (Schlei, Slien), the last of which is a narrow inlet that flows into the Baltic. There, Scandinavians later founded the town of Hedeby and the West Slavs their Sliaswich: two prominent harbour towns that merged to become future Schleswig and its castle. The draught of an ancient vessel was about a metre or less, and even in a dry year it would not have been pulled more than four or five kilometres across a drainage divide.

It was on this part of the journey, following the Jutland portage, that Pytheas recorded his first toponym: Scandia. With Mela's corroboration, we can, without risk of contradiction, now recognise it as Scandinavia (Skåne, to be precise).

Next up in the text is Dumna, or the common present-day Damme: an ancient toponym that was later adopted by West Slavs on the coastal areas of what is now Germany and Poland. Just south of the island of Rügen is the small town of Demmin, which was populated by the West Slavic Lutici tribes in the early 1st century CE. Historically, it was also called Dimine, Dymine, and Dyminium. Let us simply stick to the conclusion that Pytheas's Dumna was on the West Slav-settled southern coast of the Baltic Sea. This also confirms our inference that Pytheas reached the Baltic

SILVERWHITE

Sea via the Eider–Schlei riverway and continued along the southern Baltic coast to somewhere near the island of Rügen.

However, his journey onward to the lands of 'white nights' traced the Swedish coast.

If Scandia and Dumna lie within the Baltic Sea, then we can deduce that Abalus and Basilia are in the same region. They may be, though need not be, classified as islands for the reasons mentioned earlier. 'Timaeus [...] has given to the island the name of Basilia' (Pliny the Elder): this allows us to conclude that Pytheas described two separate islands, which Timaeus later mistook for one. Additionally, based on the vernacular and similarity of style, we can deduce that Pytheas's Abalus is Pliny's Raunonia. Let us now review the distances. Abalus/Baunonia/Raunonia is a day's sail from the Baltic Sea/Mentonomon. At a three-day sail from the southern shore of the Baltic Sea lies Basilia/Balisia/Balcia. If the Baltic Sea is a 'bay', then how are we to understand a one-day sail? The question is left unanswered. The information we have does not allow us to precisely locate Pytheas's amber islands and others, with the exception of Thule. However, an important conclusion is hidden in the little we have. After recording the Dumna toponym, Pytheas moves to the south and southeast of the Baltic Sea. Amber was Pytheas's second focus. He exhausted the topic but continued on his journey and steered northward.

What drew him to the north? We'll follow his motivation at a later point.

Scandia and Dumna are followed by Bergos. Again, we must look to later analogies.

Ancient Scandinavian legal interactions were governed by a collection of customs known as the Bjarkey (Birka) laws. It influenced the older city laws of Visby and Riga and, even before that, defined the rules of coastal Estonians. Scandinavian historians believe that the Bjarkey laws developed around Sweden's Mälaren chain of lakes, an ancient trading centre, and is a proto-Scandinavian word (Ebel 1963: 155). Björkö (Birka), an island on Mälaren, means 'Birch Island'. Around the year 1060, Björkö became the residence of Hiltinus, the first bishop to the Estonians. Björkö is first mentioned in the year 875, and even then it was a centre for missionary work. This also speaks to it being an ancient place of pre-Christian worship, where the first missionaries tended to set up camp. The interdependence of ancient holy sites and market peace has been widely documented. Several other factors point to the long history of the name. Bjarkey laws spread to Norway by the year 997 and an impressive

SHIPS TAKE FLIGHT

22 Bjarkey-adhering trade islands and harbours have been counted in the Gulf of Bothnia, the Gulf of Finland, and the Väinameri Sea (Niitemaa 1963: 201–2). This was dynamic and long-lasting dissemination. An old Mälaren toponym managed to transform into a legal concept that evolved into a new toponym in more than twenty other locations in turn. We should add that the trading island of Helgö was also found in Mälaren, and the connection between it and Björkö is, though yet unproven, beyond doubt. The oldest archaeological finds on Helgö date to the 1st century CE. Thus, the temporal gap between Pytheas's voyage and Björkö / Bjarkey as a harbour protected by market peace begins to close remarkably.

Pytheas's destination could only be a place of market peace, where foreigners were guaranteed their personal safety. The oldest written forms of Bjarkey law are *Bjärköjarett*, *Byrkös rätt* (Birch Island law), *Bjärkö*, and *Birca*. Could *Bergos* be an even older form from 325 BCE? We cannot say.

Nevertheless, Bergos (Björkö?) is not yet Thule. Bergos is followed by Berrice. So, let us continue the voyage. The sources say that Thule was only accessible from Berrice by ship. And did you notice that they don't claim Thule was large? Balcia, 'which by Pytheas is called Basilia', is immense. Berrice is large. But Thule?

Onward, the way is quite familiar from later analogy. Pytheas hugs the Åland Islands to cross the Gulf of Bothnia, navigates the coast of southwest Finland to Porkkala Peninsula, and from there steers directly south to Naissaar, no matter whether his ultimate destination was Viipuri (Vyborg), Narva, or the mouth of the Neva. There was no other route. Finland's rocky southern shore was unnavigable by ship.

'Berrice, the largest of all, from which the crossing to Thule starts.'

At that time, Estonia hadn't any counties or parishes yet, and no difference was even drawn between Estonians and Finns. We were a Baltic-Finnic people nestled among deep forests and tiny fields. We tended swine and, for some reason, herds of horses roamed the land around the fortified settlement of Iru, not far from present-day Tallinn. People lived their lives as *suurpered*, large families.

Suurpere is a modern term. Back then, we simply said *pere*, family. It was the sole social organisation of the time; a mark of one's heritage; the degree, size, and shape of our laws, obligations, and social relationships.

Pere is a word that extends back to the proto-Finno-Ugric language. For some obscure reason, it is also associated with the word *pära*: rear, remain, distant past, the beginning of all things (Collinder). If you draw a straight line from Narva in the northeast to Pärnu in the southwest, then

SILVERWHITE

to the west of it, including the islands, the word *pere* appears often in place names. However, the word *pere* sounded a little differently in Pytheas's day. Then it was pronounced more like *pereh* (SKES 3, 523), the plural of which would be *perehden*.

Linguists give final clarity to the question. I myself lack the competency to objectively evaluate the sentence that takes its form for the very first time here on this page: '[The land of] Perehden, the largest of all, from which the crossing to Thule starts.'

II

BENEATH TWO SUNS

THE FIERY ISLAND

We must seek our motivation.

In the first edition of this work, I presented my belief that the impetus for Pytheas's travels was the Kaali impact event on the island of Saaremaa. True, it did take place two hundred years or more before Pytheas's expedition. But even then, Saaremaa was well populated, and the busy Baltic seafaring of the Nordic Bronze Age was well under way. I postulated that the force of the meteorite's impact must have had a profound effect on the psyche, languages, and customs of peoples living along the Baltic Sea, and shaped their local interpretation of natural mythology. To underpin my argument, I presented examples taken from language, folklore, and beliefs. Unfortunately, examples lack the weight of an argument. In the best case, they are able to add to the hypothesis's right to life, but a hypothesis is, by definition, merely a supposition that becomes credible fact only after meticulous substantiation. At least that's how it is in the pure sciences. History is different. History's method is general sociology, the outcome of which is a statement along the lines of: 'this is how it might have been, and most likely was'. Collecting, analysing, and interpreting a mass of sources sufficient for achieving likelihood is, for the most part, beyond a writer's abilities. Literature offers not history but visions and, in the best case, historical possibilities. Recognising this might be bitter if it were not sobering. However, the exceptional appeal of the Kaali phenomenon lies, for me, in the interweaving of the pure and the human sciences, which allows the former's method to be fruitfully applied in the latter and vice versa. In the second part of this book, I present proof that Pytheas visited Saaremaa, as well as the premise that the reason behind his visit was a mythological and folkloric ripple caused by the Kaali impact

SILVERWHITE

event. The proof is mathematical. The premise proceeds from my earlier statement, which can also be expressed as: assuming that people of the time equated the meteorite's collision to the sun collapsing onto the earth, then Pytheas's reason for visiting Saaremaa could have been, and likely was, the Kaali phenomenon.

* * *

And what was that mad Greek seeking on 'the families' shore'? The Island of Fire, naturally.

To test my hypothesis, I submitted three questions to Ago Aaloe (1927–80), a senior researcher of the Kaali meteorite craters at the Estonian Institute of Geology:

1. Could you please describe the possible visual appearance of the Kaali meteorite in the sky and assess the impact of its overhead trajectory on the human psyche?
2. Can you assess the power of the explosion, the range of its effect, and its psychological effect?
3. To what degree of precision has the date of the event been determined?

Ago Aaloe

The Kaali giant meteorite entered the denser layers of the earth's atmosphere within the Jõgeva–Tartu–Kallaste triangle and ignited at an altitude of 125 kilometres. The ball of flame, scientifically known as a bolide, was brighter than the sun. It caused a ballistic wave similar to that of a supersonic flight but with a tremendous explosive force, as it plunged at an angle of just 35 degrees to the horizon. We estimate its radius of visibility to have been 700 kilometres. The sound of the ballistic wave could also be heard within a radius of 700 kilometres, and the radius of its psychological effect could have been approximately 500 kilometres, even further in mountainous areas.

The light and explosion were perceptible in the Valdai Hills, on the western shore of Lake Ladoga, in south and southwest Finland, along most of the Swedish coast including the Mälaren lakes and the mountainous interior, to the south including Skåne, within the region of Gdansk, and naturally on all islands in the Baltic Sea.

Owing to the bolide's trajectory, it would have appeared to descend vertically to anyone witnessing the event to the east or west of Saaremaa.

BENEATH TWO SUNS

Author
I'd like to interject with the following comment. Any observer to the east of Saaremaa would have seen the sun-like bolide setting in the west as the sun normally does. To anyone located on the shore of Scandinavia, however, it would have appeared to be an unnatural event of the sun-like bolide setting in the east.

Ago Aaloe
In fact, the bolide flew quite low over the entire territory of Estonia. The shock wave, in which air pressure rose by half a standard atmosphere, caused widespread devastation. Archaeologist Vello Lõugas [1937–98] hypothesises that the burning and temporary abandonment of the Asva fortress is tied to the Kaali meteorite event.

Vello Lõugas
I have no qualms with you citing that. The Asva fortress burned to the ground. It lay in the meteorite's trajectory, 15 kilometres from the epicentre. There, the ballistic shock wave was at the height of its physical and psychological power.

Ago Aaloe
At the moment the giant meteorite struck Saaremaa's soil, its cosmic kinetic energy transformed into heat energy with an explosion and appearance comparable to that of an atom bomb – without the radioactivity, of course. A split second later, eight smaller meteorites struck near the giant meteor, each with the explosive force of a huge aerial bomb. There are altogether nine craters.

Ustus Agur [1929–97], director of the Estonian Information Institute
The energy of the Hiroshima atomic bomb, if we recalculate its trinitrotoluene equivalent into ergs, was 2×10^{21}. The explosion of the first Kaali meteorite was equal to that of a tactical atomic weapon.

Ago Aaloe
As for the explosion, it could be heard throughout the entirety of continental Estonia and in eastern Scandinavia. The roar was accompanied by a minor earthquake.

In regard to your third question, the age of the Kaali meteorite craters was estimated to be 4,000 years old in the early 20th century. Later, its

SILVERWHITE

age was studied again using radiochemistry. The result was surprising. According to the new method, the event took place just 568 ± 130 BCE.

An error was suspected – a carelessly collected carbon sample or an incidental factor. The test was repeated a year later, taking the carbon sample from crater no. 2. During the Bronze Age, the area around Kaali was covered in an oak forest that incinerated the moment the event occurred. This sample was taken from one such oak and was extracted with extreme caution, ruling out any contamination from later material. The result overlapped the first, dating to 698 ± 200 BCE.

On average, three giant meteorites strike our planet every millennium. The giant Kaali meteorite's mass has been estimated at 450 tonnes. An event of such magnitude is even more infrequent.

* * *

Researchers have published new data and modified older information since *Silverwhite* was first published. I will summarise these findings later. Here, I simply wish to note that five kilometres north of Asva, the Ridala fortress was destroyed at the same time. The shock wave of half an atmosphere under the meteorite's path meant several tonnes of pressure were applied to structures that subsequently exploded. In that era, both Asva and Ridala were important harbours on the coast. The Asva fortress was never rebuilt. Although the iron-nickel-alloy meteorite had an estimated mass of 450 tonnes, only a little over a kilogram has been collected since. Iron falling to earth on Saaremaa on the cusp of the Bronze and Iron Ages must have been of unimaginable value to a person at that time. In Baltic-Finnic folklore, iron and ship nails were acquired from an island surrounded by mysterious dangers. Is this motif part of the Kaali phenomenon?

If the assumption is true that people settled along the coast of the Baltic Sea interpreted the impact event as the falling of the sun or the sky, then the Scandinavians people's assessment must also be determined. I hypothesised that from their perspective, the sun set in the east – an intolerably unnatural direction. Knowing the meteorite's trajectory, we can now sharpen those apocalyptic observations. For instance, to a person living in the area of Stockholm, the sun rose in the east and set in the east-southeast (ESE). In Karlskrona, it rose in the northeast and set in the north-northeast (NNE). On the northern coast of Gotland and even in the vicinity of continental Västervik, the sun rose in the northeast, as it always does in summer, but paused in the sky for a moment before sinking

BENEATH TWO SUNS

back from whence it came. Fifteen minutes later, the roar of the Baltic Big Bang rolled across the coast of Sweden.

MEMORY GOVERNS TIME

Time, memory, and knowledge are the keys to unlocking the mystery of Thule.

Estonians were party to an extraordinary catastrophe in Europe. Could we, as a people, remember it? And if we could, then why should we?

Exact scientists posed this question to themselves and came to a negative conclusion. The age of the Kaali meteorite craters was determined by Ago Aaloe, Arvi Liiva, and Evald Ilves. The summary of their findings contains the following statement: the Kaali craters 'cannot [...] be regarded as very young, because if they were [...], then the event should have been directly reflected in historical sources or folklore'.

We have learned to trust science in the last quarter of the 20th century. Still, I sense a mistake. Perhaps we do not understand one another or are speaking different languages.

Firstly, we must do away with the dizzying abyss before us. What do alpinists do? They rest and let their gaze drift cautiously across the landscape to the neighbouring abyss, which is even deeper. Our abyss is memory, which is a connection to the past; which is the dictation of time. The oldest attempts at creating calendars reach back to 35,000 years ago (Marshak 1972: 446–61). A time span of that length is no longer poetically perceptible. We could express it with the sterile formula 1.07×10^{12} sec. It isn't even an empty breeze, but a blank sheet of paper that is covered in invisible ciphers and can be used to write a sonnet about ecstasy or a folk haymaking holiday. It is likewise reality, has been reality, and yet both 1.07×10^{12} sec and sonnets are part of cultural history, influencing and engendering each other.

Human beings, creatures capable of learning, move from a known past into an unknown future. We can only take from the past the part that is known to us, abridgeable, and that has been given an explanation or a name, made transparent and personal. Thus, the past, which is to say historical experience, lives on within us: humans need history, and humans themselves are most likely history as well. The more unimpaired a person's tie to the past, the more capable they are of learning. These days, a connection to the past materialises through books, city blueprints, written legislation, and ploughed fields. It happened differently in the ages

53

of unwritten literature, though an active bond to what came before has always been necessary for a strong present and a vigorous future. Baltic-Finnic habitation of the shores of modern-day Estonia was outstandingly stable in comparison with the rest of Europe. The world's northernmost prehistoric grain farming occurred here. Both speak to the population's superb ability to adapt and learn – in other words, to intense feedback from the past. It is unlikely that such an extraordinary event as Kaali would be forgotten.

The destruction of the Asva and Ridala fortresses and the decimation of the local population took place at a time when Saaremaa was one of the most settled locations in present-day Estonia. As with any encouraging or cautionary experience, the event must have densified into a message that could easily be preserved and passed down to one's children. We could compare such information to a news bulletin and the length of time to the distance of the radio receiver. When the message is dispatched from far away, then disturbances are inevitable, and if the distance exceeds a critical limit, then the information may deform beyond recognition or never reach us at all. However, radio stations have various ranges and the information transmitted is of varying importance. Saying 'it snowed in winter' is certainly a message, but it lacks the quality of significant information. Yet a statement such as 'the apple trees blossomed in winter' would cause us to prick up our ears, as such a phenomenon would be unusual in Estonia's climate. The opposite would naturally apply in the French Riviera. The more plausible the report, the less information it conveys. And vice versa: the less plausible the report, the more information it contains. Utilising metaphor, one could say that an implausible message has a greater mass of information, as a result of which it penetrates deeper through time or space. An implausible message has been dispatched from a more powerful radio station. If, however, a message is entirely exceptional, then it may cross the distant past to reach us today. Upon receiving it, we must therefore account for the deforming effect of disturbances and decode it according to certain rules. Call it 'interpretative mythology'.

* * *

In a place named Saidmarreh, events of one form or another occurred. I'd now like to compare them to those that took place in Kaali. Saidmarreh, located in the Tigris–Euphrates river system, is a former British Petroleum oil concession in a desolate, rocky desert landscape.

BENEATH TWO SUNS

The Kaali impact event is Europe's largest and only; the Saidmarreh landslide is unique and the greatest to have happened globally. The Kaali catastrophe took place in a well-populated area during the Bronze Age; the Saidmarreh catastrophe took place in an area that was likewise well-populated, 11,400 years ago during the Stone Age. The effects of the Kaali catastrophe could be observed within a radius of about 1.5 million square kilometres; the power of the Saidmarreh catastrophe, both physical and psychological, was even greater: a natural gas explosion set off by an earthquake detonated 12 cubic kilometres of earth, burying the Saidmarreh and Kashgan riverbeds and causing massive flooding. Twelve cubic kilometres is merely a figure. For comparison, the 27 August 1883 Krakatoa eruption between Sumatra and Java threw 18 cubic kilometres of earth into the atmosphere. The plume of smoke could be seen within a radius of 1.5 million square kilometres, the explosion could be heard even further, and the cloud of fine ash carried by air currents eclipsed the sun in Europe. Yet unlike Krakatoa, the Saidmarreh event lasted for quite a long time. Towering geysers of petroleum spouted and caught fire while burning rivers of the substance rushed towards the sea. Based on the traces of incineration, modern petroleum geologists are able to determine the power of the eruptions, the duration of the burning, and the shape and colour of the pillars of flame. British academic James V. Kinnier Wilson compared Middle Eastern myths and the region's most ancient written sources to British Petroleum's exploration maps for incredibly practical reasons: 'our goal was to transform mythology into history' – to filter an 'exceptional historical event' from ancient folklore and reconstruct it. He came to the following conclusion: 'As soon as we learned to interpret the accounts, it was downright astonishing just how much material could be found to support and expand our theory' (Kinnier Wilson 1979: 25).

A human can only successfully function in an environment where they understand, or at least believe themselves to understand, the causal connections. The singular events of Saidmarreh placed incredible difficulties before humans of that era. They lacked any previous encounter with natural gas or petroleum and lacked the terms themselves, but those phenomena were burning in 300-metre-high pillars and blocking the horizon with a wall of flame. The lives and continued existence of those ancient humans depended on whether, and to what extent, they were capable of explaining the catastrophe's outcomes, dicing and reducing them to simple, safe bits that one's mind could wrap around, and thus passing the symbolised information

SILVERWHITE

down to future generations as a cautionary experience. Roaring towers of burning natural gas resembled lions' manes ruffled in the wind or roaring lion-headed monsters. The pre-catastrophe toponym of the Saidmarreh region was mythologised and became a symbol for peace, fertility, and prosperity in later tradition. Natural gas and petroleum do not figure in folklore per se, as one unknown cannot be described in terms of another unknown. In mythological interpretation, petroleum became 'the venom or outrage of a giant snake, and one gets the impression that this rather primitive understanding has retained its presumably Mesolithic origins with rather few changes to our present day,' as Kinnier Wilson theorises. Myths and writings of the Sumerians and Akkadians speak of the mountain Ebiḫ, a toponym meaning 'destroyed'. Petroleum geology can now confirm that the mountain was located in a place named Jebel Hamrin before the massive explosion. Ebiḫ has truly disappeared from the earth, has been decimated, 'destroyed' in the eyes of ancient humans. Here, the juxtaposition of 'decimated' and 'destroyed' is telling. Our scientific worldview shapes the question: what happened? Ancient humans asked: who did it? and presumed a causative answer. The aftermath and consequences of the catastrophe have been described in great detail, and even breccia from the explosion is recorded in myths. Against the backdrop of this myriad of details, it is striking that the event as a whole or, as Kinnier Wilson says, 'the enemy himself' is depicted much more ambiguously: 'Apparently, ancient humans had a much spottier understanding of it: they heaped names upon it without adding definitions.'

The manifestations are described by myths about primordial darkness in which 'man could not see his brother'. The 12 cubic kilometres of surface material launched into the atmosphere drifted as a fine dust and blocked out the sun for long period of time. Equally as true is the story of the Great Flood in the *Epic of Gilgamesh*. Leaning upon the latter, we could schematically follow the orderliness with which information spread. The myth of a great flood is known across the world through the dissemination of the Bible. Yet in the early 20th century, the social anthropologist Sir James George Frazer pointed out that the myth spread to North and South America independently of biblical tradition. The professor of folklore at the University of Tartu Walter Anderson made an important addition to Frazer by showing that the myth of a great deluge also reached north Siberia without biblical influence.

BENEATH TWO SUNS

Based on the geological analyses of Kinnier Wilson, we can now draw another conclusion. The Mansi, Khanty, and Samoyed peoples of north Eurasia, and the Tsimshian and Washoe indigenous peoples of the North American Pacific coast, portray the great flood as burning water or tie it to the burning of the entire world. We can imagine the spread of information as concentric rings made by a stone tossed into a pond. An ideal throw would cause them to spread in every direction at a steady speed. In reality, the speed depends on natural belts that may either facilitate or slow, amplify or suppress, the dissemination of information. In physics, this phenomenon is known as interference. The spread of the flood myth along the Arctic Sea, across the Bering Strait, and southward down the Pacific coast directs us to a narrow belt we may call a communication channel. If the information is extraordinary, the stone cast into the pond large enough, then the ripples will reach the shore. They rebound more intensely off closer banks, the reflected wave picking up local influences along the way, blending with the spreading wave, distorting and arriving back at its source, where its revised form is no longer identical to the original: the information no longer expresses contact with a natural event, but with creative psychology. There probably exists a certain limit, after which mythology remains myth, Kinnier Wilsom believes. The ring arrives at the furthest corners of the pond in its purest form. Thus, correspondences between some folkloric styles of, for example, the Sámi of the Kola Peninsula and the Kawésqar of Tierra del Fuego become understandable, having a statistical probability that cannot be written off as chance. The depiction of the flood as a sea of fire in the myths of northern Siberian people is very likely the most precise peripheral account of the Saidmarreh catastrophe. Folkloric style preserved the unity of fire and liquid, which elsewhere has been dismissed by sober experience and lost.

The energy released in Kaali was at least one unit less than that in Saidmarreh. The extent of the information's dissemination is proportionally less as well. Temporally, the Kaali event is four times closer to us. As the arc of forgetting has a parabolic shape, both events have been recorded with the same intensity in social memory and are comparable.

Kaali, now a circular pond with steep crater-like banks, caught the attention of natural scientists as early as the early 19th century. An extinct volcano was originally presumed. Folklorists did not overlook the fact that the lake carries multiple names: Kaali and Pühajärv ('sacred lake'). Hydronyms are the most unassuming of place names. Lake Peipus and Pärnu River found their way into Estonian from the Kunda language. *Püha*

57

was once used for something forbidden, untouchable; something akin to the ethnographic term 'taboo', borrowed from the Tongan *tabu*. On the Saaremaa page of his atlas (1798), Count Ludvig August Mellin noted that the lake was the gravesite of the mythical Estonian hero Big Tõll. Pastor Martin Köber romantically described the sacred site in the same way in the 19th century, imagining sacrifices being brought down to the water's edge.

An untouchable lake, the grave of a hero, and an ancient sacrificial site. The fourth characteristic of Lake Kaali/Püha is even more substantial: the name of ground zero expanded to cover the entire locality, which is unique among Estonia's nearly 110 parishes. Unique inasmuch as the source is not the name of a feudal lord or a Catholic patron saint, but an assessment taken from indigenous beliefs. Unable to subvert it, Christian missionaries were forced to adopt it. It's worth mentioning that several variations of Püha toponyms can be found in the parish: Püha and Pühattu Forest, Pih, Pyha (1818), the Pühendi and Pühattu farmsteads, and the village of Lasnama, which, according to the Institute of the Estonian Language's catalogue of place names, 'locals unofficially call Püha Village'.

As one of the lake's two names derives from folk beliefs, our curiosity also turns to the other. The oldest surviving forms of Kaali are Kali and Kall, which lead us in turn to the earliest days of seafaring. According to research by the respected academic Paul Ariste, the Estonian word *kalev* (an epic hero) developed similarly to the word *tugev* ('strong'), its basis being the specific concept of *tugi* ('support'), which became the more general and abstract *tugev*. So it was with *kalev* and the specific hero Kalev: the former, derived from *kali*, came to stand for a strong, great, and powerful individual. The various shades of meaning have been documented in writing since 1641. But what did *kali* mean? A *kali* is a lever, though an unusual and important one from the point of our travelogue: logs used for turning boat rollers. The ancient Baltic vessel was universal, travelling both on rivers and the sea and portaging on wooden drags (*lohisti*) that were rolled along with the help of levers. The *kali* was stiff and unbreakable, made from a hard wood, and was even employed as a weapon for striking. Just as the anchor today, the *kali* was a symbol for the ancient seafarers' profession. Given that the crater lake developed the name Kali in addition to Püha, it must have possessed special significance and place in the minds of ancient seafarers.

Geologist Ivan Reinwald started working on Saaremaa in the 1920s, seeking rock suitable for producing plaster. He was soon captivated by the Kaali craters and they became his life's work. In 1938, independent laboratory analyses performed in London (Spencer 1938) and Tartu (Andres Väärismaa) proved that bits of iron found from the main crater came from a meteorite. It was a noteworthy discovery that failed to gain a wide audience owing to the beginning of the Second World War one year later. Reinwald erred when dating the age of the craters, placing them in the very distant past. Even at that age, however, they would be Europe's only meteorite craters to have formed during the Anthropocene. There was no human settlement on the coast of the Baltic Sea in the era at which Reinwald dated the discovery, and therefore it did not interest anthropologists.

It seems as if tiny mistakes and trivial miscalculations helped to heap a fatal bank of semi-truths around the Kaali event. Fate has also kept hold of the baton.

In 1935, three years before Reinwald's laboratory results, Estonian linguist Andrus Saareste (1892–1964) published a twenty-page pamphlet. In it, he listed more than fifty Estonian-language words that represent, in one way or another, a fire soaring across the sky, and came to the following conclusion: 'This mythological image appears to have stemmed from a natural phenomenon; specifically, a meteor.' The terms are natively common even in areas with sparse linguistic contact. 'These islands are the remnants of a greater *tulihänd*[1] continent of their time.' Where the Kaali meteor entered Earth's atmosphere above eastern Estonia, the word *pisuhänd* (*pisu* meaning 'spark' in the Tartu dialect) is most prevalent. In western Estonia, where the meteorite exploded and burned brighter than the sun, the word *tulihänd* (*tuli* meaning 'fire') is most common.

So far, we have recreated the Kaali event with the finest scientific tools, photographed it, measured it, and recalculated it into units of energy, but we still do not know its cultural and mythological consequences. We cannot say the event *had* any consequences. Yet, to echo the words of Ago Aaloe, Arvi Liiva, and Evald Ilves: is the Kaali event reflected in historical sources or folklore?

This question in no way contradicts the idea of exact mathematical sciences. Scientists themselves do not fetishise their conclusions as the final degree of objective truth. The opportunity for subjective evaluation is never totally ruled out. Yes, the Kaali event is reflected in both historical sources and folklore, though exceedingly more so in the latter.

First, let us allow the witnesses to walk past the judge's bench. An impressive number of them have gathered, and even more will arrive

SILVERWHITE

tomorrow. Our travelogue is somewhat like a classified ad: who saw it? Is anyone able to add anything? Yet, we also understand that those who were in the closest proximity perished, rain and the years washed away many traces, and only a distant whisper and glow found their way to the 'witnesses' themselves. They are able to describe the consequences, not the cause. We know the latter, but not the former. Let us remain pedantic: we even have no idea whether the event produced any consequences.

So, let us hear from the witnesses. There are ten.

To this, popular opinion has added, that the tumult also of [the sun's] emerging from the sea [variation: falling into the sea] is heard, that forms divine are then seen, as likewise the rays about his head. (Tacitus, 98 CE)

[Our gods] are black of visage, winged and tailed, and they mount up under heaven [...]. (*The Chronicle of Novgorod*, 1071)

There are in this sea many other islands, of which a large one is called Estland [...]. Its people, too, are utterly ignorant of the God of the Christians. They adore dragons and birds and also sacrifice to them live men. (Adam of Bremen, 1073–6)

When Brother Theoderich was sent into Esthonia, he likewise endured from the pagans a great many dangers to his life. Because of an eclipse of the sun which took place on the day of John the Baptist, they said that he was eating the sun. (Henry of Livonia, 1186)

They baptized three villages within Wierland [Virumaa]. There was there a mountain and a most lovely forest in which, the natives say, the great god of the Oeselians, called Tharapita, was born, and from which he flew to Oesel [Saaremaa]. (Henry of Livonia, 1220)

At night, there often appear in that land flying, flaming serpents and other beasts of Satan. (Sebastian Münster, 1544)

And there appeared in the sky another star, and behold, a great bloody *pisuhänd* with seven heads and ten horns. (Joachim Rossihnius, 1632; Andreas and Adrian Verginius, 1680)

A gigantically large and fiery firestone a.k.a. *pisuhänd* burst above in the wind, and eight very large pieces fell. (Otto Wilhelm Masing, 1822)

Henry of Livonia was a missionary who followed on the heels of the crusaders during Estonia's long and bloody struggle for freedom, baptising the conquered. His chronicle is one of the most important sources of Estonian history. Henry arrived in Virumaa in northeast Estonia in the summer of 1220 CE, following a road that is still in existence today. The

BENEATH TWO SUNS

'mountain' from which the 'great god of the Oeselians' flew to Saaremaa lies in remarkable proximity to the area where the Kaali meteorite entered earth's atmosphere: the Ebavere esker (146 metres), which runs along the road connecting the villages of Väike-Maarja and Kutsi in the Pandivere Upland. The German traveller Karl Feyerabend traversed the same road five and a half centuries later and described Ebavere as follows:

> The highest elevation is on Thorapillade or Thorapita Hill a few miles to the west of Narva [a mile here equals 7.432 km]. To its peak, it is covered in dense primeval forest in which shadows of the past flicker, moaning in lament over their fatherland's oppression. In ancient times, a *hiis* was a sacred site to which Estonians came with prayers and sacrifices. Even now, Estonians stealthily creep to this holy place beneath the darkest arching canopies to appeal to the gods of their ancestors with gifts of thanks and appeasement.

Feyerabend goes on to describe Estonians' bravery in their visits to the sacred site, as pastors inflicted dreadful punishments for the worship of any god but their own. He writes that according to local legend, the forest on the crown of the hill was a sacred site already in ancient times. The people of Saaremaa were said to have once struggled with a powerful enemy. Just as they were on the verge of defeat, the gods came to their aid: Thor took human form at the top of that very hill and flew straight to the besieged island, driving the enemy back and giving Saaremaa's warriors the upper hand. 'From that time onward, Thor became a special guardian god of Saaremaa and the *hiis* where he first took human form was named Toorapita, which means Thor-Helper' (Feyerabend 1797).

The descriptions given by these witnesses from a range of eras possess varying degrees of precision and value, but are consistent in terms of what matters most: the Kaali event is reflected in both historical sources and folklore. In language, it still echoes today. Consequently, the human effect of the meteorite has been great and lasting. During Pytheas's day, a mere two or three centuries after the event, its impact on the human psyche must have been even greater.

Conclusion one: The Kaali event had consequences. One point of the witnesses' accounts intersects: fiery beasts move across the Estonians' sky. Estonians have a rather amiable relationship with them, though they are hostile towards the Christian witnesses.

Conclusion two: Peoples living along the Baltic Sea found their own natural-mythological interpretation of the Kaali event.

SILVERWHITE

Three giant meteorites strike Earth every millennium, but four-fifths of the Earth's surface is covered by water, which retains no trace of them. One giant meteorite strikes land every 5,000 years, and even less frequently in settled territory. Kaali was the last of these.

Conclusion three: The Kaali event was exceptional, which allows us to infer that it had an equally exceptional impact on humans and their psyche. Unfortunately, it also allows us to infer that there are no other analogies in human history.

Almost 1,700 years separate the *Chronicle of Novgorod*, the closest witness, from the event itself. Based on today's measurements, Pytheas and the meteorite are separated by a maximum of 350 years, a minimum of 100, making him five to fifteen times closer to the event. Oblivion's arc is hyperbolic. Echoes from the event were not five to ten times more vibrant, but dozens, maybe even hundreds.

Conclusion four: The exceptional event could be a direct cause of Pytheas's exceptional journey into the distant north.

But how did Estonians themselves manage the catastrophe?

SUN-TERROR

Henry of Livonia tells us (1220) of the Baltic Finns' great god Taarapita, who flew from east to west from Ebavere Hill to Saaremaa. The trajectory and destination coincide with the Kaali meteorite. It was flaming (1544), smouldering (1632, 1686), or fiery (1822). This is a significant quality. Even in the 18th century, not one serious scientist believed biblical stories of 'stones' falling from the sky, and even after the French Academy of Sciences grudgingly admitted their existence (1803), meteorites were still seen as cold 'moonstone'. We can state that the excerpts cited above are, as sources, sufficiently relevant for addressing the Kaali event, but they also do not thoroughly detail the phenomenon.

What did Estonians witness?

A long-tailed sun shot across the heavens with a deafening roar, flattening forests, igniting trees, destroying fortresses.

The blinding flame of the explosion made the shores of the Baltic Sea quake and could be seen on opposite shores.

Then came a deadly silence, perhaps an inky blackness, and the distant glow of burning forests.

The sun had fallen from the sky and was destroyed. We had no other way of describing it. Yet the next day, the sun was once again in the sky. This simultaneously fed our fears and hopes, forcing us to seek an explanation and ways to avoid danger. Our former worldview was in ruins, and from those ashes a new manner of thinking, a new language of symbols, needed to arise.

Leaning upon the four conclusions, let us attempt to reconstruct the collapse of the Baltic Finns' ancient cosmological system and the development of a new one. The stones we will use are language, traditions, folklore, and ethnography. Perhaps archaeology and literary texts as well? These materials cannot build a bridge, not even a plank into the past. They are merely a sporadic dotted line that be used, or cast aside, by a future engineer if he or she has acquired enough additional material. From the perspective of cultural history, both possibilities are equal.

Nature presented us with entirely new questions.

Based on the event, there existed a nameless force that was capable of destroying the sun. Thus, the sun wasn't a gift from nature that could be taken for granted? It could exist and could also not exist, depending on the will of the sun-devourer?

Words have great power. The sea listens, animals comprehend. Our first measure of caution was to avoid using its name so as not to summon it without cause. But who was 'it'? How could 'it' be recognised? We needed to know and remember it so as to recognise, avoid, or appease it when necessary. Language was our sole shield, and the mechanism of taboo words our strategy.

The sun had a long, flaming tail like a thousand bolts of lightning. With tremendous thunder, it pursued the sun, killing and destroying and ultimately acquiring its prey. Was this a bundle of lightning bolts with which the ruler of the world chased the sun? Or was it a completely unknown force, a snake shooting across the sky, a Long-Tail spraying sparks and fire?

Thirty-one words formed to substitute the new phenomenon, all ultimately boiling down to *pitkne* > *pikne* (in modern Estonian, 'lightning'). It's peculiar that the substitutes for a meteor, lightning, and a snake blended together. Yet, as Otto Wilhelm Masing demonstrates, other nations' *cold* moonstone still persisted as a *fiery* snake in the Estonian imagination, two and a half millennia after the event. Estonian memory has preserved other connections, also. According to folklore recorded on Saaremaa in 1872: 'The weapons with which Taara [Taarapita] pursues evil spirits are bolts of lightning [*pitkse*].' Their origin is a meteorite: 'It is believed that the bolts are "stony and smooth like a stake"' (Holzmayer 1872: 50). Let us remember this for later: Taara's weapon is compared to a stake, *vai*. It is a Bronze Age perimeter defence against lightning; self-defence against the terrible unknown. It remained invisible, but the fissures in Saaremaa's limestone surfaces, the scorched forests, and the fortresses levelled in a split second were visible and tactile. The sun as a bearer of life was no longer guaranteed. It could be destroyed. It was a novel experience, a completely different world in which we suddenly found ourselves, and that world could face an end. Such an apocalyptic line of reasoning was foreign to the balanced psyche of the shamanic era. Now, it shaped a new system of beliefs and a folk calendar with dual culminations. At the summer solstice, the sun was at the peak of its power and there was cause for joyful celebration. At the winter solstice, the sun was dying and Yuletide was thus a time of fear. On Saaremaa, people protected themselves by drawing sun wheels, specifically the rays of the sun, over their doors and windows, only adding a circle around the resurrecting star several days later when the sun began regaining its strength at a snail's pace. Light was not to be extinguished indoors during Yuletide. Straw was brought inside homes and the whole family stayed together to take turns guarding and protecting the flame.

From whom?

The sun-devourer gradually began to acquire a figurative form. Just in the century before last, it was believed on Saaremaa that slow worms were to be killed on sight: 'that animal is ravenous for the sun's rays and, because it does so very greedily, the world would otherwise grow dark over time' (Holzmayer 1872: 38).

So, did the innocent slow worm take the role of sun-devourer? Such a simple conclusion would be categorically false. Other nations' cold moonstone was, in our imagination, associated with a sun-gobbling snake. A long chain of taboo words shows us that over time 'the foe itself' took the form of a snake or a dragon. Like Adam of Bremen, Henry of Livonia confirmed the practice of sacrifice in the region, though it was unusual and tied to warfare. Whereas the exploding meteorite, 'the foe itself', was depicted as a dragon, describing the sun as fire is rather ordinary. Attempts were made to appease the foe with a sacrificial meal. Eating a sacrifice meant acquiring its strength, power, and fertility; identifying with the sacrificed animal. The party being appeased also takes part in the ritualistic meal. Fire symbolised the sun, a sacrificed animal the sun's foe. In what is now Estonia, a Yuletide barrow was sacrificed to the extinguishing sun to ensure its rebirth. Unlike the traditions of other peoples, Baltic Finns believed that swine were 'pure'. The woman of the

BENEATH TWO SUNS

household would secretly pick the animal in autumn, secretly fatten it, also secretly slaughter it, and bake the whole thing in the oven at the winter solstice. 'Then, it is set upon the table in the very same position and is left there for several days' (Holzmayer 1872: 55). According to Masing (1839): 'But in our places, our people has the tradition of baking a loaf of bread during the Christmas holiday that they call the "Christmas bread": on it is a key or sometimes a pig into which holes are poked with a bone.' According to Saareste, the Christmas barrow on Saaremaa was 'a Christmas bread into which holes were poked with a thimble. It was left on the table at Christmastime and was shaped like a pig.' In wealthier homes, an entire pig was ritually cooked. Poorer families without the means for a pig made a substitute sacrifice from dough instead, still calling it a barrow: 'In Sõrve, a two-pointed Christmas pig was cooked for Christmas Eve, which is to say an elongated loaf of bread with both ends turned upward. The loaf stood on the table during the holidays and was given to the animals on the morning of New Year's Day' (Holzmayer 1872: 56). Livestock were also able to take part in the Christmas pig's strength, power, and fertility. With one exception, of course: the Christmas pig was shared with all livestock except pigs.

Earlier, I spoke of Yuletide as being a time of fear. Families gathered around a flame kept alight day and night to symbolise the extinguishing sun, barely clinging to its strength so as to be reborn. Rituals surrounding the winter solstice – the extinguishing sun – are most directly tied to pigs. They are many and rich in significance. Swine have been important livestock for the peoples of Europe and many other places around the world, and yet it seems as if Baltic Finns' traditions, folklore, and languages emphasise its exceptionality in our ancient beliefs.

Try to solve this Estonian riddle: *the pig roots in another land, its bristles seen from here.*

Unless you happen to be a folklorist, the answer is extremely hard to fathom. Think a little longer.

In the meantime, here's another: *the pig breathes through everyone's bristles.*
Did you figure out the first? Patience!
What is the meaning of *a black pig in the oven*?

We cannot explain fantasy. All we can do is acknowledge that it exists, observe its manifestations, and use them to restore, shard by shard, the ancient understanding of the world in which the Kaali event played a central part, demanding mythological interpretation.

65

SILVERWHITE

In the first riddle, the pig represents the sun; in the second, the stones heaped atop a sauna stove; and in the third, an extinguished fire.

Riddles originally had a ritualistic or magical purpose. We attempt to explain them with a 20th-century gaze, but even that explanation is enigmatic. Metaphorically, a radio operator working in a hostile environment must communicate information that is unintelligible to the enemy and understandable to their own side. The code is known by those who should know it, carried within since childhood or, at the latest, initiation celebrations: the earliest Martinmas traditions; customs that required youth who reached adulthood to withdraw for meditation and perform tests of strength and masculinity. Only afterward were they entrusted with the family secrets. It was something of great cultural importance before the spread of literacy. Initiation celebrations ensured the continuity of experiences and traditions; a living connection to past generations or, to return to our comparison, the adoption of a code of understanding the world. This may seem absurd, because we now have a much better grasp of causal ties. The label of absurdity is given to the analyser, not the analysed; absurdity is a white flag of surrender in a writer's grip. The pig as a symbol for the sun, a stove, and fire is not absurd. It is obvious from the similarity of the solutions. The degree to which we can remember our former code and are capable of restoring the riddles' associative connections is another thing entirely. Still, let's try.

It is difficult for us to delve (or rise?) into the wondrous world of Bronze Age poetry. 'If poetry's possibility wasn't an indisputably established fact, then one could quite convincingly prove that it cannot exist,' the academic Juri Lotman once said. The sun can be destroyed. The idea seems fantastical even today, but it was a reality to our ancestors, an experience that needed to be recorded and passed down as information of paramount importance. To do so, the cosmic event and its cause needed to acquire sufficiently familiar contours. Life and death? Apparently, the pig and fire were the best fit for personifying two possibilities of such a contradiction. Swine are one of the oldest forms of livestock and were very likely present in what is now Estonia by 2000 BCE at the latest. A symbol is a sign, its sole task being to reference that sign's use. Now, let us add an important ethnic and temporal criterion: Latvians and Lithuanians lack several elements of Baltic Finns' swine worship. This allows us to draw a line between the dissemination of these rituals and the Balts, thus dating the former to the age before the latter even arrived at the shore of the Baltic Sea.

BENEATH TWO SUNS

The *tulihänd*, fire-tail, is an event; fire is the sun; and the Yuletide barrow is a sacrifice of appeasement and perhaps also a symbol of the sun-destroying fire-snake. How does this tie into the folk belief that slow worms eat rays of sunlight and could leave the world in darkness?

Firstly, let us recognise that the snake is common to the mythology of every nation. That of Estonia stretches back to the Finno-Permic era. The archaeologist Harri Moora described a sculpture found in Tõrvala, near Narva, which dates back to 5000 BCE (Moora 1957: 226). So far as we know, that is also the beginning of Estonian art history. The ancient artist carved a slithering snake out of a moose antler. A hole bored near its head allowed it to be carried atop a pole like a flag. Snake worship was first recorded in writing during the transition from the Stone Age to the Bronze Age: 'And Moses made a serpent of brass, and put it upon a pole, and it came to pass, that if a serpent had bitten any man, when he beheld the serpent of brass, he lived' (Numbers 21:9). A bas-relief on Trajan's Column, a ten-minute walk from the Roman Colosseum, depicts a barbarian of Central Europe. The man is holding a two-metre-long pole topped by the head of a wolf, the rest of its body waving. It was no longer a wolf and wasn't yet a flag. The transportable symbol could act as a totem. Baltic Finns' totem animals were most likely bears, moose, wolves, swine, snakes, stoats, weasels, dogs, or birds – the full list would cover the majority of animate nature. A totem animal is seen as the mythical ancestor of a family or tribe, the hunting of which is subject to strict limitations and launches a long chain of ritualistic acts. In a funeral ritual that lasts for days, the Mansi people sing and dance to ask for a slaughtered bear's forgiveness and convince it that the culprit was actually the member of another family. The Itelmens of Kamchatka refuse to kill wolves despite the harm they may cause to their reindeer herds. Totem animals were given supernatural powers to strengthen what was forbidden. In southeast Estonia, lizards were regarded as the most venomous reptiles, which were not to be touched, much less killed. Inasmuch as totem animals were conduits for ancestors' connection to the present, they could, depending on one's behaviour, unleash illness on a person or call the sickness away. Baltic Finns' relationship to snakes is conspicuously contradictory and does not seem to fit the global myths in which a snake represents evil, downfall, and death. Instead, Baltic-Finnic folklore reveals a friendly and tenderly caring attitude towards the reptiles.

Snake worship persisted in Estonia until the 20th century, and even in the 1950s it wasn't uncommon for a lucky 'house snake' to be brought

SILVERWHITE

indoors and fed milk in some areas. Perhaps Adam of Bremen's words 'they adore dragons' should be cleansed of religious prejudices and we should ask: what was adored, how, and why? Yet we shouldn't jump to the conclusion that Adam of Bremen meant Estonians' 'dragons' were necessarily snake-shaped or reptilian in nature. Dragons were a fruit of fantasy and usually patched together from several different animals – the more outlandish, the more frightening. They could fly, move on land, and, for the most part, swim. However, their outer appearance always reflected a local imagination wherever they were found; the opportunities for experimenting with local nature.

Is this so important? Yes.

Estonia's Stone Age Tõrvala sculpture could be called a worm, snake, dragon, *vexillum*, crucifix, or flag, depending on the given era and the observer's fantasy, ideology, and language. If we were to add a few more Estonian-language names to the list such as *tulesaba* ('fire-tail'), *pisuhänd* ('spark-tail'), and *valgemees* ('whiteman'), then we may grasp the reasons why a Stone Age custom has remained alive and well in Estonia to the beginning of the present day.

Dragons were not uncommon tribal totems. Thus, the symbol of a dragon may hand us one end of an important thread in the history of Baltic exploration.

To recap: the dragon, or *tulihänd*, can be interpreted as an event, fire as the sun, and the Yuletide barrow as a sacrifice of appeasement.

At the summer solstice, the sun reaches the pinnacle of its power, and there are no known swine-associated customs on the occasion. Perhaps we may thus draw a tentative line between the sun and the sun-devourer. Again, fire as the sun and the pig as the sun-devourer. This would make the foundations of the ritual quite transparent: the sun-devourer is appeased by a constant flame in one's hearth, the pig and the sun reconcile, and the pig is consumed as a traditional meal to acquire some of its strength and might. Both the former and the latter required a symbol; its own sign. The sun was depicted as spokes with a ring drawn around it. The force pursuing the sun was depicted as a pig, which over time was simplified to its head: a symbol of strength and destruction, but also of power and protection.

Let us also observe this from a religious perspective. 'Dragon' is an ecumenical term that evokes overly refined associations. In Catholic art, it appeared as a fantastical serpent with bat wings, claws, and the head of a Galapagos land iguana. According to Church ideology, the dragon

BENEATH TWO SUNS

embodied 'paganism and evil'. All these elements could have nothing in common with Estonians' flying serpent-dragon. Barbarian dragons were simpler.

The first Christian missionaries to reach what is now Estonia were forced to work with ancient beliefs and adapt Christian traditions to local customs. When Henry of Livonia arrived at sacred Ebavere Hill, a priest 'cut down the images and likenesses which had been made there of their gods. The natives wondered greatly that blood did not flow.' The missionary cut down a dragon, thereby identifying himself as a dragon-killer. In the pantheon of Catholic saints, dragon-killers include Margaret, George, and Michael. If the first step of missionary work was to replace the local dragon/*pisuhänd* tradition with that of a dragon-killer, then it would be reflected in the renaming of sacred sites, the erection of chapels in those places, and the development of new toponyms.

In Estonia, this speculation is valid. According to Estonian onomastician Edgar Rajandi, Margaret the dragon-killer has been adapted into Estonian as Mareta and Karuse. On the Feast of St Margaret (*maretapäev* or *karusepäev*), it was the custom on Saaremaa for people to gather around a 'sacred pillar' on Mustjala Cliff to sacrifice beer and vodka (as well as bulls and humans, as an earlier author alleges). In a way, the custom is also associated with fire: stoves were sealed on that day (Holzmayer 1872: 69; Kruse 1859: 7; Vilkuna 1963: 54–6). In southeast Finland, *mareta/karusepäev* was called *madontappajaiset*: snake-killer day. The Feast of St Margaret is celebrated on 20 July in the Catholic world and on 13 July in Estonia and southeast Finland. The exception is worth noting. According to M.J. Eisen, 'The day was marked in old calendars with a pig's head' (1922: 31). Naturally, the pagan sacrifices on the Feast of St Margaret could not have been done in the name of a Catholic dragon-killer. It was an appeasing sacrifice to something that was killed but was still in rather good health. Those taking part in the custom identified more with the killed than the killer.

SUN JOY

Beginning with the *Chronicle of Novgorod* (1071), Christian sources described in hostile tones fiery dragons that flew in the Estonians' sky. Our own relationship with the *tulisaba* or *pisuhänd* was trusting; even friendly. Yet, considering the Kaali event, the total destruction of the fortresses in Asva and Ridala, and a fear of the sun in general, the question arises: why and how could fire falling from the heavens transform into a symbol of wealth and power in the eyes of Saaremaa Estonians?

We find the answer in a taboo manner of thought. In Estonian, other words for the Kaali crater and the surrounding parish include the word 'holy' (*püha*), which is to say off-limits. Ebavere Hill, which Henry of Livonia said was the point of departure for the chief Estonian god's flight to Saaremaa, was still sacrosanct ground in the late 18th century. 'Even now, the Estonians would clearly be prepared to exact the most dreadful retribution if anyone were to resolve to obliterate the sacred site of their ancestors here,' wrote Feyerabend in 1797. However, the Kaali event was visible and audible across the entire Baltic Sea. If taboo notions still persisted so vibrantly in the 18th and 19th centuries, then it is extremely difficult for us to accurately assess their extent and influence in the Bronze Age. The catastrophe could have caused large-scale migrations and would explain curious folkloric ties between western Estonia and southwest Finland. In the eyes of neighbouring peoples, the glow of inviolability could have spilled over into the island of Saaremaa, where the *axis mundi* had collapsed. Saaremaa could have become the centre of Baltoscandic cosmogony and its inhabitants the guardians of the grim secret of the sun's demise and rebirth. Inviolability granted rewards and protection. In the mythical accounts of the event, fear is interwoven with hope, dread with trust, and fear with fearlessness and prosperity.

The conclusion of Henry of Livonia's chronicle appears to somewhat support these postulations. After a long and bloody conflict, Saaremaa, the last pagan outpost of northern Europe, fell to the joined forces of Emperor and Church. 'Thus did [Riga] now water [Saaremaa] in the middle of the sea,' the preacher proclaimed as he watched, from the Soontagana fortress, the Livonian Brothers of the Sword return.

'When this is finished, when it is done, when all the people are baptised, when Tharapita is thrown out, when Pharaoh is drowned, when the captives are freed, return with joy, O Rigans!' (Henry of Livonia 2004). In the original Latin: *Tharapitha eiecto, Pharaone submerso, captivis liberatis*. In this sentence, the Estonian god Taarapita is elevated to represent 'paganism' in its entirety. This was also noticed by Uku Masing (1939: 1–16), though he failed to ask why Saaremaa likewise rose to become the Olympus of the mythical worldview. Now, we likely know the answer. Despite time and space, the Big Bang that was the Kaali event found its way into written sources and its soft echo reverberates in epic national poems.

BENEATH TWO SUNS

Nevertheless, another question arises from Henry of Livonia's chronicle: why and whence the reference to the Pharaoh at the culmination of a long and punishing struggle? In baptismal liturgy, Egypt symbolises the heretical kingdom of evil and the Pharaoh its ruler, i.e. Satan. Metaphorically, *Pharaone submerso* would thus mean the victory of 'Christians' over the 'pagans'.

However, Henry of Livonia had already described the military achievement earlier. The enemy was crushed, the land totally conquered, and the militaristic theme complete. The baton was handed from the soldiers to the preachers: Estonians were being baptised, Taarapita 'thrown out', and 'the Pharaoh' drowned in the sea. We should recognise specific missionary tactics in the list: sprinkling the conquered with water while seizing and blessing ancient holy sites as sacral politics. In the preacher's eyes, this equated to casting out Taarapita and, ultimately, destroying totems, just as had been done on Ebavere Hill seven years earlier and on Rügen island one hundred years before. In short, we should interpret the Pharaoh as the destruction of idols associated with his name and their being thrown into the sea. In this way, individual elements in the list are put into a logical context which was missing from the earlier interpretation and which lead to an entirely new question: wasn't the Pharaoh a stand-in for the sun, even its symbol and living manifestation? Wasn't Taarapita the sun falling upon Saaremaa? Henry of Livonia himself wrote of the god flying to Saaremaa. Consequently: could the sentence also express the Kaali event, the sun falling onto an island, and a sun cult that, when the chronicle was written, was associated with the Egyptian sun god? One argument against this line of reasoning is that a 13th-century missionary might not be familiar with Egyptian history. Another argument in favour is that Henry of Livonia detailed the successful completion of missionary work elsewhere, but only in Saaremaa's case did he write that 'Pharaoh is drowned'. Here, we must also conclude with a question mark.

* * *

Henry of Livonia's phraseology, which has been regarded as formal stylistic metaphor, suddenly acquires a meaning just as substantive as the Aesti people's Isis for Tacitus as he held his stylus a thousand years earlier.

Let's imagine walking past us the bards who passed myths and true names down from generation to generation; the tanners who made parchment from calf hide (known in Baltic-Finnic as 'writing skin');

SILVERWHITE

the seafarers and swordsmen who conveyed to the earliest chroniclers their folklore and names in God knows what language and with what pronunciation; the chronicler himself with his goose quill, ink, and non-standardised handwriting that stumbled over both folds in the parchment and the unintelligible pronunciation of foreign names. Let's imagine the diligent procession of scribes who, in the silence of monastic chambers, contributed a personal error for each earlier mistake out of ignorance or, even more frequently, on purpose, for, as always, the mistake-maker sees a flawless improvement, or at least a more precise interpretation, in their own mistake, just as I do here. In short, let us transfer our precise Information Age formulas that erase folklore and budding literature from the world onto a hazy path. When we do so, a fascination will spark not over the mutation of *Tooru* and *Taara* into *Taranis, Turupit, Taarapita*, and *Tharapita*, but over that ancient knowledge reaching our hands in the first place.

The sun god Pharaoh was drowned in the sea? The priest's jubilation was premature.

Taara's mythical fire-dragon had ensured the Aesti people's untouchability for millennia and continued to snake through our leaden skies for many long centuries more.

During the period 1640–2, the Pärnu county court sentenced fourteen witches to death, the most noteworthy of which was 'Pig-Mart'. In 1682, before the same court, Jürgen Sacke admitted to communicating with a dragon, among other crimes. In 1723, terrified Tartu judges faced a peasant named Ado Wielo who was said to have learned 'witchcraft' in Sápmi, commanding a dragon to set foes' houses alight, though also employing its aid to heal people and stop bleeding. In the writings of Pastor Martin Büchner, who lived on the shore of Livonia in the mid-17th century, we find the following coded warning, which was decoded by the linguist Karlis Draviņš (1960: 89–90):

> If any witch who keeps a dragon
> happens to speak these words –
> Schmiz poerfilene – then his
> house will turn to ash above him
> at that very moment.

Yes, it all seems more ridiculous than meaningful at first glance.

* * *

BENEATH TWO SUNS

Diverging from the main road is a boulevard that leads to the gleaming two-storey Salli Manor, repurposed as a school in the early 20th century. Extending beyond the athletics field is the former manor park. Maples, oaks, lindens, elms, all a verdant mountain rising amidst the level fields of Kõljala. Some paths are maintained, others lost in tall grass or choked with hazelnut thickets. Gnarled old-growth trees hunch over a little lake at the base of a steep declivity, their crowns entangled, making the still ring of water seem even smaller. The thunks of a ball and children's squeals echo from PE classes under way. Late in summer, a drowsy silence settles over the park. Mosquitoes buzz in the lush, dim greenery, and at midday sunlight only filters down onto the banks.

Early spring is the best time to visit Kaali. Budding trees don't shroud the reflective surface of the lake and there is space for one's gaze to wander across the crater walls and heaps of dislodged limestone. The earth displays its bones and a deep wound. Not long after *Silverwhite* was first published, archaeologist Vello Lõugas and I took samples from the site. His shovel entered the soil on 8 May 1976. The morning sun was shining, the smell of humus pervaded the park, and the world was still and windless. As he laboured alone, I had time to doubt everything and harbour all hopes. The blade struck our first findings after fifteen minutes.

Stable habitation developed on the narrow crater mound not long after the natural catastrophe took place. The ancient settlement can be traced from the late Bronze Age to the days of Henry of Livonia. Archaeological finds at the site include shards of ceramics, amber, tools, a flax comb made from an animal's shoulder blade, and a cupola furnace for crafting bronze jewellery. Metal fragments have been preserved in the pond. The prehistoric settlement was located on the northeast edge of the crater, directly in the meteorite's path of descent.

The entire lake and crater are surrounded by an orderly ring: a piled embankment of large stones. Who would think to defend a lake without fish? Who would come up with the idea of settling on the shore of such a body of water? Who would be capable of building such a wall? By Lõugas's calculations, it would have taken 36,000 wagonfuls of boulders to build.

A defensive earthwork? To defend against whom? Were the islanders trying to separate the grave of the sun from the world of the living?

Lake Kaali/Pühajärv hasn't always been so still, brooding, and lifeless. A few years ago, an attempt made to pump the lake dry failed owing to intense groundwater activity. The water table was even higher twenty-five

SILVERWHITE

centuries ago. After the meteorite's explosion, water would have shot up from the crater as a geyser. Would it have been hot?

The Kaali meteorite weighed about 450 tonnes. To date, only a little over a kilogram of shards have been collected. Where did the iron go? In the early Iron Age, it was a priceless commodity not yet mined in Scandinavia. There are signs of an ancient iron bloomery at the northern foot of the crater. Was the tiny prehistoric settlement capable of using every last trace of the meteorite?

Items crafted from meteorite iron mysteriously began to spread in northern Europe in the later Iron Age: perhaps its source isn't so mysterious after all.

The dig is finished by midday. Our finds are laid out and the hole filled in again. The earth is a book that can only be opened once. I walk down to the village store and buy fresh bread, milk, and cheese, amazed to find it open at all. The street is still deserted. I knock on the door of the post office, where I buy a postcard with a photograph of Lake Kaali taken by Rein Maran. I address it to my son Kristjan, who is eight and stayed home. However, my mind blanks on what to write next. So little time passed and so many questions arose. So, I wrote: 'We're in Kaali, where we found Pühajärv. Mart and Dad.' I stamped it myself, which they allow at a rural post office, and dropped it into the mailbox. Later, I reckoned, I'll slide it between the pages of his copy as a keepsake. Then, suddenly, I remember that Kaali's proper name is Kali. And then a children's song comes to mind. Children's songs are, for the most part, very old. Generally, everything cultural that adults no longer need is tossed to children as playthings, and those children then carry on playing hide-and-seek, war, or environmental protection until they grow up in turn, though some continue even later. One suitable answer to the question of where Kaali's iron disappeared to lies on that discarded song that kids caught.

> Mina kakku Kalevile,   I give bread to Kalev
> Kalev mulle rauda   Kalev gives me iron

Children in Mustjala used to sing that, fewer than a dozen kilometres from Kalijärv (Lake Kali). Were that song not recorded in several dozen other places across Estonia, the detail would be meaningless. Nevertheless, the songs are not gone, and neither is the iron.

BENEATH TWO SUNS

KAALI IRON AND KALI BLACKSMITHS

In the eyes of a first-hand witness, was it the sun or fire that landed on Saaremaa?

Folklore studies treats these as separate motifs, but here we should consider whether they aren't two different branches of a single phenomenon. What I have in mind is the grandiose generalisation of the Estonian linguist Andrus Saareste (1892–1964): the meteorite, which approached Saaremaa as a glowing falling star from the east, exploded in a gigantic ball of fire on the island. The Estonian language is spoken over a relatively small territory. Therefore, it's surprising that the meteorite is overwhelmingly referred to as a 'spark' in our eastern dialects and a 'fire' in the west. The distance of a few hundred kilometres was enough for eyewitnesses to evaluate the same event differently. The divergent words for meteorite – *pisuhänd* in the east and *tulihänd* in the west – are the simplest, most basic, and consequently the most archaic ancient words for the Kaali event.

The dirge for Finland's Bishop Hendrik of Uppsala (killed in 1156) contains the following curious lines:

Caswoi ennen caxi lasta,	Two children grew up before,
Toinen caswoi *caalimaassa*	One of them grew up in Kaalimaa,
Toinen Ruitzis yleni …	the other in Sweden.

Caalimaa? Kaalimaa? Kaali Land?

Bishop Henry is believed to have been English. Scandinavians popularly believed that *kaali* ('cabbage' in Finnish) was a staple in England. Thus, the word has been interpreted as a reference to the bishop's native land, which is rather plausible. The Latin liturgical text composed when the bishop perished was read out in every church in Finland. The fourteen written copies of the dirge are thus folkloric echoes of the liturgical text. Without significant risk, we can presume that if the missionary's homeland was mentioned in the Latin text, then it was done so in a form and a way that required a folk-etymological explanation. The Latin text could have gone without mentioning Bishop Henry's land of birth. The bard needed to emphasise that whereas the crusade's military leader Erik Jedvardsson came from Sweden, the man of the cloth hailed from another, further place. In different variations of the song, Sweden is pressed together with *maalla wierahalla* / *maalda wierahalda* ('a foreign land'; Th. Reinus's copy). In Estonian folklore, a 'foreign land' or one 'beyond the sea' was overwhelmingly equated with Sápmi. It would be sensible to presume a symmetrical phenomenon in the Finnish tradition. There has also been real substance to the statistical frequency of Sápmi/Kaalimaa earlier. To later users of the toponym, it simply referred to a land that the dirge's audience would recognise as beyond the sea. The recorder of the dirge seems to have understood this as well: *Ruotzi* is capitalised, *caalimaa* is not.

Do we dare link Kaalimaa to Lake Kaali, Lake Kali, and Saaremaa?

We find ourselves in an unusually favourable position, knowing the event that caused the original impetus. Metaphorically speaking, we've been handed an extremely old and rare cast. Now, we must search museum cupboards for ornaments that match it as closely as possible. Only fragments remain of those ornaments. Even worse, they've only been preserved if they remained in perpetual use. Consequently, they have been mended, patched, and melted with other ornaments to be fit for adorning one's neck, belt, or blouse. True, the ancient mender was a sparing wordsmith. If they did use a new cast, then they searched through older fragments to find ones that filled it best and then poured bronze or silver into the gaps. Hopefully jewellery artists will not fault me for imagining an impossible technology as possible. All I wish to express is that two similar fragments, two similar verses or metaphors, often contain more history than two long and perfectly overlapping texts.

The event itself was extraordinary but short-lived in every manner in which it was expressed. The consequences of the event were less extraordinary but more enduring in human consciousness: the stone bulwark around the crater is evidence of a long-term undertaking driven on by flattened forests, the smoking ruins of fortresses, dead landscapes, charred wood, the whispers of the deceased, and the memory of an explosion that split heaven and earth. The sky healed, the land was left with a scar, and a wall was built around it. Then, it was discovered that although the event wrought destruction, it also gave the gift of iron. Hence, the Kaali event must also echo in motifs of an island of the dead, the birth of fire and iron, and smithing. Saaremaa adds seafaring. The *Kalevala* dimension dominates in songs of the Estonian, Votian, Finnish, and Karelian peoples; it is missing in those of the Sámi and the Vepsians. Therefore, its genesis is thought to lie somewhere in the late Baltic-Finnic era. Metre is nevertheless just a tool for recording information. It reshaped and rendered mythical, magical, and shamanic visions, the most archaic elements of which developed towards the end of the Early Stone Age.

BENEATH TWO SUNS

A travelogue does not present proof, but possibilities.

Are we able to separate the Iron Age from the Bronze Age with the help of folklore?

First, fire descends from the heavens, sowing death and terror. Then, iron is discovered upon a devastated island.

In the *Kalevala*, compiled in Karelia by Elias Lönnröt in the 19th century, the mistress of Pohjola tracks Väinämöinen for three days before confronting him to reclaim the magical Sampo artefact. Let us see in the mythical character of Väinämöinen a man who lives along a strait and has acquired shamanic powers. Folkloric names given according to local geographic features was a widely documented practice. People settled along the Daugava River (Estonian 'Väinajõgi', Livonian 'Vēna') were called *Lyvones Veinalenses* in Henry of Livonia's chronicle. Väinämöinen escapes his pursuer with the aid of fire (SKVR I_1 441, lines 134–43):

Iski tulda Ilman ukko	Ilmataati threw the fire
välähytti Väinämöine	Väinämöinen made the lightning
Kolmella kokon sulalla	at the eagle's three feathers
viijellä vivutšimella.	at its five tail feathers.

But oh, woe was he! The spark slipped through his fingers and fell to earth. What ensued was catastrophe as the fire injured and slew:

Läksi sielda Kyböine	Then the spark spread outward
läpi moan, läpi Manulan	over land, over Manula,
läpi reppänen retuizen	over the black smoky abyss

läpi lapsen kätkyöista,	over the babe's cradle,
parmoiset emolda poltti	burning her mother's breasts
tissit pissyt piikaizilda.	the lass's plump tits.

The fire incantation, recorded by Matias Aleksanteri Castrén, belongs to the very oldest layer of Baltic-Finnic folklore. It was recited during the ritual lighting of a fire when beginning slash-and-burn practices or at the summer and winter solstices. Although the next incantation partially coincides with the last, it contains greater detail and a significant geographical foothold for our travelogue (SKVR I_4 250, lines 1–20 and 36–41):

SILVERWHITE

Iski tuita Ilmarinen	Ilmarinen threw the fire
välähytti Väinämöinen	Väinämöinen made the lightning
päälä taivosen kaheksan	over the eighth sky
ilmalla yheksänellä.	over the ninth world
Kirposi tulikpuna	A spark of the flame charged
läpi maan, läpi Manalan,	over land, over Manala
läpi reppänän retusen	over the black smoky abyss
läpi lasten kätkyettä,	over the babe's cradle,
rikko rinnat neitosilta	ruining the lass's breasts
poltti parmahat emolta.	burning even the mother's breasts.

It's striking that the fire which fell into Manala found *living* people before it. The mythical Mother saved her family by casting the fire into water.

Tuo ange *Aluenjärvi*	The immense *Aluejärvi*
kolmitse kesässä yönä	swelled over three summer nights
kuohui kuusien tasalla	rose as high as the spruce
noissa tuskissa tulosen	in them the fire caused pain
valkiaisen voakahissa.	the white glowed in the agonies.

Let's keep in mind those five dramatic lines and the lake named Alue. We might recognise post-catastrophe Saaremaa in the incantation's scorched landscape:

Palo ennen maita maljon	Much land first burned
pahana palokesänä	in that wild fire-summer
tulivuonna voimatonna.	in that ruinous burning-year.
Jäi vähä palomatonta	Little was left unburned
Ahin aian kääntimillä	Ahti's orchard in the far corner
Hirskan pengeren perällä.	Hirska's garden in the back reaches.

Ahin, aka Ahti, similarly named Hirska, was the Baltic-Finnic water god who, among other things, was responsible for successful fishing. He was mentioned by Mikael Agricola in his 1551 list of local gods. In this context, Ahti's 'orchard' and Hirska's 'garden' are symbols for the shore. Thus, the notion of a 'burning-year' is connected to a coastal area. Ahti's most frequent epithets are those of an islander, which allows us to envision him as living on an island, just as Väinö resides on a strait.

BENEATH TWO SUNS

The famed Latvajärvi bard Arhippa Perttunen described Ahti's island to Elias Lönnroth as follows (SKVR I$_2$ 759, lines 22–68):

Maki on taynnä seipähiä	The hill is filled with stake fences
ne on taynnä miehen päitä	the stakes are filled with men's heads
[...]	
Matallas tulinen koski	Your path to the cascade is fiery
kosessa tulinen koivu	the birch in the cascade are fiery
[...]	
Suet on pantu suitsisuuhun	The wolf awaits, gnawing the bridles
karhut rautakahlehisin.	the bear is rattling the chains.

In another song, the islander Ahti lives near a fiery eagle next to a fiery birch on a fiery island in the fiery cascade of a fiery river. The eagle has slain a hundred men and a thousand heroes (SKVR I$_2$ 716, lines 97–8):

sada miesta on siiven alla	a hundred men beneath his wing
tuhat hännän tutkamissa.	a thousand tails under the tip.

Our goal is to glean the overall context, atmosphere, and details from the descriptions of dwelling places, which may be influenced by the Kaali event. A fiery eagle in the crown of a fiery birch is poetry, a fiery cascade on a fiery island has been a geographical reality. That landscape existed and remained. Heroes lived in them, blended together, switched names, were killed, and awoke anew. We must take into account the erosion and interruption of folklore as well as its mixing with new or older traditions, which had the same blood type in terms of atmosphere and plot and enabled the implants without leaving a single scar. The notion of a dangerous island appears in several folk songs of varying substance: it was an intact whole, convenient, sparking the imagination, and probably favoured by the listeners, weaving into the stories whenever and wherever there was a need to emphasise the protagonist's masculinity, strength, adventurousness, or supernatural abilities.

Shamanic Väinämöinen's companions are usually the characters Ilmarinen and Ahti. Ilmarinen's name derives from a word that can be traced to proto-Finno-Ugric and originally meant the god of all the world, winds, and skies. Later tradition added blacksmithing, forging the

SILVERWHITE

sky and the stars, transforming him into a cultural hero. Väinämöinen's environment, on the other hand, was originally tied to water: a broad, deep, slow-flowing river; a strait; or even the bottom of a river or the sea, thus associating him with fishing, humankind's most ancient practice. Contrary to his partner, he appeared in our ancestors' world as a water god, though as mythical systems developed he was also promoted and the watery job was given to the islander Ahti. The outlines of Ahti's character blur and fade. In several pieces of folklore, he blends with Lemminkäinen, no matter that the latter is Väinämöinen's traditional opponent and adversary in ritualistic spells, singing competitions, and other contests. Based on a number of attributes, we may regard Väinämöinen, Lemminkäinen, and Ahti as offshoots of a single trunk and presume that they were only made to counter one another in later interpretations, especially when influenced by the dramatic adventures of Viking sagas. Lemminkäinen's name can only be translated to a limited extent, the root being the Baltic-Finnic word *lemppi* ('love, warmth'), which also sprouted into the name of Estonia's ancient military leader Lembitu, who in Henry of Livonia's chronicle is referred to as *Lembite*, *Lembitus*, and *Lembito*. Here, I would also like to note that the Votian word *lemmüz* means 'spark', *pisuhänd*, *tulihänd*, 'meteorite'. In the best case, the blurring of archaic lines allows us to formulate a question: did the effect of some external event cause the formerly single water god to split into two different directions which we now know as the rivals Väinämöinen and Lemminkäinen?

We could add yet another question. The enemies of the heroes of ancient folklore were those who dwelled in the dark, cold North. A true chain of events glimmers through the back-and-forth forays — clashes between the Baltic Finns who migrated into present-day Finland and its indigenous inhabitants, the Sámi. The antithesis of the North – i.e. a combined notion that comprises home and homeland, perhaps even the homeland from which the Baltic Finns departed – is Päivölä: a mythical land, whose name comes from the word *päev*, which in present-day Estonian means 'day' and in that era meant 'sun'. The Land of the Sun is the homeland of our ancient local folkloric heroes, but, unlike the North, it is described much more ambiguously. The gods and heroes gather in the Land of the Sun to feast and drink. A parallel name for the Land of the Sun is *Saraja*, also *Sariola* (SKVR I 759, I$_1$ 362) – a word that has disappeared from Estonian but in Ob-Ugric languages means the 'sea' and in Komi-Zyrian stands for the southern region to which migratory birds fly for the winter. Our Olympus is anything but a pleasant place: features of the Land of the Sun

BENEATH TWO SUNS

in Arhippa Perttunen's folkloric memory include heads on stakes and a fiery birch next to a fiery cascade. When Osmotar starts brewing beer and the whole island is covered in smoke and flames and entire forests burn to the ground, Lemminkäinen gets the urge to join the party. His mother warns him that three deaths lurk in Päivölä. The first is an eagle by a fiery cascade, the second is a snake thicker than a beam, and the third is

aid on rauvasta rakettu	the fence is fused from iron
tšitšiliuskoill on sivottu	binding the lizard
kiärmehill on keännyteldy	wound round the snakes
muasta suate taivahahe	from the land up into the sky
	(SKVR I$_1$ 362, lines 138–41)

Naturally, the hero overcomes all perils, makes it to the islanders' feast, competes in song, and fights the master, whom he ultimately kills. Competitive singing is meant as two shamans testing each other's strength. The slain master, however, is the water god Ahti, simply referred to as an islander. In this context, he appears as Lemminkäinen's adversary, consequently Väinämöinen's ally, and perhaps shouldn't even be differentiated from the latter. Päivölä's iron fence fits nicely with Ahti's own. In the fire incantation just cited, the spark or *lemmüz* fell into Lake Alue. I owe a debt of gratitude to Matti Kuusi for informing me that Lake Alue also appears in Estonian folklore under a different name. Fire fell from the heavens straight into Lake Kaleva:

Tuli tuhmalta putosi	The foolish fire crumbled
läbi karsun rautaharkon	fell from the spread iron prongs
polvelta pyhän jumalan	from the sacred god's knees
aita parran autuaan	to the barn helper's beard
keskelle Kalevan järven	amidst Lake Kaleva. (1818)

Is Kalevanjärv, i.e. Lake Kaleva, 'Kali' (Kaali) and Saaremaa Päivölä's iron fence?

The extraction of bronze from iron went on for centuries along the Baltic Sea. Bog and lake iron were used for ore, smelted to iron over the course of a long and painstaking process. The iron contained in the Kaali meteorite, about 450 tonnes, was of immense value, totalling more than the entire world's annual iron production. We might assume that attitudes towards Saaremaa changed with the dawn of the Iron Age:

SILVERWHITE

notions of terror and catastrophe were complemented with fairy tale-like themes of wealth, ultimately blending and overwhelming the earlier reputation. Instead of a cosmic blast, we should hear the steady clanging of blacksmithing hammers from the area of Saaremaa. The 'sun and moon' suit well for a poetic line between the Bronze and Iron Ages. The sun vanished into a cliff and darkness flooded the earth. Väinämöinen, born and raised in darkness, works in a smithy. A flying serpent appears on the doorstep and states the reason for his presence as follows (SKVR XII$_1$ 99, lines 14–18):

Tuota minä lienen kynnyksellä	For I stand here on this threshold
sanomatta soattamassa:	so I may speak a message:
jopa nyt kuu kivestä nousi	the moon has risen from the stones
päivä peääsi kalliosta	the sun escaped from the cliff
takohissa Väinämöisen.	in Väinämöinen's hammer blows.

Thus, the sun starts to shine once again as the result of smithing. Against the backdrop of the developments on Saaremaa, we may interpret this as a new assessment in its final form. The weaving of images of catastrophe with smithing could thus belong to an even earlier time and illustrate the formation of a new view of the island – overcoming fear with apotropaic magic. Perhaps the legend of Väinämöinen's journey into Toonela (the Baltic-Finnic land of the dead), recorded in Karelia in the last quarter of the 17th century, also belongs here. We interpret the shaman's odyssey primarily as his falling into a trance, during which it was believed the soul could soar into the netherworld. Features of the song draw it closer to the globally prevalent theme of visiting the land of the dead, but also permit a more mundane interpretation. The iron-fingered daughters of Toonela, weavers of iron belts, ask Väinämöinen's purpose for his request to be ferried across the Toone River. The shaman replies that he has been sent to Manala and brought iron to Toonela; iron-death by the sword. As the man's clothing is not bloody, the daughters refuse to grant his wish. He names a second reason: fire, fire-death, but his clothing is not singed. Upon being asked a third time, he admits he has come to Toonela to acquire iron bars and ship's nails. They ferry him across the river in a small boat. Transforming into a brown snake, the shaman escapes home and warns the iron-weavers not to visit. Compared with other ancient, global legends, the magical power of iron stands out: a power so great that Toonela's defining feature is of a metallurgical nature. Fear had to be

BENEATH TWO SUNS

overcome and an island of the dead visited to acquire the metal. Is this perhaps the line between the Bronze and the Iron Age?

During another trip to Toonela, Väinämöinen faces a young opponent whom he defeats with the power of words (SKVR II 161, lines 134–7):

lauloa lambihi kalattomaha	he sang the lake into fishlessness
aivan ahvenettomaha	fully into perchlessness
kynzin kylmänä kivehe	that in his clutches into cold stone
hambahin vezihagoho	that in his teeth into a soggy stake.

Can we interpret this as Lake Alue? Lake Kaleva? Lake Kaali, and thus Saaremaa? The question justifies the reason for journeying into Toonela. Väinämöinen planned to borrow a wise man's words from the great shaman Antero Vipunen, whose name also appears as Vironi, Virunen, Viroine, and Virokannas. As we demonstrated earlier, Emil Nestor Setälä treated Virokannas as the *axis mundi* and the name's connection to Estonia is beyond doubt. The ancient blacksmithing culminates in singing about the creation of a zither and a gold woman. Both motifs are tied to Estonia (SKVR VII$_1$ 547 a, lines 1–5):

Teki kauko kanteloista	Kauko made a little zither
Viron seppä vinkeloista	Viru's blacksmith a little spiral
eikä puusta eikä luusta:	not from wood, nor from bone:
sapsosta sinisen hirven	from a shoulder that blue moose
poropeuran polviluista.	a reindeer from the knees.

Here, also, the Viru blacksmith is seen more as a world blacksmith, i.e. a parallel form of the weather god Ilmarinen. It replaces one unknown with another and leaves unanswered the question: why did the name of the world, Viru, borrowed from Scandinavia, become affixed to Estonia and the Estonian blacksmith? How can we explain the equivalence between *Viru blacksmith* and *world blacksmith*?

The Estonian word for blacksmith, *sepp*, comes from the proto-Finno-Ugric language and originally meant 'skilled, masterful, handsome'. It is over 4,000 years old. In Estonian, it has led an unusually fruitful life and produced a family of over 75 compound words, which include couplings with beads, silver, swings, beer, and words. 'Reflected in this is the extraordinary age of handicraft traditions in the area of Old Estonia,' commented linguist Julius Mägiste, also noting that similar compound

SILVERWHITE

words in Livonian and Finnish are mostly borrowed from Estonian, and are lacking entirely in more distant kindred languages (Mägiste 1970: 43).

In the following lines, the toponyms support one another, deepening the conviction that associating blacksmithing with Saaremaa is an ancient practice of Estonian folklore (SKVR III$_3$ 4033, lines 1–10):

Saaren maat saroin jaettu	Saaremaa's lands are smoothed
Viron maat viipin vaapin,	Viru's lands are sliced,
pellot on piusten mittaeltu	the fields measured by stick
ahot on vaaksoin arvaeltu.	the harvest viewed by vaks.[2]
Jäi sarka jakamatointa [...]	One plot was left unparcelled [...]
tuohon seppo seisattaise	that's where the blacksmith stopped
takojainen pani pajaa.	the hammerer put down his pot.

Life posed questions to us and we responded as well as we could. Not one could be left unanswered. The ambiguous echo of the Kaali event in older Baltic-Finnic myths points to the catastrophe being divided into pieces, separated into causes and effects, and being gradually joined to a world that had become home over the course of millennia.

And yet. From a letter written by Professor Kuus: 'The Saaremaa meteorite spurs my imagination as well: Finnish folklore contains a wealth of material that can be tied to Saaremaa and its role as a central site for ancient Finnish seafarers. One folk song of possible interest is *Saaremaa põleb* ["Saaremaa Is Burning"] ...'

And here they are, the opening lines which match in every version:

Näin mie Saarenmaan palavan	I saw Saaremaa burning
ja tulen runnista tulevan	fire rising from the spring
lehmuksesta leimahtavan.	the linden tree in flames.

Here, the event as remembered by the bards of Harju-Jaani and Kuusalu:

Nägin Saaremaa põlema,	I saw Saaremaa burning,
tule loogeti tulema	the fire winding closer
pikki välja, põiki välja.	going straight, going sideways.

Worded differently (Estonian Literary Museum, EÜS VIII 705/7, 288):

Oi imeta või imeta	Oh wonders, oh woe,
Mis nägin mina imeta	What wonder did I behold

BENEATH TWO SUNS

Õuessagi käiessagi	What did I witness outdoors?
Nägin Saaremaa põlema	I saw Saaremaa burning
Tule luugista tulema	I saw the pyre rampaging,
Sood süütsid, järved põlesid	The fen caught fire, the lakes blazed,
Kivid keereldi ujusid	Stones swam in the whirlwind,
Tähed lõivad tääringida	Stars were tossing dice,
Mõõga otsad mõõringida	Swords were howling thunder.
Pühi nutab Saare maada	The blessed weep for the land of Saare,
Teine Saaremaa mehida	Others for the men of Saaremaa,
Nutan Saare neidisida	I weep for the girls of Saare:
Vöö kudujaid, kooga loojaid	Braiders of belts, weavers of fabric,
Rahaskirja kirjutajad	Writers of laced scrolls,
Laia raamatu lugejad.	Readers of broad script.
	(Kurrik 2013: 251)

The image of a burning county is exceptional and unparalleled in Estonian folklore. For this reason, it's unlikely that the motif is meant to describe 13th-century battles that rolled across the entire territory of Estonia and left every county ablaze, but rather an extraordinary and unrepeating event that we believe we have now identified.

In spite of time and space, is it possible that a distant message has reached our ears?

THE DEATH OF THE GODS

Language has delivered to us a shock that struck us, as well as a natural defence mechanism: the smouldering crater was taboo, the scorched surroundings were taboo – could Saaremaa and the entire western coast have been taboo as well? Was it also the case for our neighbouring tribes? Virokannas raises a question: was the Kaali event recorded in Scandinavian mythology as dramatically as it was in ours?

The catastrophe occurred during a time of active seafaring – in the so-called First Viking Age. The Estonian word for Swede, *rootslane*, developed at about the same time. 'Ship' was a taboo word in Scandinavian languages. Placeholders were used instead: 'wave-ride', 'sea horse', or simply 'horse', *rhos* (Lönnroth 1963: 12–19). Rowers (sails hadn't been invented yet) were named *rodskarlar*. At times, it meant a person of any tribal affiliation

SILVERWHITE

who had arrived by boat from the west. When a Karelian of Rukajärve or Paatene took a Finnish wife, she was called a *ruotšakko*. Over time the word's meaning narrowed, became more specific, and came to refer to western neighbours in general: rowers of wave horses became *rootslased*, Swedes. Our small sea had become a body of water that joined peoples together and was a conduit for trade.

But when the Kaali event made coastal Baltic Finns mysterious and terrible in the eyes of their neighbours, perhaps even taboo and untouchable, then it must have had an entirely positive effect on Baltic Finns themselves or perhaps at least the residents of Saaremaa. Fear was one thing, but fear held by neighbours could be a better defence against invasions than the Baltic Finns' own weapons – given, of course, that they were recognised as inhabitants of the sun-grave. Consequently, the Baltic Finns' own relationship with the sun-devourer had to become dualistic, directly contradictory. Terror must have woven with trust, dread with gratitude, fear with mercy and courage. So it seems to have been.

The meteorite was seen in eastern Estonia as a spark and in western Estonia as a flame. What about in Sweden? The visual phenomenon extended across the sea, which means its cultural echo must have crossed the water as well.

Seen from Scandinavia, the sun set in the east. What could be more shocking than a world turned upside down?

I believe that here may lie the key for understanding the fearful symbols in Scandinavian mythology. Estonia's western neighbours observed the Kaali-sun setting in the east, an unbearably unnatural direction, falling and being destroyed. Perhaps that is why the sun-eater Fenrir acquired the form of an uncompromising wolf – a contrast to Estonians' dualistic notion of a flying serpent that sometimes sows destruction and other times gives life and prosperity. Those responsible for the sun's doom resided east of Scandinavia. Fenrir's father was Loki, a god of the *Edda*, and his mother Angrboða, a giantess of the threatening Ironwood to the east beyond the sea:

> In the east lives the old one, in Ironwood,
> and breeds there Fenrir's kind

Relations were much more amiable in earlier times. This is acknowledged by Loki when he addresses his father, the god of war and poetry:

> Remember, Othin, in olden days

BENEATH TWO SUNS

> That we both our blood have mixed;
> Then didst thou promise | no ale to pour,
> Unless it were brought for us both.
>
> *(Poetic Edda* 1936)

But the past is the past. The *Edda* knows only to hate and fear Loki and his Muspell, people of the realm of fire. Loki boasts of this challengingly:

> But when Muspell's sons through Myrkwood ride,
> thou shalt weaponless wait, poor wretch.

The following lines are believed to have been written in the 10th century:

> From the east comes Hrym with shield held high;
> In giant-wrath does the serpent writhe;
> O'er the waves he twists, and the tawny eagle
> Gnaws corpses screaming; Naglfar is loose.
>
> O'er the sea from the east there sails a ship
> With the people of Muspell, at the helm stands Loki

To tie these distant echoes to later folklore, we should add that the sun-eater Fenrir's sisters include the giant serpent Jörmungandr, who was cast into the sea, and Hel, ruler of the underworld. From the same family comes the singing giant Starkadr, the Baltic Sea's daring Roland, an embodiment of masculine virtues. Starkadr was said to have hailed from Estonia. As Saxo Grammaticus wrote in the late 12th century: *in ea regione que Sveciam ab oriente complectitur, quamque nunc Estonum [...] sedibus tenet originem duxisse.*

The older parts of the *Edda* were recorded only around the year 850 or later, at a time when the mythical understanding of the world was dying, only to be reborn as poetry. Are we capable of restoring and understanding even more ancient ties?

> A daughter bright Alfrothul bears
> ere Fenrir snatches her forth ...

The connection between the sun-eater Fenrir's name and the Sámi/Finnish people is beyond doubt. Before our recorded history, Sámi populations extended much further south in Scandinavia. They were the immediate neighbours of Sweden's ancient Scandinavians and resided just across Estonia's northern shoreline, extending in a wide band to the southern banks of Lake Ladoga and Lake Onega/Äänisjärvi. Scandinavians' direct

87

contact with the Sámi, who, because of active seafaring, they knew to live also in the east, may have led to identifying eastern peoples as the sun-eater. To an observer in Scandinavia, the Kaali meteorite ('sun') rose in the northeast and set in the east, northeast, or south-southeast, depending on their location.

If Manala, the underworld, tends to blend with Lake Alue in Baltic-Finnic myths, and Kalevanjärv/Lake Kalevan with post-catastrophe Saaremaa, then might we find something similar in Scandinavian mythology?

Strangely, we do, and, strangely, it coincides with *helmekuu* ('bead month/moon'), which in archaic Estonian meant the first month of the year, and in Finnish the second (after the beads of ice frozen on bare branches). The word *helm* is borrowed from the Kunda language, as Paul Ariste has determined, and originally meant 'amber' in Estonian. The richest excavations of that odd organic mineral are located in what is now the coast of Russian-controlled Kaliningrad/Königsberg. It was historically home to the Sambians, an Old Prussian clan that was gradually assimilated. The term *glaesum* offers no anchor: it is a Germanic word with the same roots as 'glass', *Glas* (German), and *klaas* (Estonian). Before amber was mined, but was gathered, the Courland coast probably produced equal yields. Amber has also been found on the Kihelkonna coast of western Saaremaa. In archaeological sites in Estonia, it has been found in layers dating to late 3000 BCE. In Valma, on the shore of Lake Võrtsjärv, the archaeologist Lembit Jaanits found an amber boar figurine in the grave of a woman buried in a settlement. Amber's magnetism could not have gone unnoticed by an amulet's wearer. The mysterious phenomenon fascinated peoples of the Baltic Sea and developed into a belief that amber possessed magical power.

Based on archaeological finds, amber's ancient tradition was transported along waterways to the east, northeast, and north. Therefore, it is difficult to write off as pure coincidence the fact that in the German historical tradition, Central Europe's ancient amber route is called *Hellweg*. Germanists acknowledge that the word hasn't been satisfactorily unpacked yet. Generally, it is seen as a compound word formed by *heller Weg*, or *via regia*, which is a false national-etymological interpretation. In Westphalia and the Groningen province of the Netherlands, *Helwege* and *helwegen* (respectively) refer to a path of death and peril, both having developed from ancient taboo ideas. A somewhat more concrete example is a strophe from the Scandinavian chronicler Snorri Sturluson's *Prose*

BENEATH TWO SUNS

Edda: *helvegr ligrr nidr ok nordr* – Hellweg led 'downward and northward', in other words to the realm of the dead. I took my suspicions to Ariste and am grateful for his written reply: 'German *Hellweg* – the first part of the word could be connected to *helm*. There's nothing linguistic to disprove it, but facts and fantasy allow them to be tied. One should take into account the fact that all kinds of extraordinary changes happen with place names. [...] For example, one highly credible linguist has connected Saransk to Saaremaa. Someone like me doesn't believe in that connection, of course [...]. If you were a linguist, then I'd have a much more hostile attitude towards your conjectures. [...] Nevertheless, I like your interpretation of the Hellweg problem' (Paul Ariste to the author, 17.2.1974). The Russian linguist Aleksander Popov also proposes that the amber-acquiring Aesti should be identified as ancient Estonians, justifying his belief with the much broader range of Baltic-Finnic settlements during Tacitus's day (Popov 1973: 67–8).

Fenrir the sun-eater lives on the eastern shore of the Baltic Sea. Fenrir the Sámi's sister is Hel, ruler of the underworld. And the ancient trade route to the eastern shore of the Baltic Sea, including that to Saaremaa, is the death-road Hellweg, which may have originally meant the Amber Trail.

Taking this as truth, questions regarding the origins of Thor inevitably arise. He crowns the Nordic pantheon as unshakably as Zeus sits upon Greek Olympus. He is too inherently a part of Germanic mythology to even arouse any suspicion. Except among linguists. For in terms of linguistic history, it is impossible to prove Thor's Germanic origins. One would have to compose a new linguistic history that proceeds from irregular phonemic shifts. Not to mention the geographical questions which are charged to the brim: myths leave no doubt that Thor is the parvenu of the frosty Nordic Olympus. A newcomer, one could say, if you delve deep enough into the past, having come from a place in the east called *Glyssiwalla* – also ancient Scandinavians' name for the amber coast. In this ancient name, we can recognise Tacitus's *glaesum* – amber.

The world is home to outstandingly unpleasant characters. One tries to look past them. They're addressed only when extremely necessary. They must be made to reply with a yes or a no even when they'd otherwise wish to tell their tale. But because their long-heard tales are rather unsavoury or make it hard to sleep, one leaves them as soon as possible and hurries to turn the nearest corner. Thor is one such figure.

SILVERWHITE

For the Khanty of the Ob River, the bear is an ancestor and totem deity, and also the son of the paramount god *Num-Torem*. The Mansi people knew a similarly sounding god, *Thor*, written down thus in 1730. For the Chuvash peoples, in whose development the Volga-Finnic tribes played a significant part, the word *torum* appears to signify a god or the sky. 'We remember you, we regret nothing for you, we pray for you, Tori,' they spoke, appealing to ancestors in ancient times. In those kindred Finno-Ugric languages, the symbol for the sky, the weather, and the world is pronounced *taarem*, *toorem*, *tuurem*. Sámi expands the geographical reach, its *dier-mes* used for thunder. And, finally, closing the circle is Johannes Scheffer's comment about the female Sámi thunder god, which was written earlier than 1673: 'Beside these greater, the *Pithenses*, *Luhlenses*, and their neighbours have some inferior Gods, as the *Tornenses* likewise have, though they worship them all under one name, excepting only that which they call *Wiru* [Viru] *Accha*, signifying a *Livonian* old woman, which *Olaus Petr.* with some alteration calls *Viresaka*' (Dalin 1756: 97; Scheffer 1674).

The relationships of the Baltoscandic pantheon are too complicated to scratch more than superficially here. Let us limit ourselves to the conclusion that, based on ancient beliefs, Thor came from a region that lay to the east of Scandinavia. For some ambiguous reason, the dissemination of funeral rituals is associated with the lord of thunder.

Unexpected weight is given to this relatively inarticulate summary by the dissimilar prevalence of *toor/thor*-type place names in Estonia and Scandinavia. According to data collected by the Estonian Language Institute, there are around 124 toponyms with the root *toora-*, *tuuri-*, *taara-*, and *taari-*, nearly half of them (58) clustered around a quite small area to the west of the Tallinn meridian. They are noticeably prevalent on the islands of Saaremaa and Hiiumaa, along coastal Lääne (Western) County, and in Keila Parish near Tallinn. To a lesser extent, they follow the Pärnu–Emajõgi river waterways to Viljandi and Suure-Jaani Parish. In north Estonia, the toponym follows the coast until the city of Jõhvi.

In Scandinavia, however, the toponym is far less prevalent: six Torskaker-like place names in East Sweden, which date back to the beginning of the Common Era; a single village named Torsager in Jutland (to which, in truth, can be added fields of the same name); and ten Thorshof toponyms mainly in the vicinity of Oslo, which do date back to a much later period. Jan de Vries (1970) has deemed it necessary to presume that 'the renaming of old sites of worship' provides a satisfying explanation for such a remarkable anomaly. But, strikingly, Sweden's oldest Thor-type place names are clustered around the Lake Malar area and larger river deltas. It appears they had a direct connection to ancient waterways during the First Viking Age.

We may at least regard the Finno-Ugric origins of *tuur/toor/taar* as probable. This allows us to date the supreme world god to before the splitting of the Baltic-Finnic tribes, which means a far older time than the Kaali meteorite event of the Bronze Age. This latter point is of exceptional importance to our travelogue. An unprecedented natural phenomenon required above all an explanation and interpretation within the bounds of existing cosmogonies. That was obviously done. The crumbling of the heavens, the death of the sun, and the unbearable power of a supersonic shock wave were fitted to ancient notions of the world/lightning god named *tuur/toor/taar*.

If this inference holds true, then the myth and the Finno-Ugric *tooum* could both be prevalent on other Baltic coastal areas to varying degrees of mutation and development.

One intermediate stage has been preserved in the *Knýtlinga Saga*: in the 12th century, the Danes conquered the island of Rügen and destroyed the 'pagan' holy site there. Idols were chopped to pieces. The chronicler recorded the names of these gods, listing *Turupid* in second place. Already in the 18th century, researchers saw it as a version of *Taarapita*, the 'chief god' of Saaremaa. The interpretation, belonging to Aleksandr Kotlyarevsky (1870), failed to gain broader traction because 'Taarapita' was decrypted as a cry for help: *Taara, avita!* However, an idol erected on the trading island of Rügen could not carry the name of a cry for help. Nevertheless, this decryption of Taarapita is questionable. Various transmutations of the name appear in the chronicle of Henry of Livonia: *Tharaphita*, *Tharapita*, *Tarapitha*, *Tharapitham*, *Tharapitha*. The first can be read as *Taara, avita* if the missionary used the letter *v* for *ph*. He does not do it elsewhere. Henry of Livonia's four latter versions end in *-pita/-pitha*, which more likely than not derive from the word *pikne*, 'lightning'. The ancient *pitkne*, *pitkä*, *pitša*, *pitkäne* meant both a taboo snake and lightning. Even in the 19th century, it was recorded that 'Taara's weapon' was as smooth 'as a pike'.

Ancient Baltic-Finnic fire spells place the words 'fiery cascade' next to '*tuur*'s [Tuoni/Turi's] new room'. Fire plummeting from the sky struck the people of Tuoni next to a lake named Alue aka Kalev. Folklorists interpret the Tuoni people as subjects of the Scandinavian thunder god

SILVERWHITE

Thor. Perhaps it is time to see Thor as the creation of the Tuoni people instead?

Let's not evaluate these bits and pieces while sitting in an armchair cast in the soft glow of an electric light bulb. Let us perceive time, the cruellest adversary of heritage, the most attentive gardener of culture; let us perceive the abyss that separates us from the clear-watered spring of all these complex phenomena. On our journey, we've arrived at the mouth of a majestic river (if you'll allow me to drift into sentimentality), a self-aware current, the oily waters of which carry white ocean-going vessels, giant tankers, and busy tugboats, framed by a modern harbour landscape and a noisy forest of iron cranes. This is our time and our reflection. But for a moment, let us awaken the thought that even this river begins with a palm-sized spring nestled in the quiet bosom of distant trees. It is up to us whether we find the time and the desire to walk upstream and reach the site beneath those rustling leaves, which hides the genesis of all and from which our next step – a single, arbitrary step! – takes us further from the edge of that spring, over a generation-high watershed towards another river mouth and another sea.

Let us perceive the distance of the spring while standing amidst our rumbling harbour landscape, and let us embellish the comparison: we, here and now at the mouth of the river, wish to behold and taste the pure water of the spring so far away. It is difficult but not impossible: it requires the same precise analyses that schools of salmon are able to make when they return to the waters of their birth to spawn; a precision that humans are yet unable to achieve. Nevertheless, that degree of precision already exists in nature and we ourselves are a part of nature, too – by learning from nature, leaning upon it to evolve and better perceive ourselves, we nourish the hope of someday moving from hypothesis to truth or from hypothesis to its disproval, which, unlike in sports, is of equal value and depends on the time and our own interests and wishes.

* * *

'Dad, what's the point of history?' Who could say?

ISLANDS OF THE HEART

The fiery sun descending into Püha (Sacred) Lake; the oak forests of Saaremaa igniting and the eruption of a searing cascade; a sacred site settled

BENEATH TWO SUNS

by humans for a millennium, meteoric iron, and amber discoveries; and finally, an Amber Trail that led 'downward and northward' and was called the path of death – these and other myths, folklore, and archaeological finds were a living reality in Pytheas's day. Extraordinary reports spread like a powerful wave from an island located far in the northeast. The information they conveyed was diametrically opposed to the scientific worldview of the time. Could Pytheas have not visited Saaremaa?

This, let it be emphasised, we do not know yet. We have posited reasons why the Kaali meteorite crater could have attracted men of science from Pytheas to Alfred Wegener, and it certainly did so. We have shown that Pytheas visiting Saaremaa is more logical than not. In spite of all this reasoning, we have only reached the outer limits of probability: it may and may not be true.

But if it was ...

If Pytheas's work sparked a wave of fierce debate that lasted centuries, then wouldn't his descriptions have stirred up both ideas and emotions, touched the mind and feelings? Based on this, could we assume that descriptions of the catastrophe clashed with the scientific worldview and were discarded, though they found grateful welcome in the spacious world of folklore and poetry? That Kaali settled in science and built up in literature?

Pytheas's new travelogue captivated the minds of educated men in the time of Apollonius of Rhodes (295–215 BCE). The Library of Alexandria, which Apollonius directed for a time, was one of the most famous cultural hubs of its era. It is there that he began working on his epic, *The Argonautica*. Leaning upon the structural example of Homer's *Odyssey*, Apollonius has his seafarer journey instead towards a holy amber island in the Far North (Apollonius of Rhodes 1912: IV):

> Thence they entered the deep stream of Rhodanus which flows into Eridanus; and where they meet there is a roar of mingling waters. Now that river, rising from the ends of the earth, where are the portals and mansions of Night, on one side bursts forth upon the beach of Ocean, at another pours into the Ionian Sea, and on the third through seven mouths sends its stream to the Sardinian sea and its limitless bay.

Let us attempt to define the features of this place. In translation, lines 504–6 read: 'And quickly they entered the ship, and toiled at their oars unceasingly until they reached the sacred isle of Electra [*electrum*/ *ēlektrōn* = amber], the highest of them all, near the river Eridanus.' There, they

93

SILVERWHITE

came across a small, deep lake, the grave of the sun, and 'even now it belcheth up heavy steam clouds from the smouldering wound' (lines 599–600).

In the Finnish epic *Kalevala*, the land of the Tuoni lies:

> To the island in the ocean,
> To the meadows rich in honey,
> To the cataract and fire-flow,
> To the sacred stream and whirlpool. (XV)

Ovid, among others, addressed the same topic three centuries later. 'May man know his limits and not demand the rights of the gods' could be the maxim of the *Metamorphoses*, in which he used over two hundred Greek myths. Phaethon, son of the sun god Helios, demanded to share his father's rights and is punished: he veers through the sky like a shooting star that appears to come from out of the blue without falling to earth. Struck down by lightning as a consequence, Phaethon turns to dust and the river Eridanus swallows his smoke-blackened face. If we are to believe myth, Ovid adds, then there passed a day without sunlight; fires were the sole source of illumination. The sea ebbed, revealing sandy dunes and ridges, fish seeking escape in the hollows, seals floating belly-up, lifeless. It is psychologically refined literature, intellectual literature, and, above all, literature. Let us take a moment to appreciate the fantasy that composed a realistic catastrophic landscape from secondary details. The Sun's wailing daughters stretch their arms out towards the grave, but the arms do not obey; they freeze and transform into branches while the daughters' flowing robes stiffen and crumple into bark. The Heliades' tears stream down and congeal in the sunlight to become amber, dropping into the clear water, washing ashore, and ultimately adorning genteel Roman women.

These motifs all appear in Apollonius of Rhodes's *Argonautica*, one of Ovid's sources, in a more lapidary way. Associating the death of the sun with amber is very significant. During the time of Apollonius, and especially Ovid, amber symbolised the world map and Europe's northern periphery, if not the Baltic Sea itself. However, we find in Apollonius's text an important nuance that is missing from Ovid. He, quite unusually, also references its source: according to the Celts, the amber tears may have come from the Sun himself when the heavenly body 'came to the sacred race of the Hyperboreans' (verse 614).

When we studied Baltic-Finnic myths, we asked ourselves: did fire or the sun fall upon Saaremaa? Now, we are faced with the same question.

BENEATH TWO SUNS

According to Greek myth, Phaeton fell from the sky. The Rhodanus aka Eridanus is now the Rhône. Since the days of Herodotus, 'Hyperboreans' was used to refer to people of the Far North. What caused Apollonius to shift the catastrophe from the Mediterranean northward? Referencing Celtic myth, he adds only that the Sun himself, not Phaeton, died in the North. But what was the impetus that turned his gaze from the Mediterranean to the Baltic?

The reason I am returning to Pytheas lies in the blacksmiths of Kalev.

In the fourth book of *Argonautica*, Apollonius names seven islands including the Ligystians, which are called Stoechades, the dwelling of Hephaestus. Hephaestus embodies the blacksmith and metalwork, and is symbolised by the hammer. Commentary composed during Apollonius's lifetime shows that the islands' names were borrowed from Mediterranean tradition, but their descriptions from the work of Pytheas. The island booms with the roaring of fire and the rhythm of blacksmiths' hammers. This all characterises Hephaestus, of course, but not the following: in days of old, Pytheas asserts, seafarers would land on the island, deliver raw iron to the smith, and receive their ordered item the following day for a fee. The names of the islands have been poeticised. *Lipaino* means 'oil' in Greek, thus Ligystian. For, as commentary confirms, the iron would turn 'as soft as oil' in its forge.

Have we now found the first literary flecks of our 450-tonne iron meteorite?

Seafarers come, seafarers go.

> I give bread to Kalev,
> Kalev gives me iron ...

Is it really so? The memory of song extends even further into the past:

> I give iron to Rebu,
> Rebu gives me gold ...

In north Estonia, children hoped to get gold from Rebu; in Viljandi County, from Rebu; in southeast Estonia, from Räbo. We should keep this in mind, along with a question: what did 'Rebu' mean in the era of the Kalevs?

But now we've come to a fork in the path: one leading to science, the other to literature.

Strabo quoted Pytheas above all to add weight to his own worldview while ridiculing the latter.

SILVERWHITE

Literature, unlike science, prefers imaginative poetry over unsmiling probability. Literature received Pytheas with open arms, and there he lives a nameless life to this day.

Measure twice, then cut straight. Perhaps this also applies to travelogues, especially so?

Why rummage through classical literature when we Baltic Finns have our own common epic, the *Kalevala*, the oldest songs of which remember a time millennia ago?

It was a dark night. The supreme deity Ukko, 'the great creator of the world', decided to cast a little light.

> Lightning Ukko struck in darkness
> From the edges of his fire-sword;
> Shot the flames in all directions,
> From his blade of golden colour,
> Into heaven's upper spaces,
> Into Ether's starry pastures.
>
> (*Kalevala*, XLVII)

He entrusted its nurturing to a young maiden named Imbi so

> That it might become a new-moon,
> That a new day might follow.

But alas!

> Long the fair and faithful maiden
> Stroked the Fire-child with her fingers,
> Tended it with care and pleasure,
> Till in an unguarded moment
> It escaped the Ether-virgin,
> Slipped the hands of her that nursed it.

Catastrophe ensued. The details dispel any doubts that it might simply be a distant echo of biblical myths. The events, seen through the eyes of indigenous hunter-fishers, hold rich metaphoric significance: a ball of fire fell to earth.

> Quick the heavens are burst asunder,
> Quick the vault of Ukko opens,
> Downward drops the wayward Fire-child,
> Downward quick the red-ball rushes,
> Shoots across the arch of heaven,

> Hisses through the startled cloudlets,
> Flashes through the troubled welkin,
> Through nine starry vaults of ether.

The fireball killed and injured.

> In the cradle burned the infant,
> By the infant burned the mother,
>   [...]
> First it burned the fields, and forests,
> Burned the lowlands, and the heather;
> Then it sought the mighty waters,
> Sought the Alue-sea and river,
> And the waters hissed and sputtered
> In their anger at the Fire-child,
> Fiery red the boiling Alue!

And, suddenly, an unexpected parallel with the Roman poet Lucan.

Marcus Annaeus Lucanus (39–65 CE) received a first-rate education from his uncle, the philosopher Seneca the Younger. Yet he did not venture into science but into literature. Metaphors, myths, and facts were equal building blocks of his writing. Even so, he is regarded as a credible author; a representative of documentary belletrism. 'Lucan does not deserve a place among the great poets, because he has not written poetry, but history,' wrote a scathing critic of his day. Petronius was even more categorical in his critique of presenting facts in poetic form.

Nevertheless, there are facts galore in Lucan's *Pharsalia*, or *On the Civil War*. His geographical erudition was flawless, demonstrating familiarity with the entire Mediterranean, including the shores of the Black Sea, as well as northern Europe and distant China. His earth is naturally a sphere, and if you dig a deep enough hole in it, the Antipodean sky would be revealed. Listing the possible causes of the tides, he bases his arguments on Pytheas's theory and references the moon's effect. To this, he adds two more explanations, more poetic than scientific. 'Be it the wind which thus compels the deep from furthest pole, and leaves it at the flood' (Lucan 1896). The author's gaze is undoubtedly focused northward. The third possibility is also phrased as a question: 'Or else, in search of fuel for his fires, the sun draws heavenward the ocean wave?'

If we wish to describe the Kaali event's tsunami with words from the *Kalevala*, the following verses might do:

SILVERWHITE

> Three times in the nights of summer,
> Nine times in the nights of autumn,
> Boil the waters to the tree-tops,
> Roll and tumble to the mountain,
> Through the red-ball's force and fury;
> Hurls the pike upon the pastures

It is the last line that particularly entrenches my belief in the *Kalevala*'s veracity. Lucan's third explanation is not borrowed from the clichés of classical geography. It is listed alongside Pytheas's theory, possibly even coming from Pytheas himself. The earthquake accompanying the meteoric disaster produced a gigantic wave – a tsunami-like phenomenon inundating Baltic coastlines, unprecedented and extraordinary at our latitude. Fish flung onto dry land? Let us consider what a deep scar that unnatural experience must have left on the balanced psyche and memory of ancient hunter-fishers.

The consonance between the *Kalevala* and Lucan's descriptions are important: far in the North, fire fell from the heavens down to earth and could cause a rise in sea level that 'hurls the pike upon the pastures'.

Can we use the *Kalevala* to pinpoint the event's location?

Väinämöinen and the blacksmith Ilmarinen rush to the site to determine

> What the kind of fire that falleth,
> What the form of light that shineth
> From the upper vault of heaven,
> From the lower earth and ocean.

Here, we should recall that 'Väinämöinen' originally symbolised a person who lived on the shore of a broad river or strait. Ilmarinen's name extends even further into ancient Finno-Ugric history, being used for the supreme deity or god of all the world. The sky and the *axis mundi* Sampo are his handicraft. Ilmarinen is present in the Udmurt language and myth as Inmar, and the same stem means 'sky' or 'sky-ruler' in Komi.

The divine pair carve a dugout and row

> On the Nawa [Neva]-stream and waters,
> At the head of Nawa-river.

Meeting Ilmatar, Väinämöinen's mother, they learn that the fireball fell

> From the plains of the Creator,
> Through the ever-moving heavens,

Through the purple ether-spaces,
Through the blackened flues of Turi,
To Palwoinen's rooms uncovered.

Thus, the catastrophe struck the Turi people and their new home. Where to seek it? Where to place it? Väinämöinen and Ilmarinen rowed to the site by boat. This doesn't necessarily mean the destination was an island. Waterways, we may recall, were the primary, if not the only, transportation routes. However, the *Kalevala* gives us a definite foothold to identify the Turi settlement as an island, and specifically in the 15th canto.

Lemminkäinen's mother is joining her slain son's bones and members together again and asks a bee or honey-birdling to bring life-restoring honey, a magic balsam. The bee first flies to retrieve honey from the realm of the forest ruler Tapio,

But the balm is inefficient,
For her son is deaf and speechless.

Next, she dispatches the bee to the Turi people, instructing him as follows:

Little bee, my honey-birdling,
Fly away in one direction,
Fly across the seven oceans,
In the eighth, a magic island,
Where the honey is enchanted,
To the distant Turi-castles,
To the chambers of Palwoinen

And the bee sets off:

On the verdure does not settle,
Does not rest upon the flowers;
Flies a third day, fleetly onward,
Till a third day evening brings him
To the island in the ocean,
To the meadows rich in honey,
To the cataract and fire-flow,
To the sacred stream and whirlpool.

Thus, the bards who sang the *Kalevala* remembered that the Turi people lived on an island, next to a fiery 'cataract', and along 'sacred', i.e. taboo, waters.

SILVERWHITE

Yet, to what extent can we trust the resilience of the epic's 'memory device'? It has endured the test of time astonishingly well. There are no oak forests in Finland or Karelia, something which Professor Kuusi has noted. Only in southwest Finland can lone oaks be found, relics of a more ancient epoch. For the last two millennia, Estonia's shore has constituted the northernmost limits of the species. Even so, oaks appear in Finnish folklore more frequently than birch or spruce. The roots of Finnish folklore irrefutably extend deeper into a more ancient geographic environment; into the time before the territory of Finland was settled. We may trust the memory device.

Indeed, we may conclude that the Turi/Tooru/Taara people lived on an island.

The *Pharsalia* describes the civil war between Julius Caesar and Pompey the Great. The former, preparing for battle, gathers his cohorts beneath a single flag and summons garrisons from the furthest marginal fortresses. Barbarians see this as liberation from the yoke of Rome. Listing the barbarian tribes, Lucan flies northeast from the Rhine, even mentioning the Sarmatians between the Vistula and Volga rivers. He contrasts the tribes beyond the Rhine with those in Nordic Europe. This gives us ample probability to separate the Nordic peoples from the Celts and the Germanic tribes. The poet's rhetoric is straightforward. He speaks in the second or third person about barbarian tribes rejoicing over the Romans' departure, calling them by name. But then, for the first time in the whole 52-line passage, he abandons the structure because he's reached indeterminate distances where the tribal names are unknown. Instead, he employs mythical names, one of which is particularly noteworthy. Among those rejoicing are

> those who pacify with blood accursed [...]
> and Taranis's altars [...].
> And you, ye Bards [...]
> pour forth in safety more abundant song. [...]
> Your dwelling-place, and forests far remote.
> If what ye sing be true, the shades of men
> Seek not the dismal homes of Erebus
> Or death's pale kingdoms; but the breath of life
> Still rules these bodies in another age –
> Life on this hand and that, and death between.
> Happy the peoples 'neath the Northern Star
> In this their false belief; for them no fear

BENEATH TWO SUNS

Of that which frights all others: they with hands
And hearts undaunted rush upon the foe
And scorn to spare the life that shall return.

Taranis belongs to Celtic mythology.

The Celts, in addition to Pytheas, could have carried reports of the catastrophe on Saaremaa to the Mediterranean region. Their magnificent expansion reached its climax shortly before the days of Pytheas and Apollonius. On the southern shore of the Baltic, they became the Baltic Finns' neighbours across the sea; to the west and the south, they reached England and Italy, pillaging Delphi in 279 BCE. When Alexander the Great asked them before a decisive battle what they feared most, they replied with an enchantingly frank statement: nothing except the sky falling.

Taranis, whose altar is sprinkled with human blood, carries the epithet 'Thunder'. The Celts' Taranis is the same god of thunder as the Baltic Finns' Tooru/Taara and Ukko, the Germanic Thor, and the Roman Jupiter. Unlike the last-mentioned, Taranis is symbolised by a wheel – a mark representing thunder or the sun. A spiralling wheel depicts a thunderbolt. The meaning of this wheel is the same across the entire Baltic Sea, including on Saaremaa. Celtic folklore adds embellishing details to the human sacrifices that Lucan mentions. The ritual is established in the global motif of a hero's triple death: to complete a seemingly impossible task, he must overcome death by wounding, fire, and drowning. In Baltic-Finnic folklore, it directly corresponds to the wound struck to Väinämöinen's knee, Väinämöinen striking fire or visiting the iron-weavers of Toonela more ambiguously, and Väinämöinen's drowning for seven years. These common threads in folklore appear to point to long-established cultural contact between the Celts and Baltic areas. Scandinavian archaeology does not support this belief, pointing out the noticeable lack of Celtic artefacts in archaeological sites. A Scandinavian vacuum is mysterious, because both earlier and later eras show intense trade between the peninsula, continental Europe, and the Mediterranean.

Paul Ariste believes it likely that a common proto-European substrate exists in Baltic-Finnic and Celtic languages. If this is true, then it is hard to dismiss the similarity in sound between our *tooru/taara* and the Celtic *Taranis*. *Tooru/torum* can be traced back to the proto-Finno-Ugric language. Consequently, the direction of borrowing could only be from east to west. Do we see a concentric circle expanding out from the Kaali crater once again? Taranis is prevalent in the Central European areas of

SILVERWHITE

Celtic settlement but is totally absent on the island of Ireland and is only represented in Welsh by two questionable linguistic myths (MacCana 1973: 12–23). The Celts only invaded the British islands late in the 4th century, slightly prior to Pytheas's journey. Does this allow us to conclude that the Baltic-Finnic *tooru/taara* developed its features earlier and spread to become the Celtic Taranis and Scandinavian Thor after the given period? Tarand-type graves (communal burials within rectangular above-ground stone-wall enclosures) typical of those in Estonia have been discovered north of Sweden's Mälaren and near Täby and Skålby; the word *tarand* has become a historical term in Swedish. Mårten Stenberger, professor of archaeology at the University of Uppsala, has characterised the period 300–150 BCE as such: 'There are almost no archaeological finds and it is therefore the darkest and most difficult-to-penetrate period in ancient Swedish history' (Stenberger 1977: 239, 245). Could this be the darkness following the death of the sun?

Pliny the Elder, with whom Tacitus was in close contact, adds another *tara*-stemmed name. In the distant Scythian north, the humanist and near-democratic aristocrat remarks, there lives an unusual four-legged creature, about the size of a cow, with branched antlers and a deer-like head. The animal is said to be called *tarandus*. Do we dare to identify this as the Sámi's reindeer, a domesticated animal that was previously unknown to the Romans but had to be given a name somehow associating it with the mysterious North? A pedantic reader might scoff at such a bold assumption but would grow more thoughtful if they were to glance at the word *tarando* in the modern Spanish dictionary: reindeer, *Rangifer tarandus*.

Natural myth drifted through areas of Celtic settlement and cast a password onto the shores of classical literature.

Unexpectedly, we have reached the literary sea where continents are starting to dissolve on the horizon. Sailing in Pytheas's wake, in addition to Apollonius and Lucan, were Virgil, Seneca the Younger, Antonius Diogenes, Statius, and Avienius – just to name the poets. Poetry comprises fact, of course, but requires different rules in reading. Perhaps Adam of Bremen knew these rules better than we do. When he began composing a church chronicle of northern Europe ten centuries after Lucan's lifetime, he used the *Pharsalia* as one Baltic Sea source.

What drew poets to Pytheas's path?

A distant island that receded further with each passing century, losing its true outlines and acquiring new ones in relation to humans' hopes, longings, and ideals. Non-existence transformed into utopia, but utopia

BENEATH TWO SUNS

crystallised into a model. I have in mind, of course, Thomas More's Utopia and envision Campanella's City of the Sun; I understand Rousseau's yearnings and see a rising bourgeoisie through the eyes of Robinson Crusoe spying on natives singing their songs on the sea instead of drilling for oil. There have been many such islands, and many more are to come. When the horizon begins to shrink and turns into a small glowing blue disc at the height of satellites, they will find their final shelter in hearts, hopes, and illusions; on an arena split by the thirst for justice and lust for gain. In moments of weakness, we all carry islands within us, overlooking the fact that the continents carry us.

Helios!

A FINGER BRUSHES THE SUN

Are we able to prove that Pytheas visited Saaremaa?

The fragments of proof can, to summarise, be concentrated into five lines (Mette 1952):

> (a) On 21 June, the ratio between a gnomon and its shadow in Massalia is 120:41.80, and the length of a day is 15.25 hours. (Pytheas > Hipparchus > Strabo)
> (b) In the place marked with the letter B, the height of the sun on 22 December is 9 cubits. (The same; and Strabo)
> (c) In place C, the height of the sun on 22 December is 6 cubits. (The same)
> (d) In place D, called *helios koimatai*, the height of the sun on 22 December is 4 cubits and the length of the day on 21 June is 18 hours. (The same; and Geminus)
> (e) In place E, the height of the sun on 22 December is less than 3 cubits and the length of the day on 21 June is 19 hours. (The same)

Pytheas used a gnomon, a cubit called a *pechys*, and an hourglass that measured time with 15-minute precision. A Greek cubit is 0.444 metres. One cubit contains six fists (*palaiste*, 0.074 metres), i.e. 24 fingers (*daktylos*, 0.0185 metres). In addition to these, he uses a ship's day, the length of which is unknown and depends on the speed of travel or the destination, which we are still determining.

Let us begin where Pytheas did.

A gnomon is a simple staff that casts a shadow. The direction and length of the shadow change over the course of a day and a year. The

103

SILVERWHITE

changes vary in different places, depending on the distance from the equator. Therefore, one can use a gnomon to determine their location or movement in a north–south direction. The shadow's daily changes are easy to understand. The earth rotates towards the sun, which we perceive as sunrise. As the sun rises in the east, the gnomon casts a shadow to the west in the morning. Summer shadows are shorter than in winter. The shadow's changes over the course of a year depend on the earth's axis, which is slightly tilted. When the upper end of the axis and the northern hemisphere are angled towards the sun, it is summer. The sun rises higher, shadows are shorter, and days are longer. The longest duration of sunlight is experienced on 21 June when the angle reaches its greatest. The tilt is 23° 27′ 8″ and is called by the Greek loanword 'ecliptic'. Pausing for thought, we could consider how much we owe to this phenomenon. Winter starts to recede, the dripping of snowmelt fills the air, buds swell, and poets reach for their pens. If there were no tilt, there would be no awakening. Identical days would shuffle past monotonously and disappear from sight. Every day, every season, every year would be a replica of the one before. The Earth's tilt is one of the most important factors in the environment's incessant changing and in species' adaptation to change. Classical authors had a better grasp of this and it was last praised in Ovid's *Metamorphoses*.

The equator faces the sun year-round, days and nights there are equal, and the gnomon's shadow walks its dull path day after day. But the further away you move from the equator, the shadow shrinks more remarkably as summer approaches and grows towards winter. By observing the shadow's change, one can, with simple rules, determine one's geographic latitude, which is to say one's distance from the equator. If observations are taken in two different places, then the opportunity arises to determine the sites' relative and absolute distance. The former could be worded this way: Öland is two times further north from Massalia than it is from Cornwall. To determine absolute distance, one must know the earth's circumference. The length of a gnomon's shadow naturally depends on the height of the sun. In order to compare these lengths in different places, measurements must be taken on the same days and at the same times. Solstices are most opportune for this, as the gnomon can be used as a calendar and a precise clock simultaneously. For example, the gnomon marks astronomical midday on 21 June by casting its shortest shadow of the year. And vice versa: on 22 December, it casts its longest shadow, which at the moment of the astrological midday is also the longest midday shadow of the year. Having an exact date of measurement matters most when otherwise

104

BENEATH TWO SUNS

lacking a sufficiently precise calendar system. The height of the midday sun gives one the geographic latitude at which the measurement is taken. At the summer solstice, the ecliptic tilt gives that much greater longitude and the angle should be subtracted from the sun's height. At the winter solstice, the sun shows a lower latitude and therefore the ecliptic angle should be added to its height. In the absence of a gnomon, geographical latitude can also be expressed simply by the height of the midday sun or the length of the day. This, as we can see, is what Pytheas did. In Massalia, he measured the gnomon's shadow and the length of the day. He took a simpler approach when travelling, measuring the height of the sun at winter solstices and the length of the day at summer solstices. This apparent inconsistency will, we'll later see, turn out to be a consistent pursuit for precision.

On 21 June, the ratio of the gnomon to its shadow is 120 : 41.80.

Two figures: 120 and 41.80. The former evokes credibility; and the latter, questions.

Other pairs could have expressed the same ratio: 30 and 10.45 or 600 and 209. The latter would particularly have provided a simpler, more general overview. We could, for example, say that a room is 8.64 metres long and 4.32 metres wide. It could also be put in more straightforward terms: the room's ratio is 2 : 1. It all depends on what information we wish to convey. The first pair contains double the information. Why did Pytheas prefer the clumsier option?

The reason is obvious: in addition to the ratio of the gnomon to its shadow, Pytheas also provided the actual length of the gnomon and its shadow. He did so using a measure that we do not know at first. Determining it is crucial. It will give us the key to mathematically interpreting Pytheas's measurements of sun height and transferring the outcome to a map.

Pytheas began his scientific journey by painstakingly ascertaining Massalia's latitude. Why? It points to an intention to determine the relative distance from Massalia of more important regions on his journey, and perhaps even absolute distances. Proving the latter hypothesis would require a new chapter in the history of mathematics and is not a part of our travelogue.

In order to pinpoint Massalia precisely, Pytheas needed a tall gnomon that would cast a sufficiently long shadow even on the summer solstice. The longer the shadow, the easier it would be to measure and the smaller any probable degree of error – up to a certain point. And where is that point?

105

SILVERWHITE

The ratio of the gnomon staff to its shadow is the tangent of the sun's angle. In order to determine Massalia's location with a precision of *up to* 10 kilometres, the sun's position would have to be measured with a precision of more than 5 arcminutes. Such accuracy could only be achieved if the length of the shadow was measured in increments of a fifth of a *daktylos* (finger). That fifth, about 3.5 millimetres, is the lowermost limit of what Pytheas could accurately measure.

Let us suppose that Pytheas measured the height of the Massalia gnomon and the length of its shadow in fists. One-fifth of a fist is about 14 millimetres. Error at this amplitude would result in a quite inexact latitude.

Of course, we must ask ourselves whether the height of the Massalia gnomon could have been 120 fists or 120 cubits. The first would translate to 8.88 metres, the second to 53.28 metres. The shadows would also lengthen accordingly. Earlier, we deduced that a longer shadow would provide a more exact measurement. Now, we must add that the claim only stands up to certain limits. The gnomon's precision depends on the verticality of the staff. The taller the gnomon, the greater the probability it will deviate from a perfect right angle. Thus, the gnomon could not be too short because that would increase the likelihood of errors in shadow measurement, but it also could not be too tall because that would increase the likelihood of erecting the gnomon at a 90-degree angle.

We have arrived at our first conclusions:

1. Pytheas used the *daktylos* to measure length: a finger measuring 18.5 millimetres.
2. The Massalia gnomon's ratio contains both the height of the staff and the length of its shadow in absolute values, which Pytheas presented in *daktylos*.
3. Pytheas presented the Massalia gnomon's readings as tangential functions.
4. The height of the Massalia gnomon (120 *daktylos*) was 2.22 metres, and on 21 June it cast a shadow with a length that did not exceed (41.8 *daktylos*) 773 millimetres.
5. When necessary, Pytheas measured to the fifth of a *daktylos*, i.e. 3.5 millimetres.

These conclusions can be partially checked.

Crediting Pytheas with such accuracy arouses doubts, and justifiably so. Coincidentally, one literary critique written by Hipparchus, the most famous classical astronomer, has been preserved – the only text out of his prolific scientific research that was not destroyed. In it, he challenges the astronomer Eudoxus to achieve Pytheas's precision: 'Eudoxus is mistaken in regard to the polestar. [...] There is not a single star at the pole, but it is an empty space near three stars that, together with the pole, form more or less a rectangle, as Pytheas of Massalia has demonstrated.'

These lines reveal a flash of character: keen observational skills paired with impeccable vision. The celestial pole was indeed empty 23 centuries ago. The head of a pin held in an outstretched hand would cover the space. Pytheas's level of precision to which Hipparchus alludes here is even a degree greater than his accuracy with the Massalia gnomon.

Hipparchus and Strabo convey to us a line that was recorded with Pytheas's stiff fingers in early winter: 'Hipparchus says, at all events, that at the B. [...]'. Strabo writes out the toponym, but let us abbreviate for now: 'at the B. [...] at the winter solstice the sun ascends at most only nine cubits [...]. But Hipparchus, trusting Pytheas, puts this inhabited country [...].' Let us leave Strabo to argue with Hipparchus and Pytheas while we attempt to unravel the meaning of this absurd claim: the sun rises to a height of nine cubits.

The position of a celestial body in relation to the horizon is measured by altitude. If a scientist encounters an unexpected natural phenomenon while away from the observatory, they may use their hand to make a measurement. First, they determine its distance from the pole (azimuth). To do so, they extend their right arm, create a right angle at the wrist, and lift their eyes so that the arm aligns with the horizon. When the body lines up with the index finger, then its altitude in relation to the horizon can be determined: 9.5 degrees, a contemporary observer would murmur without realising that they had awakened the ancient Greek cubit, fist, and *daktylos*. But so it is. An arm slightly bent at the elbow is one cubit away from the eyes of the observer. The width of a hand from the last or even second-to-last joint of the pinkie is 74 millimetres. Under these conditions, the altitude genuinely is 9.5 degrees.

As no simpler method exists, we may assume it is the one that Pytheas employed. To achieve greater precision, he used a finger instead of his hand. Pytheas's measuring finger, *daktylos*, is 18.5 millimetres wide at the joint.

Pytheas used the tangential function to describe the Massalia gnomon and its shadow. It sounds complex but is simple: he presents the ratio of two catheti (adjacent legs) of a right-angled triangle. In other words,

SILVERWHITE

altitude. Judging by Hipparchus's quote, we can conclude that Pytheas was a professional astronomer and projected triangular coordinates onto the sky: 'together with the pole', three stars form 'more or less a triangle'.

* * *

The winter solstice crept up as we busied ourselves in deliberation.

Pytheas extends his right arm, bends his hand at the wrist to form a right angle, and aligns the base of his index finger with the horizon. He ensures that his hand is exactly one cubit away from his eyes. Perhaps he double-checks with a cubit-long measuring string, one end held between his teeth and the other tied to his thumb. Arabs practised a similar method with a tool called a *kamal* several centuries later. A fair amount of space is left between Pytheas's measuring finger and the sun. Without moving his right hand, he reaches his left arm out to the distance of a cubit as well and rests his left index finger on his right. 'A second full cubit,' he says to himself. A gap still remains between both fingers and the sun. Keeping his left finger still, Pytheas repeats the action with his right hand, still making sure the fingers remain exactly a cubit away. 'A third full cubit,' he says to himself. He is frozen in place, his gaze unmoving, his arms stretched toward the sun as if in a curious prayer, and the barbarians we may conjure in his company gradually fall silent as well: they've never seen anything like it. Today, 23 centuries later, we can understand those worried barbarians. They bore witness to the birth of the three-dimensional world, and it was full of foreboding. There is no space for barbarians in a three-dimensional world.

Nine times, Pytheas placed one finger next to another a cubit away from his face. Nine times, he increased the angle between the horizon and the sun until his right index finger finally brushed the upper edge of the solar disc. He had done so the previous days and did so on the following: he needed the solstice. But what if it happened to be cloudy? Pytheas wasn't a gambler: Pytheas was a scientist.

He starts to get cold and ducks into a shelter, where he warms his hands by a fire. He is satisfied with the results, records them. Perhaps he's already thought up the title of the work: *About the Ocean*. We do not know. We aren't even aware if he ever finished it or where, and how he died.

'Here,' he writes, 'at the B., the sun rises nine cubits on the winter solstice.'

* * *

BENEATH TWO SUNS

These words were also put into Pytheas's mouth, but, unlike in previous instances, this has the weight of argument and is no longer speculation: the claim can be mathematically checked.

Pytheas increased one cathetus of the triangle nine times until his index finger brushed the upper edge of the sun. The upper edge? Day dawns with the first ray, ends with the last, and the upper edge of the sun's disc is day's envoy. Nine *daktylos* is 166 millimetres. The second cathetus of the triangle remained the stable length of a cubit: 444 millimetres. With this ratio, the sun's altitude from the horizon is 20 degrees and 30 minutes. The sun, as we now know, is a sphere, and its movement should be determined by its centre, not its edges. The sun's circumference is 32 arcminutes. This means the altitude of its centre must have been 20 degrees and 14 minutes from the horizon. At the winter solstice, the sun lies low on the horizon in this region. The northern hemisphere has tilted away from the sun and its ecliptic dips exactly 23 degrees and 27 minutes below earth's equator. If we combine both angles, we get 43 degrees and 41 minutes: the distance of the observation point from the North Pole. Latitude reads the opposite way on our maps, i.e. from the equator to the pole. We must do one more equation: $90° - 43° \, 41' = 46° \, 19'$.

Pytheas took his measurements somewhere at a latitude of 46° 19′.

Shedding the abbreviation, Strabo's quote reads: 'at the Borysthenes [...] at the winter solstice the sun ascends at most only nine cubits [...]. But Hipparchus, trusting Pytheas, puts this inhabited country [...]' The Borysthenes is the Dnieper, but it is also a Greek city-colony also known as Olbia, which was located at the mouth of the southern Bug. Both rivers flow into the same bay 50 kilometres apart. Factoring in the minute changes that have taken place in the ecliptic over the course of 23 centuries, we get a minimum error of 20 kilometres (11 arcminutes), a maximum of 80 kilometres (43 arcminutes), and an average of 50 kilometres (27 arcminutes). The Massalia gnomon's ratio places Pytheas's hometown at a latitude of 43° 13′. Modern nautical charts place Marseilles at 43° 18′, though this reading centres on a sea mark at the southern tip of the breakwater at the city's fore-harbour. Pytheas's gnomon most likely stood in the centre of the market square on the Vieux-Port de Marseille, a natural inlet nearly 700 metres south of the present-day sea mark. He erred by 8,600 metres, 4⅔ arcminutes, when ascertaining Massalia's location. This is an outstanding level of precision that would be unsurpassed for several centuries, and marks the birth of modern navigation.

SILVERWHITE

PYTHEAS ANSWERS

However, we are interested foremost in Saaremaa. We'll get there in a moment.

Did you notice Pytheas's quandary? In two places, he tries to determine latitude using two different methods: he measures the sun's altitude at the winter solstice and the length of the day at the summer solstice. Why this inconsistency? The length of the day would also have provided a precise-enough latitude.

The sun is on the horizon. Let us suppose it is a Nordic sun. Unlike the Mediterranean region, it is not preceded or followed by night: something simple enough for us northerners to understand, but the first experience of white nights he and his scientific world have ever had. The day begins and ends with the upper edge of the sun's disc aligning with the horizon. Does the definition of a day still apply when it fails to end? Even where no night follows? Pytheas had to first determine the method and then the latitude. The criteria for defining a day were uncertain and, in addition to its length, he measured the altitude of the sun in winter.

Why not vice versa? For two reasons. Measuring altitude when the sun is high in the sky in midsummer would increase the probability of error. Solar altitude is much easier to measure in winter than in summer. Secondly, measuring the length of a day requires one to observe the sun twice over a 24-hour period: at sunrise and sunset. One must follow the disc. Sunny weather is highly improbable in northern Europe during the month of December. Nevertheless, the sun's zenith at midday can be determined even with dense cloud cover. And the length of a day is significantly easier to measure in summer than in winter.

Apparent inconsistency is becoming a consistent pursuit of precision.

Rather grudgingly, I must now employ the terms 'civil' and 'nautical' twilight and explain what they mean. At the equator, the sun rises and sets at a right angle to the horizon. Day turns to night like switching off a light. The further one is from the equator, the longer twilight grows at the limits of day and night, and past the Arctic Circle there is a belt where the midsummer sun never sets: it slants up from the horizon, reaches its zenith to the south, and returns to the horizon by midnight without dipping below the field of vision. Contrary to the equator, the sun's daily path runs at a slight angle to the horizon. An array of transitions develop as one crosses latitudinal lines and moves closer to a polar circle: the angle of sunrise and sunset decreases, days lengthen, and the sun remains hidden behind the horizon for decreasing amounts of time. When the angle of the

BENEATH TWO SUNS

sun's descent is less than 12 degrees, the horizon stays clearly perceptible even at night, and seafarers can measure the altitude of celestial bodies and set their vessel's course accordingly. This is called nautical twilight. Even further north, the sun's angle of descent is less than 6 degrees, its light scatters across the land at night, the sky remains bright, and dawn and dusk blend together. The time is colloquially known as white nights and astronomically as civil twilight.

However, the shifting ratios of days and nights are not evenly distributed from south to north. For instance, twilight lengthens thirty times faster from the north coast of Saaremaa than it does on the path beginning from Massalia. Pytheas, who had never before encountered such twilight phenomena, could have had serious difficulties defining the criteria for day and night. This assumption is proven by Geminus, who wrote that the longest day lasts 17 and 18 hours, and night lasts only 2 and 3 hours, meaning 'the sun began to rise again a very short time after setting'. This is a highly important quote that has nevertheless remained unappreciated. Let us rearrange the numbers: day lasts 17 hours and night 3 hours, day lasts 18 hours and night 2 hours: in either case, this amounts to a 20-hour day, not 24. Where did those four hours go? The answer is obvious: Pytheas had trouble judging twilight. We're able to phrase an even more accurate conclusion: when measuring the length of a day at home in Massalia, Pytheas applied astronomical criteria from sunrise till sunset; his hourglass was off by only a minute. Northern white nights forced him to abandon this method.

Does the scant material that has been preserved permit us a more exact interpretation?

It seems as if Pytheas anticipated this issue and therefore presented geographic latitude in two ways: solar altitude in addition to the length of a day. He strove to measure the latter as accurately as he could. Judging by Geminus of Rhodes's quotation, the four lost hours should be attributed to twilight. Therefore, we could reword Pytheas's findings as follows: the day, including twilight (dawn and dusk), lasts $17 + 4 = 21$ hours, and night is 3 hours accordingly. Alternatively, day, including twilight, lasts $18 + 4 = 22$ hours, and night 2 hours. The two measurements could be based on two different places, though not necessarily. Pytheas could have devised his own twilight-defining method that is unknown to us. In this case, the two calculations would be from different places: day and twilight last 21 hours in the more southern location and 22 hours in the more northern. Previously, we discussed the length of summer days increasing

SILVERWHITE

exponentially as one moves towards the pole. In the zone where twilights last four hours, it takes a mere distance of about 80 kilometres to lengthen daytime (dawn + astronomical day + dusk) by an entire hour. In other words, no matter whether our definition of civil twilight coincides with that of Pytheas, both areas could not have been further than 80 kilometres apart. Pytheas may have been uncertain about defining our northern summer days. In that case, the quote should be understood as an approximate estimation: day (including dawn and dusk) lasts 21–2 hours. In either case, the length of Pytheas's day can be mathematically interpreted and placed on a map with an accuracy of up to 80 kilometres.

A summer solstice day that lasts 21 hours is positioned at the latitude 58° 30′. One lasting 22 hours is at 59° 20′. A summer solstice day lasting 21–2 hours is positioned between those two limits.

Now, we can finally compare the latitudes that Pytheas determined on the basis of the length of a summer solstice day, the altitude of the sun at the winter solstice, or both. Additionally, we can factor in the knowledge that Pytheas tended to err towards the south when taking his solar measurements: he positioned Massalia (43° 18′) 4.6 arcminutes, and Borysthenes (46° 45′) a full 26 arcminutes too far south. In short, due to reasons of which we're now aware, Pytheas erred by 6 arcminutes or 10 per cent to the south on average with every latitude, beginning from that of his hometown.

In figures, Pytheas's journey begins to look something like this:

	1	2	3	4	5	6
Massalia	15¼	43°	120 : 41⅘	43° 13′	43° 18′	43° 09′
Borysthenes	–	–	9	46° 19′	46° 45′	46° 45′
'Keltike'	16	47° 19′	6	52° 31′	53° 34′	47° 19′
						53° 34′
'Helios	21	58° 30′	4	57° 05′	58° 29′	58° 29′ 30″
koimatai'	22	59° 20′				58° 54′ 30″
'Even further'	23	60° or more	>3	>59° 26′	>60° 36′	>60° 18′

The measurements and results presented in the chart are as follow:

(1) the length of daytime on 21 June in hours, (d) and (e) including civil twilight;
(2) latitude calculated from the latter;
(3) altitude of the sun on 22 December in cubits; the Massalia gnomon on 21 June;
(4) latitude calculated from the latter without adjustment;

112

BENEATH TWO SUNS

(5) latitude calculated from the latter with adjustment of 6′ latitude;
(6) average latitudes calculated using both methods.

The latitudes of Massalia and the Borysthenes are known to us, can be checked, and turn out to be outstandingly precise. Although the last two regions are unknown, the latitudes calculated using the different methods (columns 2 and 5) are remarkably close; too close to assume coincidence. The variation for the 'Keltike' region (columns 2 and 5), on the contrary, is too great and allows us to conclude that the observations were taken in two different places. Sixteen-hour days occur at the latitude 47° 19′. At the winter solstice, the sun rises to a height of 6 cubits at the latitude 47° 19′. The first points us to the Loire and the second to the mouth of the Elbe, perfectly coinciding with Pytheas's presumed route.

Day-length readings (column 1) and the word 'Keltike' were conveyed to us by Pliny the Elder. Solar altitude readings (column 3) were passed along by the astronomer Hipparchus, who was renowned for his precision. Pytheas via Hipparchus via Strabo also gives us the absolute distance of Keltike from Massalia: 6,300 stadia (1 stadium = 185 metres). Placing Keltike at the mouth of the Elbe with a latitude of 53° 34′ gives us a distance of 1,170 kilometres from Massalia. Thus, our calculations diverge from those of Pytheas and Hipparchus by a mere 30 kilometres. Therefore, we may trust the calculations within the limits of this error, bid farewell to Keltike at the mouth of the Elbe, and enter the final stretch of our mathematical marathon, which leads us to the oldest mystery in exploration history.

The tip of the Sõrve Peninsula extending southward from Saaremaa has a latitude of 57° 54′. Geologically, it was a separate island in Pytheas's time, only connecting to the main island later. The latitude of Saaremaa's northern Pammana Peninsula is 58° 39′. Lake Kaali's latitude is 58° 22′ 21.82″. Details conveyed to us by the astronomer Geminus concerning a place that 'Pytheas of Massalia apparently visited', one that we have so far referred to as *helios koimatai*, point us to the latitude of 58° 29′ 3″ (row (d) in the chart). The difference between the latitudes of Lake Kaali and our calculation of *helios koimatai* is 13 kilometres.

In addition to the distance between Massalia and Keltike, Hipparchus gives us another: that of Massalia from an area where the winter sun rises only 4 cubits, but a summer day (without twilight) lasts 18 hours. According to Hipparchus, the distance from Massalia to Pytheas's 4-cubit-high sun (58° 29′, according to our calculation) is 9,000 stadia, i.e. 1,700

SILVERWHITE

kilometres. Modern maps give us a distance of 1,688 kilometres. Pytheas erred by 12 kilometres to the south when determining latitude. If he took his measurements on the shore of Lake Kaali, then he was 9 kilometres off.

Precision is anything but a strength of sovereigns.

This forces us to point out that up to this point, we have only discussed latitude. That of Lake Kaali, as exact as it may be, also cuts through Sweden, the Norwegian coast, Labrador, Alaska, the Kamchatka Peninsula, and Lake Peipus. Latitude alone does not allow us to determine a definite geographical point. For that, we need another coordinate: a meridian intersecting the latitude. Determining a meridian requires one to compare the time at the harbour of origin with local time in the unfamiliar region, i.e. precise chronometers; and such a chronometer was not built until 1775 at the hands of watchmaker John Harrison: a saucer-sized spring-driven clock that would fall 15 seconds behind on a 156-day voyage. Pytheas possessed no chronometer, and thus his *ultima Thule* lacks a meridian. Without that measurement, it is practically impossible to pinpoint a place, no matter how exact the latitude.

And, gently as dawn, a miracle rises over the Baltic-Finnic cultural landscape. Let us interpret this 'miracle' as the confluence of improbable chances. The reference to Pytheas in Hipparchus's writing is a miracle as well, as all the astronomer's other works are lost. But for those very lines to be preserved … The miracle is the fact that Pytheas also gives us a meridian.

Earlier, we noted that the linguistic origin and meaning of the word 'Thule' remain a mystery. The *Oxford English Dictionary* claims so, and it is wrong.

Tuli, 'fire', is an Estonian word with roots that reach back to the proto-Uralic language. It is also *tuli* in Finnish, *tul* in Mordvinian, *tul* in Mari, and *tyl* in Udmurt. In Komi, *tyl-kort* is a fire-striker. The word can also be traced to further-removed kindred languages (Collinder). It was common along the Baltic shore already in the Neolithic period. In the year 350 BCE, its nominative case was *tule*. The root can also be found in Celtic languages with the meanings 'pillar, nail, or phallus' (Pokorny). Let us remember, for a moment, the ancient Baltic-Finnic shaman Virokannas. For the Goths of Gotland, *tyle* was, somewhat blurrily, a spell-whispering shaman or oracle portrayed more clearly by a strophe in J. de Vries's *Altgermanische Religionsgeschichte* (De Vries 1970: II, 389).

BENEATH TWO SUNS

Another astronomer specifies the meridian. What does Geminus's *helios koimatai* mean? *Helios* is the sun. The verb means 'to lie down to sleep', 'to doze off', and metaphorically 'to slip into the sleep of death'. Here, I'd like to thank Professor Tuomo Pekkanen of Helsinki for helping me to delineate those nuances. We should also show appreciation to Geminus, who, over a distance of millennia, has handed us a single authentically preserved line from Pytheas, which from this point forward should read as such: 'The barbarians showed us where the sun fell into the sleep of death [earlier read as: where the sun goes to sleep].' And Cosmas Indicopleustes adds that far in the North, barbarians showed Pytheas the 'coffin of the sun [earlier read as: the cradle of the sun]'. Meridian crosses latitude.

When determining the latitude of Lake Kaali, Pytheas erred by 9 kilometres. Or did he?

Cultural history accredits the first exceptionally precise measurement of longitude and earth's circumference to Eratosthenes of Cyrene (282–202 BCE). Eratosthenes, who headed the Library of Alexandria after Apollonius of Rhodes and was an outstanding geographer (and average mathematician), was nicknamed Beta. Among other details, Pytheas's unusual precision in practical observations and theoretical generalisations demonstrate that he managed to measure his longitude even before setting off on the voyage, likely by travelling along the Rhône's longitudinal valley and observing the changes in the heavens. This possibility, as tempting as it may be, no longer has room in this work, and even if it does, then only to ask: Eratosthenes was Beta, the second. Does that mean Pytheas was the first, Alfa?

Tin led him to Cornwall, amber to the Baltic. Here, he heard an outlandish story about a ball of fire that landed upon a sacred island, astonishingly similar to the Greek myth of Helios, a god who charges across the sky in a gold sun-chariot and lands in the hallowed waters of Oceanos in the evening.

Pytheas was a scientist above all. How could he have refrained from journeying to the Isle of Fire?

He saw, he detailed, and that sealed his fate. Pytheas wandered far ahead of his time, which is just as tragic as missing your time by a day. *Helios koimatai* and *helioy koiten*: the sun, according to every rule of mythology, descended into the Kaali crater for slumber, falling into the sleep of death.

This burden was far too great for a travelogue, and in the eyes of scientists – not poets, mind you! – Pytheas's days were numbered. There was no longer space in the scientific world for a place where the sun 'fell into the sleep of death' or 'went to sleep' It was absurd, evoking more pity than humour. Strabo wrote that everything Pytheas reported about Thule was imagined and brazenly concocted, though he also added in recognition: 'And yet, if judged by the science of the celestial phenomena and by mathematical theory, he might possibly seem to have made adequate use of the facts as regards the people who live close to the frozen zone.' Strabo might have been envious as a fellow adventurer, but was fair as a scientist. He remarked in Pytheas's account of Thule a mythological anachronism that stood in striking contrast to the exact description of the dawn-and-dusk phenomenon. Hipparchus and Geminus must have held Pytheas's measurements in even higher regard: both were astronomers. No one from classical civilisation had ever ventured as far north as Pytheas. His accounts confirmed later theoretical deductions in practice. It was unfathomable. Despite his ridiculous 'cradle of the sun', Pytheas turned out to be the only professional scientist whose observations confirmed the alternation of long summer days and long winter nights. It was as clear as day to Geminus that Pytheas was not lying; that it was impossible for him to have lied with such mathematical accuracy 250 years earlier. 'The barbarians showed us a place where the sun fell into the sleep of death.' To rescue Pytheas's prestige, researchers have interpreted this as the northwest. Pytheas could hardly have needed barbarians' explanation of where the sun set, and it is highly unlikely that Geminus read it as 'northwest'. For why, and how, did he happen to quote that very line – perhaps the most important of any in Pytheas's manuscript, even if it has been preserved in its entirety to our day? It continues: 'for around these places it happens that the night becomes very short, two hours for some, three for others' (Evans and Berggren 2006: 6, 8–9). And so it was: Pytheas sailed through the Åland Islands ('even further', 60° 18'), skirted the coast of southwest Finland, and reached the Estonian shore (58° 54' 30": Hiiumaa? Muhu Strait?). Even so, we should grant ourselves the right to surmise something more.

The ship came to shore. He disembarked. The walk was a mere hour – at that time, the coastline cut four or five kilometres inland. He passed a cemetery. The time of arrival was fortuitous. He was shown the place where the sun collapsed: a gaping crater, charred chunks of oak, bowels of the earth turned inside out, and glinting iron that fell from the sky. It was summer, meaning white nights. But he would also see dark days, as he measured the altitude of the Christmas sun there as well. He describes the heavy breathing of the freezing sea, a mixture of land, water, and air,

BENEATH TWO SUNS

calling it 'sea lung'. He describes the slushy salty water that can no longer carry a ship but also cannot yet bear a man, calling the ice-imprisonment a shackle around the world – something that rings with desperation. Ultimately, he also manages to describe firm sea ice: it begins a day's sail away from Saaremaa/Thule, corresponding with the isomer that gives a 0.7–1.00 likelihood of ice formation at the mouth of the Gulf of Finland. He measured, he observed, he questioned.

Estonians showed him 'the sun's endless deathbed' because the nights were light when the curious Greek arrived by sea and the midsummer bonfires were burning brightly.

What did we Estonians sing 2,300 years ago? Who may peer into such a distant past, crack open long-closed doors, and touch fates, stones, charred earth, and budding wisdom, a smooth oar creaking in the rowlock?

> I remember yet the giants of yore
> [...]
> Of old was the age when Ymir lived
> Sea nor cool waves nor sand there were
> [...]
> Winters unmeasured ere earth was made
> Was the birth of Bergelmir;
> This first knew I well, when the giant wise
> In a boat of old was borne
> [...]
> Whence comes the sun to the smooth sky back,
> When Fenrir has snatched it forth?
> Daughter bright Alfrothul bears
> Ere Fenrir snatches her forth;
> Her mother's paths shall the maiden tread
> When the gods to death have gone.
> [...]
> The sun turns black, earth sinks in the sea,
> The hot stars down from heaven are whirled;
> Fierce grows the steam and the life-feeding flame,
> Till fire leaps high about heaven itself.
> [...]
> Now do I see the earth anew
> Rise all green from the waves again;
> The cataracts fall, and the eagle flies,
> And fish he catches beneath the cliffs.
> [...]

SILVERWHITE

From below the dragon dark comes forth,
Nithhogg flying from Nithafjoll

(Poetic Edda)

VIROKANNAS'S LONG JOURNEY TO THE SAMPO

We are like travellers without a compass, zigzagging haphazardly in both time and space. However, we cannot ignore the mystery of our common Baltic-Finnic epic's Sampo – a magical mill that was forged by a blacksmith and that brings good fortune. There is little new in the following pages. Or is there?

The first time I laid eyes upon a wondrous Sampo was in 1962. We were riding horseback across the forest tundra of Kamchatka, our destination the reindeer herder Äiteki's winter camp. I've written about it in my book *Tulemägede maale* (To the Land of Fiery Mountains):

> The sun peeked through the clay-coloured clouds, making colours smile melancholically. A final stream, already in dusk. The growling of sled dogs, the jingling of chains. The tundra is black from reindeer tracks though none is in sight. Two or three sleds, poles for drying fish, turf piled up into a shell, all clumped close together like castaways in a lifeboat, and in the very centre, a lodge. It is a furry mountain bleached silver in the sun, crowned by poles jutting from the tip. Now I understand why it has shimmered so unattainably before us since midday: the conical dwelling is a good four metres tall. Smoke rose between the branches. A fire! At that moment, I dreamed of nothing more.

My comparison to castaways on a lifeboat seems formal and misleading in hindsight. As I am a city man, the warm dwelling was like a lone island amidst a forbidding sea of tundra. The over-dramatised attitude dissolved after sharing Äiteki's home with him.

> Bright fire, the tangle of sooty poles, and red strips of meat hung to dry like yarn. Everything was alien and surprising – the colours, scents, and dimensions. The lodge turned out to be much more spacious than I could have ever believed. Various household goods and firewood were piled up at the furthest edges, drowned in darkness. The arm-thick supporting post was worn to a shine from long use. The dangling strips of meat rattled dryly whenever brushed while leaning over. Hanging over the fire were chains with iron hooks, a pot of bubbling reindeer

BENEATH TWO SUNS

meat, and an old-fashioned copper kettle. Yellowed tundra turf served as a floor. Äiteki offered us a seat in his *yoronga*. Yes, for the lodge, *yoranga*, is merely a hide-covered courtyard, a work room protected from the gusting winter winds, large enough to also shelter family dwellings. Here, there are two. The *yoronga* is more than a simple room: it is a house within a house with its own floor, ceiling, and thick walls, warm enough to sleep naked in when the temperatures outside are frigid. Äiteki's *yoronga* is positioned across from the door, facing the fire with its baldachin-like door hanging open. Now, we will also stay here on his left, in the traditional place of honour.

I would gladly use a more laconic comparison or equivalent to present my attitude towards the Sampo, but ethnography and folklore are not exact sciences. The exclamation mark, pause, and dead gaze of stopping time play their part. Distances, both spatial and temporal, must be experienced if we wish to identify with the ancient fisher or hunter and evaluate the world through their eyes. This is not fact, of course, but interpretation: literature, if you will, that has its own rules. On that first night, I journalled for quite a long time:

> I'm sitting on a soft bundle of hide at the tea table and watching the coals cover themselves with a thickening blanket of ash for the night. Tossing and turning under the second baldachin is a sick child, its mother singing a monotone song like the wind. Soon, they fall asleep. Polaris shines through the sooty poles towering above the smooth support post, motionless. For a moment, the skittering of snow; for a moment, the whispering of tundra stalks buffeted by the wind; and then, it is again so quiet that one can hear the candle burning and the embers extinguishing.

The Sampo mystery was the furthest thing from my mind in that moment. It did not exist. It came to me only later, after returning to the literary world. On that night, I was facing a long, straight, tactile post, one end in the ground, the other in the dark smoke opening, where it disappeared from sight and reappeared as the glimmering point of Polaris or, rather, as a hole it had worn into the heavens. If you lie next to a dwindling fire and stare past a glowing post into the sky for hours, then your eyes and imagination will capture dual movement. For a while, it may seem as if the sky surrounding Polaris is unmoving, and that as night takes hold, the lodge is what starts to slowly spin around its axis, around the world's axis, for there is nothing else in the void. Then, in turn, it may seem as

SILVERWHITE

if the heavenly arc (which imitates the lodge's conical shape so closely!) is spinning around your home axis along with the entire universe, and that egocentric perception is perhaps even more profound. Yet either way, the lodge's hairy arc and the cosmic field of stars blend together: if one has been crafted by a human hand, then why not the other, its greater reflection revolving around the home-pillar just as obediently? The Kalevala also equates the Sampo's creator to that of the heavens:

> Since I forged the arch of heaven,
> Forged the air a concave cover,
> Ere the earth had a beginning

This gradual revolution found a later parallel in the rudimentary farmer's magical mill and quern-stones, though long before that, for millennia, it was the poetic symbol of life, work, and prosperity. The post was also much more than the world's egocentric axis, around which life revolved: it was a connection. The post connected the infinitesimal present moment with the bottomless past and boundless future. Strangely enough, the connection of past, present, and future was as strong at that time as it is for us in the era of the Gutenberg Galaxy. That is at least how I believe I can interpret an experience I had on a later expedition to Chukchi lands, which I describe in my book *Virmaliste väraval* (At the Gates of the Northern Lights).

On that trip, I met a man named Attata, to whom I owe my knowledge of many fascinating Chukchi fairy tales. He had an outstanding memory. Attata refused to show me his writings. Those people are trustworthy because they are not vain. His father was a sailor on Roald Amundsen's *Maud*, and the two later went on long hunting trips together. Attata and I were lying on the beach. 'Polaris is called *Umpener* in Chukchi, which translates to Star-Post-Rammed-into-the-Ground,' he told me. 'The stars move around it like reindeer on a halter.' This was nothing new, of course. The surprise came elsewhere. Above Polaris is believed to be a hole through which one passes from one world to the next. A week earlier, Attata watched the making of a drinking staff. The tool was 25 centimetres long, 2 centimetres wide, and only hollowed at either end. 'Just to fulfil tradition,' Attata said in excuse for such carelessness. The dead are believed to use it to drink from the afterlife ('like from a well') when they aim straight downward from the hole that was worn into the sky by constant turning. Thus, the world pillar acquires a rather fundamental significance in the upper latitudes of the North: it ties the visible world to the invisible,

120

BENEATH TWO SUNS

the person to their ancestors, the ancestors to yet-unborn generations, the past to the future. It's possible that refusing to tolerate time's barriers may be one of humankind's most charming characteristics, but that is talk for another time. The Palaeo-Arctic Chukchi and Koryaks shaped their world almost like concentric cones are shaped in geometry. A hole worn into the heavens by Polaris could be confirmed by everyday interaction with a bow drill. The apparent abstraction of an axis and revolution ceases to bother us as soon as we cease to regard ourselves and our civilisation as the world's navel and the point or reason for the course of history to date. Naturally, such a plain and large-scale worldview exists in direct dependence on high latitudes.

However, to what degree are these Palaeo-Arctic conceptions adaptable to the Finno-Ugric world? And, secondly, is it possible to project the *tepee/lavvu/chum/yaranga* world post into an even more distant past and transform it into a basic, original form that is free of mythological burden? On the Taimyr Peninsula, halfway from the Bering Strait to Scandinavia, I had the opportunity to meet a Nganasan shaman named Demnime Kosterkin. The Nganasans belong to the Samoyedic branch of the Finno-Ugric linguistic tree and are the earth's northernmost people. Until just recently, they survived off hunting wild reindeer roaming their harsh, level terrain. Like other Samoyedic tribes, they migrated to northern regions from the south, likely during the 1st century of our Common Era, and developed a culture marvellously adapted to the Arctic environment over a relatively short amount of time. Demnime set up his lodge in a sheltered bend of the Avam River. His lodge lacked a post, which may have been due to the rarity of wood in that region. The Taimyr, almost as big as the country of France, is unforested. A small number of Siberian larch, *Larix sibirica*, sometimes even 2–3 metres in height, may grow on taller hillocks, but such sites are rare. Demnime appeared highly educated. He was incredibly knowledgeable about his people's history and his own mission. 'The songs come from your head on their own,' he explained. 'Everything comes from your own head.' Defining history, he said: 'The lives lived of old have now become stories.' According to Nganasan belief, the connection between the past and the future forms as follows: At the centre of the *chum* is a firepit, the dwelling of ancestors in both actual and metaphorical terms. Archaeological finds confirm that in earlier times, the Nganasan cremated their dead in home fires. The smoke of a lodge's fire is the breath of ancestors, especially when one sacrifices reindeer fat which bursts into a crackling blue flame. The breath of living souls mixes

SILVERWHITE

with the fire's smoke and rises as a pillar from the lodge. The heavens are supported on this very pillar. In olden days, Demnime explains, 'the sky was only two to three kilometres above the top of the lodge. People were few in those times. [...] Because the people grew in number, the sky kept on rising and swelling.' The sky is the Nganasans' history and rests upon their present; their life-pillar. A scarcity of wood has replaced the actual ancient post with a powerful poetic metaphor. In this grand vision, the breath should be equated to the human soul: 'In our language, the film that is born of the steam that is created by a person's breathing is called *ngohüto*,' Demnime explained. *Ngohüto*, which in direct translation means 'that formed by a person's breathing', is also the Nganasan Milky Way. 'Human life depends upon it. Human death depends upon *ngohüto*. All know *ngohüto*.'

The Nganasan world pillar is rooted in ancestors, forms an axis of the present in their lodge, rises into the sky, and folds against the heavens to become the Milky Way – in Estonian, *Linnutee*, the Bird Path.

In the tundra near Mount Obseda along the lower Pechora River, I watched Nenets (Yurak Samoyeds) erect their winter lodge. The modern day with its primus stoves and nylon jackets has made its mark on summer conditions while leaving the strict rules of surviving winter relatively untouched. The Nenets call their lodge *mja*, close to the Baltic-Finnic *maja* (house). First, they lifted two poles bound together at the tip and rested a third against them. About thirty more were then added to the equilateral pyramidic construction, making it conical. The Nenets call the first two poles *makoda*, house-spruce. The second-to-last step was erecting the sacred central *siimsi* pole (*simzõ*). Finally, the frame was covered with two layers of reindeer hides. I was Aleksandr Taleyev's guest that evening. The door, fire, and *siimsi* pole divided the dwelling into three: I was hosted on the men's side, to the left of the pole. The women's place was on the right. The segment behind the sacred pole, smaller than the right and the left, was a holy place. There, I eyed a transistor radio in place of the totems of old, though the women still refrained from treading that ground. This is nothing new, either. All I wish to do is emphasise the spruce's importance. As I mentioned, the first two poles were called house-spruce. I can make that statement more efficaciously: house-spruce are the first building blocks of the (Nenets') lodge. In the forested area of the central Puri River, the Nenets also use the name *mjad hasava*, home men, for the first two poles. Spruce and home, spruce and man, spruce and human? The most modest conclusion would speak of a reverence for the spruce or

BENEATH TWO SUNS

rather the house-spruce, for reasons we do not yet know. The sacred *siimsi* pole speaks for itself. Horizontal poles were attached to the post, and from those hung chains and pots on hooks, though the pole itself has no function in the lodge's structure. This fact naturally does not reduce but rather increases its magical burden. The name 'spruce' is not used for any other elements in the lodge but is applied solely to the dwelling's first poles, even the cut of which is unlike the others – four-sided.

My suspicion that the spruce has played a special part in the history of Uralic peoples deepened while spending time with the Ilych Zyrians and especially the Kazym Khanty, to whose lands we organised five expeditions while filming *The Waterfowl People*. We were lucky enough to photograph and later also film a Khanty *hlōnhot* – an ancient 'post hut' for ritual purposes. It was hidden in deep overgrown forest near a slow-running tributary of the Kislor River. The structure rested upon four sedulously chosen birch trunks, whose branching roots spread out across the ground like chicken's feet. Within the hut, we discovered several well-tended shamanistic items. A lodge, be it for summer or winter purposes, is not a permanent structure. The oldest permanent structure of the taiga should most likely be deemed the post-hut (Palaeo-Arctic people's post platform), which dates to the Finno-Ugric era (Paul Ariste) and stands as a memorial to semi-nomadic fishers' and hunters' transition to permanent settlement. Thus, we stood before the oldest type of permanent structure in the taiga, which at the dawn of written history already jolted Eastern Slavs' imagination and calcified in later Russian folklore as the 'hut on chicken legs' or witch's dwelling. There wasn't the slightest doubt about the fetishisation of spruce. The eaves were attached to nice-looking birch-root pot-hooks, which the Khanty and Zyrians still use to this day, and the birch-root posts gave the structure its crucial link to the living tree, i.e. to the hypothetical taiga proto-dwelling. Hence the spruce – the lodge's predecessor. Out of every species of tree found in the taiga, the spruce is the most human-friendly with its broad branches and natural shelter. As Zyrian Prokop Bogdanov – whose winter hunts we followed along the upper Ilych River in the northern Urals – confirmed to us, a hunter who stays in the forest overnight will first seek shelter beneath the boughs of a spruce even today. To do so, they rest smaller fallen trees and larger branches around the cone of a standing spruce, mimicking a lodge itself. Old Finnish and Karelian sayings remember this practice. The spruce (as well as the branches rested against one) was likely the ancient gatherer's first shelter.

123

SILVERWHITE

Such a reconstruction would explain the Nenets' enduring memory of the house-spruce.

The prospects of finding proof to back up these inferences are as good as non-existent, of course. We are speaking about extremely ancient times and the taiga retains no trace. Trees decay into soil, soil transforms into trees. We are only able to read stone-carved cultural history, and more often than not, we tend to regard the pyramids as the only history worth reading. Yet the taiga has a more enduring memory – perhaps because it didn't need the art of writing, memory's walking stick. Writing was invented by the tax collector, and the tax collector invented creative literature for his own defence. The history of the taiga survives in unwritten literature, customs, and rites. Conceptions of the human afterlife are the most lasting – better hunting grounds where one continues one's everyday activities with one's bow, arrows, boat, and favourite dog. Burial customs are extremely conservative all across the globe and preserve overall structures, acts, and texts that have been in practice for millennia.

We had undeserved fortune. There on the Taimyr Peninsula, Demnime's son Dülsimaku Kosterkin showed us the way to the nearby Düdete sacred mound. He himself didn't dare to go, having no reason to disturb the peace of the dead. A handful of larches grew atop the sacred mound. A few trunks and branches leaned against them to create cone-like lodges just like the ones Zyrians and Komi hunters make when spending the night beneath a spruce. However, Düdete's lodges were the dwellings of the dead – *matalir-ma* in Nganasan. The corpses lay in the lodge on their sleds, their feet facing north and their heads on the rear part of the sled pointing towards the door, which, as in all Nganasan lodges, is south- or, more precisely, southwest-facing. Rising in the centre of the lodge was the larch. Although slender, it stood as a striking, living pillar amidst the flat, two-dimensional tundra, adding a third dimension to the landscape and embodying the fourth dimension of time. I believe the *matalir-ma* is the missing link between the ancient gatherer's life-spruce and the Eurasian lodge.

The names given to a few details of its construction appear to confirm this hypothesis. Opening to the southwest, as I mentioned, is the door. Why southwest? In Nganasan, the direction is called *mungka*, which today means 'forest' and which earlier was used to describe a dull arrow used to hunt smaller fur animals, primarily squirrels (Collinder). The word can also be traced to other Finno-Ugric and Baltic-Finnic languages, it belongs to the Proto-Uralic language, and consequently comes from the distant

BENEATH TWO SUNS

times when the Nganasan still lived in the taiga, not having migrated to the tundra. It is a memory of a friendly forest, a former home whence 'our own people', *nja nganasad*, i.e. Nganasans, originated.

Although the limits of the taiga depend on local conditions and do not follow latitude anywhere in the world, it still lies to the south as a whole. Thus, one would expect the Nganasan to call the south the 'forest direction'. So why this shift? The answer is simple. Memory has been forced to adapt to the conditions of the newer environment. Storms generally blow in from the southwest on the Taimyr Peninsula, which means a lodge door would need to submit to the wind. We can refine this course of logic even further and say that if a person were to live without memory, then they would have placed their door opposite to the prevailing winds, meaning to the northeast, but the Nganasan did not. Leeway was limited by memory, tradition, or an accustomedness to the south, and the northeast direction represented sterile purpose. Nature forced humans to choose, and out of all possible options, humans chose that of least compromise, thereby preserving their history and adapting it to the new environment with a minor concession. The antithetical roots and crown, anchor and sail, brake and engine, weave into a seamless whole which we call history, and which we should call human.

The opposite side of the *matalir-ma*'s door faces northwest. This direction is called *djangur* in Nganasan, though the word also means 'tundra': obvious evidence of my earlier claim that the lodge's structure and terms for its positioning were in use during the days of Proto-Uralic at the latest – a time when the tundra lay entirely north of the Nganasan. Behind the burial chamber, to the northwest and thus on the axis of the door and fireplace, was a cargo sled carrying some tools and goods for the deceased. Holes were made in the kettle and the pot. This was a widespread tradition that we'd observed and filmed a few years earlier while among the Khanty, for example, and that is also common in Baltic-Finnic archaeological finds. Still, I'd like to cite Demnime's explanation: 'Let's say that after death, a woman left her children behind. Their spirits are not closed. But a pot is closed. In order for the spirits to breathe, a hole is made in the pot and the kettle. The spirit may not be locked away.' Reindeer are also slain in a different manner when sacrificed for a funeral. In everyday life, Nganasan use a lasso to kill for meat. The animal is hung and the blood is used for nourishment. When sending the deceased on their journey, however, reindeer were hitched to the sled and then stabbed in the back of the neck, allowing their spirits to accompany the

125

SILVERWHITE

human to better hunting grounds. A sacrificial dagger is a man's knife with an extremely fine blade, which is snapped in half after the ritual is performed.

Every tradition tied to the *matalir-ma* served one purpose: sending the deceased off on their journey to the mirrored world where they would continue to hunt and fish eternally. Humans had entered through the 'forest-side' door and consequently would have to leave in the opposite direction: towards the 'tundra'. The deceased's legs, as I mentioned, extend to the northwest, and, what's more, a small hole is left in the wall opposite the door. 'A ventilation hole,' Dülsimaku reckoned. 'Our lodge has one, too.' However, the Nganasan word for it is *sieja*, meaning 'word, speech, tongue, language'. Apparently language, tongue, speech, word, and also spirit and breath formed a meaningful string in our ancient beliefs. 'Word' is seen as one form in which the human spirit reveals itself, making it humankind's most defining quality. One echo of this ancient notion lives on in Estonian and Finnish to this day: *(juttu) puhuda*, literally 'to blow (a story/conversation)'. So, across from the *matalir-ma*'s door is a tiny *sieja* door through which the soul of the deceased embarks on its long journey. We encounter a similar concept among the Sámi and several other peoples.

Nevertheless, burial rituals only interest us in so far as they help us understand the development of the lodge's predecessor and the spruce or coniferous tree's role as the oldest dwelling of the taiga. We believe the *matalir-ma* is the most ancient transitionary form from living beneath spruce boughs to building the lodge. Let us add a few terms to confirm this. The central point of the burial chamber is a living tree, which is simultaneously the axis of the conical lodge and a pillar rising into the sky, given that it is much taller than the lodge itself. The burial sled positioned between the door and the *sieja* door is the lodge's diameter, though only approximately: as a tree grows amid the space, the sled is pushed up against the trunk. At the centre of Demnime's lodge was a cooking fire. Demnime viewed its smoke and the occupants' breath that mingles with it as a pillar that rises to the heavens. Consequently, we can presume that what was once a real trunk, i.e. pillar, was replaced with a metaphorical one. The fireplace, more specifically the hearth, is called *tori-faa* in Nganasan, though in direct translation it means hearth-tree: a linguistic monument to the distant house-spruce. The Sámi language has been able to preserve ancient associations to an even greater extent. In the Vefsn dialect, the same word is shared for the lower part of a trunk shorn

BENEATH TWO SUNS

of branches, the broadest lower part of a *lavvu* or lodge, the arced horizon, and ancestors (SKES 2, 333).

Nganasan lodges have a diameter of up to six metres and can be erected with as many as forty poles. The most important of these, to which women attach the home totem, are named *kungüj*: a compound word, the latter part of which means 'roof pole' and the former 'conifer', which on the Taimyr Peninsula naturally refers to larch.

So, we have come full circle to the pillar, which was originally the trunk of a spruce. The Estonian word for spruce, *kuusk* (also when used in reference to larch and spruce), extends back to Proto-Ugric and is shared by all Finno-Ugric and Samoyedic peoples today, just like our memory of the world revolving around a pillar. In the Khanty 'Bear Song', their ancient world spins on its axis just like the multicoloured Arctic sky-cover of the Chukchi and Koryaks:

> One mitten from my pair
> Out, it watches and listens to
> the spinning of the world's round ring,
> Watches and listens to
> The spinning of the fairy world's round ring.
> …
> One of my two eye-stars
> Watches and gazes upon
> The spinning of the fairy world's round ring,
> Watches and gazes upon
> The spinning of the spirit world's round ring.
> …
> The second of two ear-stakes
> Hears and listens to
> The spinning of the fairy world's round ring, etc.

The spruce – the lodge post – Sampo?

Of all our historical eras, the Uralic was the longest; and of every type of dwelling, only that which preceded the lodge was in use for longer. Thus, we may settle for the memory of the pre-lodge house-spruce only after we have recalled the lodge itself in sufficient detail.

Many complex elements of the lodge survive and are surprisingly recognisable in the 20th-century Karelian farmhouse, having withstood the increasingly aggressively transforming temporal landscape for over forty generations.

SILVERWHITE

But the house-spruce? Since time immemorial, it has combined the construction of the home hearth and poetic metaphor. The metamorphosis of the first did not entail the erasing of the second but rebirth as a mythological idea. And yet the grand simplicity of both makes them difficult to recognise and decode.

Still, despite centuries of Russian Orthodox tradition, the Karelian dining table has not been moved to the 'holy corner' in the manner of the Russian *izba*, but juts out from the outer wall and divides the room as in ancient times and just as in the present-day Nenets lodge: to the right is the *soppi*, the women's side, and to the left is the *perä*, the men's side. The dining table, the 'place of sacred sacrifice', is thus located in the segment behind the Nenets' *siimsi* post. In most cases, the mouth of the oven faces the outer steps: fire's magical glow protected the house's occupants more effectively than any Christian saint. The spatial distribution of the lodge has been transferred to the farmhouse with pedantic precision, and in this time-defying faithfulness, the loss of the sacred post, the Nenets *siimsi*, seems unlikely: the post was older, simpler, and more meaningful than other maintained elements. And so, while filming a large handsome oven on a beautiful July evening amid the forests of Ylä-Kuittijärvi, I was startled to realise that the house-spruce, our world post of old around which everyday life revolved along with the heavens themselves, had transformed into the Estonian *ahjupost*, a modest support, and, along with its name, imparted all its ancient shamanic power to Sampo without relinquishing its original fireside location on the same line as the door, the hearth, and the division of men's and women's sides.

Piä patsas muistossasi	Remember the ancient pillar.
Eläkä orsia unoha … (23: 209–10)	Do not forget the rafters …

It evokes the *Kalevala* without suspecting that it is forgetting the ancient connections between things that were perhaps even further from that day than the *Kalevala* is from ours. Nevertheless, our ethnography knows the chimney well.

Causal connections are the easiest to sever as time passes. Metaphor is more durable, though once it has lost its former roots, it turns into a mystery and incites us to seek new explanations and, therefore, also new names. The house-spruce has survived to our day more as a symbol than a concrete word. Metamorphosis has changed the word into a post, pole, stake, or pile. Nevertheless, behind every lexicological tunic lies a common original form: a life-pole with one end rammed into the earth and the

BENEATH TWO SUNS

other aiming at the heavens. It has withstood the test of time somewhat more intact in Baltic-Finnish ornamentation, for example in Karelian and Saaremaa embroidery, which we can appreciate as art without also being able to read it as code or interpret it as a pre-alphabetical vehicle for information, which is precisely what it was. Nevertheless, the *Kalevala* has given us the most vivid notion of a house-spruce: the Big Dipper rests upon its branches and the moon on its golden crown, while the spruce itself brushes the multicoloured dome of the heavens:

> Ilmarinen, full consenting,
> Straightway climbed the golden spruce-tree,
> High upon the bow of heaven,
> Thence to bring the golden moonbeams,
> Thence to bring the Bear of heaven,
> From the spruce-tree's topmost branches. (10: 145–50)

The ethnography of other Finno-Ugric peoples both near and far adds facts to this interpretation (spruce > house-spruce > post), primarily the customs of the Sámi, which were preserved more intact than those of other Baltic-Finnic tribes thanks to their late Christianisation. Sámi erected their post at a slight angle so it would point directly north to Polaris, which in their language is called *Bohinavle*. A travelogue by Uno Harva and Kustaa Karjalainen speaks of an Ob Ugrian post topped with a carved wooden bird. Siberian shamans used such a post notched in seven places to climb through the seven spheres of heaven while in a trance, similar to Ilmarinen scaling the golden spruce 'high upon the bow of heaven'. In the ritualistic Mari summer lodge, we encounter a stake in place of a post, though it is also notched for use as a ladder and rests against the rear wall where sacred dishes are kept on a platform. Whereas the *matalir-ma* is a transitionary structure between the house-spruce and the lodge, the Mari summer lodge – *kude* – is the next transition from the lodge to the four-cornered farmhouse of the following era. Thus, we may recognise in the Mari people's notched stake the Nenets' sacred *siimsi* post and a relic of the ancient spruce trunk. Language also confirms this. The Mari word for the sacrificial-dish platform, against which the notched post leans, is *kude-vodež toja*, which translates as 'lodge-spirit stake' (Holmberg 1914: 31). The ancient post – the world axis of a tribe, village, or family, around which both quiet mundane life and the rumbling world revolved – therefore connected the distant past with the equally distant future. The post was a link between two worlds.

SILVERWHITE

Let us now attempt to envision the post in its ordinary environment and landscape.

The Sámi sacrificed animals to their supreme deity *Waralden Olmay*, the 'world man'. A post was sprinkled with the blood. The post was called *Maylmen stytto*, the 'world support', though it simultaneously performed an axial function: driven into the tip was an iron nail that should be seen as the post's notional extension, a kind of rivet, a point upon which the imagined world axis rests. The latter naturally entailed an opposite point – the end of the axis in both an abstract and definite sense. This dilemma was resolved by the Sámi with incredible ingenuity. Another nail was driven into an idol depicting the world god Tooru along with a shard of flint, which was meant to symbolise his lightning: *in capite clavum ferrum, cum silicis particula, ut si videatur ignem Thor excutiat*, as one missionary noted (Scheffer 1673; see also De Vries 1970: I, 402). If we recall the Palaeo-Arctic Chukchi's belief that the world axis had worn a hole into the heavens above Polaris, then their ancient understanding of the cosmos is revealed to us in all its simple majesty.

Astrology adds an important temporal dimension to this discussion. At the beginning of the 1st century BCE, Alpha Draconis was the central point in our heaven instead of Polaris. It is part of the constellation Draco, which was then known as Thuban, 'head of the serpent', and even today is referred to in Estonian as *Põhjamadu*, 'the Northern Serpent'. According to the Estonian Literary Museum, it has also been called *Looja Rist*, 'the Creator's cross'. We should see the latter as a mark of missionary work and confirmation that crusaders equated Estonians' Northern Serpent with the dragon of biblical myths. What's more, the Beelzebub was expelled with an *equivalent* Beelzebub. Estonian names for constellations are generally regarded as young, as they reflect the masochistic '700 years of slavery' narrative – peasants' dull, grey world. This treatment betrays the incredible tenacity of inferiority complexes endemic in Estonia's first educated generation more than it does knowledge of the subject. Homer sometimes referred to the Big Dipper as the Bear, othertimes the Wagon. It was the Wagon to the Scandinavians, Icelanders, and seafaring Frisians. To the Anglo-Saxons, it was the Plough 'cutting through waves' in a metaphorical context. The key to understanding all these names is the 4,000-year-old Frisian tradition of depicting the constellation as a funeral wagon. We should take a similar approach to our own Estonian constellation names, which may shed light on their long-extinguished functions in ancient cosmogony.

130

BENEATH TWO SUNS

For centuries, the Northern Serpent was known as the Dragon in German folk astronomy without irritating the heads of Christianity. Thus, the dual Estonian monikers for Ursa Major reveal its central place in ancient cosmogony and its mythological gravity, which hadn't been dispelled early on in the crusades. In the Stone Age, the Northern Serpent was in the middle of the sky and the sky rested upon the house-spruce. In the Bronze Age, Polaris took over the central point of the heavens and the heavens rested upon the lodge's *siimsi*-pole, the stake, and the chimney.

To what extent is the worldview of more distant peoples applicable to Baltic Finns? To Estonians? Archaeologists have long been intrigued by the regularity of ship nails in the stone burial sites of coastal Estonia, but also further inland. They have been found at dig sites on Saaremaa, in western Lääne County, in central Harju County, and in central-eastern Viru County, and their association with burial customs is obvious. Is the phenomenon parallel to the Sámi world support – to the tall stave with an iron nail driven into its tip? In the village of Ülendi in Reigi Parish on the island of Hiiumaa, there were trees with trunks driven full of blacksmith's nails as recently as 1928. Architectural sample digs at the site produced many more nails underground (see Eerik Laid's 1928 report in the Topographical Archive at Tallinn University's Institute of History). All across Estonia, there are *ristikuused* (*virvepuud* in Võru County): spruce or other trees with crosses carved into them, at which funeral processions pause. The word's meaning has tarnished in Estonian, but in Livonian it meant a stave, stake, rod, or spear; in Finland, trees used as stakes; and, borrowed into Sámi, a tall and flexible spruce. Here, we might recall Procopius's (527 CE) description of an ancient sacrificial rite: 'And the manner in which [the people of Thule] offer up the captive is [...] by hanging him to a tree, or throwing him among thorns.' Thorns are naturally to be interpreted as sharp stakes. And Procopius's claim is not a fact, but a question.

To this day, the pillar and stake have maintained several cosmogonical and mythological shades of meaning in Estonian. A päevapost ('day post') means a bright ray of sunlight; ilmasammas and maasammas ('world pillar') are associated with eternity in the Estonian expression 'they can't turn into a world post'; some midsummer fires are lit at the top of a tall post; and, according to ancient belief, a pillar would appear in the sky when someone burned to death in a fire: a superstition that no doubt contains a memory of ancient sacrifices. The following observation was made sometime between 1635 and 1637: 'they select particular trees

SILVERWHITE

in various places, especially on hilltops, from which they lop branches up to the crown before weaving red ribbons around it and voicing their pagan wishes and prayers beneath it' (Laugaste 1963b: 55). Estonians' world pillar was documented on medieval geographical expeditions. In 1557, British explorer Anthony Jenkinson undertook a long trip through Muscovy to the Bukhara oasis in Central Asia and published his famous map based on what he found. In the northwest corner of this map, we can read Rougodive vel Narue, i.e. Rougodiv, Rongodiv, Narva (1562). Mikael Agricola published that version of the name a dozen years earlier and added it between the sons of Kalev and Virokannas on his 'list of false gods': Rongoteus Ruista annoi … The clergyman associated Rongoteus with the Karelians, as Russian historical tradition, from which he acquired his knowledge, tied the name to Estonians and it had long been used as an alternative for Narva. Rongoteus, which in folklore also appeared as Runkoteivas, Rongotus, Rõngutaja, and Rõugutaja, was the Baltic-Finnic god of the rye crop, and in Virumaa he also protected fertility, cradles, birth, and infants. Even around the start of the Common Era, Virumaa, which has rather stunted forests, developed into a quite prosperous agricultural region (in addition to Saaremaa) – the northernmost in the world. It is believed that the original form of the fertility god's name was Rukotivo. The first half of the compound word means 'rye' – a cornerstone of life and prosperity – and the second half 'pillar', originally also 'sky' and 'god', as the long path of the ancient loanword's transformations in both sound and meaning shows: ancient Indian deváh ('heavenly, divine, god') and Indo-European *deiuos, which branched into the Greek Zeus, the Latin deus ('god'), the Proto-Baltic *deivas ('god'), and the Proto-Germanic *tiuz, from which the Germanic god Tyr derived. The word had time to solidify into a deity in Proto-Baltic, which means it must have been borrowed earlier, perhaps in the era of Finno-Volgaic when the word meant 'sky'. It is stored in our cosmogonic conceptions of the house-spruce, pillar, or seven-notched stake connecting the past to the future and the land to the sky. As a loanword, it also began to mean the world pillar, the sky tree, and the pole: sacrificing to it meant sacrificing to the sky.

The vertical line, horizontal line, right angle, and circle originate in nature. The conception of the world pillar is common among all tribes, and on different continents and in different eras it manifested as the totem pole, the pagoda, the pyramid, the minaret, and nowadays the television mast. Our goal is to prove that within the taiga biome, spruce was the

BENEATH TWO SUNS

original form of the Baltic-Finnic world pillar. The hypothesis appears to be supported by the Nganasans' *matalir-ma*, the Nenets' *siimsi* pole, the Baltic Finns' Ilmarinen, who forged the heavens and the pillar and could touch the stars from the crown of a spruce, and, finally, the Estonians' cross-marked spruce, but these are not enough to act as full proof.

The missing link in the chain of evidence is the Christmas tree. Celebrating the winter solstice by igniting a Christmas (Yule) tree is a young tradition. In Berlin, the first spruce was lit in 1780; in Stockholm, at the beginning of the 19th century; in Helsinki, in the 1820s; in the Soviet Union, as a New Year's tree in 1940; and in England, the spruce only became widespread after the Second World War.

In Estonia, the first documented Christmas tree was set alight in Tallinn in 1441 (Amelungen and Wrangell 1930), though as part of an ancient ritual with roots reaching back to prehistoric folk religion. In the strictly regulated medieval city centre, only the Brotherhood of the Blackheads had permission to perform such an act. The Brotherhood would celebrate the end of shipping season with Yuletide feasts which began on the Friday before St Lucy's Day and lasted until 3 January. A limited number of guests, twelve in total, was invited to the opening feast, at which the finest beer was drunk. Dried salmon was offered and fried herring heads (!) were served an hour before midnight. These feasts culminated in the carrying of two spruce from the House of the Brotherhood of Blackheads on Pikk Street to the market square before the Town Hall one day after the winter solstice, meaning at least one day before religious Christmas. The entire Brotherhood would take part in the long procession, which was preceded and followed by dancers and musicians. Members of the city council were required to attend (or pay a one-mark fine) and treat members of the Brotherhood to a joint feast. The trees were lit in the centre of the square and, as the chronicler Balthasar Rüssow complained (1578), 'the brothers took one another by the hand and leapt and danced in pairs around the tree and the fire'. Summer solstices were celebrated in similar ways in Tallinn and elsewhere across Estonia. Rüssow wrote: 'At night in the cities, towns, manors, and villages, you saw only, without exception, gleeful bonfires across the entire land. People danced, sang, and leapt with absolute gaiety, and were not reserved in their playing of large bagpipes, the likes of which were abundant in every village.' As a pastor, he wars against the pagan tradition: 'And although pastors chided them for it and called it Moses's calf dance, the people showed no regard for the scolding.' Celebrating the winter solstice with spruce branches was

133

SILVERWHITE

a widespread custom even in the early 19th century. The Russian writer Aleksander Marlinsky travelled to Tallinn late on Christmas Eve in 1820 and described his impressions as follows: 'At first when I arrived, I was startled by the great death of Tallinn's citizens when I saw spruce branches laid out upon the streets, the stairways, and the entryways, for I believed that it meant funerals here, just as in our country.'

Spruce – house-spruce – Christmas spruce? And why in Tallinn?

Several of the prehistoric, pre-Christian Estonian fortress town's rituals survived in medieval Tallinn with outstanding vitality. The reason seemingly lies in the rigorous regulation of urban life, though also in more frequent documentation of customs and events and, above all, the continuity of old traditions. Tallinn's city council regarded itself as the legal successor of the Estonian-governed Revala fortress town. This alone ensured the preservation of ancient Estonian customs and common law or their adaptation to Lübeck city law. It also, to a lesser extent, held true in rural areas, in the manor courts of which, according to Rüssow: 'the noblemen on the court did not make a ruling after the complaint was heard, but sat in silence because, in accordance with the land's ancient tradition, the oldest peasants had the right to make judgments and determine punishments.'

We tend to hold a simplified notion of conquering in which one historical era ends and another begins. This is not the case. Rather, we should see it as a tense balance that neither side was able to sway, and then evaluate its duration. For example, the Tallinn city council gathered for a ritual All Souls' Day feast in autumn. The bills still exist. All Souls' Day is recognised by both Estonian folk religion and Christianity, so it would be strange to seek ancient folk relics among the council members' rituals. Even so, the expense of one occasion was recorded in the council ledger under the Estonian word *Hinkepeve* ('All Souls' Day' in archaic form) and it leaves no doubt that the council was carrying out, 'in accordance with the land's ancient tradition', a tradition that began in the Estonian-language environment before the region was conquered. Given this fact, maintaining the spruce in winter solstice celebrations might be more understandable.

The Brotherhood of the Blackheads was active only in Estonia and Latvia. Its first mention in chronicles dates to 1399, though researchers believe the Brotherhood's history stretches much further into the past. It apparently began as a colony of unmarried merchants, seafarers, pilgrims, and missionaries who spent winters in the Revala fortress town after

134

BENEATH TWO SUNS

navigation season. The Brotherhood's oldest surviving statutes emphasise members' obligation to aid one another, but also the body's duty to defend the city as an independent partisan force in the event of war. Members were required to leave the Brotherhood upon marrying. Their original patron saint was St George, a prince of Cappadocia who, according to Christian belief, saved a princess from a dragon and was later martyred. For reasons lost to the sands of time, the Blackheads switched patrons and St George relinquished his place to St Maurice, commander of the Theban Legion of Rome. As church legend goes, Maurice refused to fight against fellow Christians, was executed by order of the Roman emperor, and was later made a saint. Maurice was black, ergo the name Blackheads and a black head on the Brotherhood's coat of arms. Do the different patron saints reflect the Brotherhood's altered function after the Revala fortress town was conquered by the Danes in 1219?

In connection with the Tallinn council's All Souls' Day rituals, historians Paul Johansen and Heinz von zur Mühlen state that such intertwining of ancient folk belief and Christianity must date back to the days of Bishop Fulco of Estonia and Finland (c. 1171–7). Similarly, an explanation could be found for Tallinn's medieval spruce-burning custom, which the pious chronicler Rüssow equated to idol-worship. This tradition has only been documented in Estonia, and the Blackheads could only be found in Estonia and Latvia. Consequently, the tradition was not imported but developed here, and the spruce's roots have drunk life-giving water from our stony soil: a memorial to a dizzyingly distant era when a tree grew in the centre of our ancestors' dwellings – the future world pillar, the beam supporting the falling sky, Sampo.

But back to the topic at hand: the world axis that holds up the heavens near Polaris is consequently a full magnitude younger than the house-spruce, as it could only have developed after the star moved to the centre of the night sky and bronze became the bearer of culture. Lodges were a thing of the past by then, and house-spruces were ancient relics. In the context of such rapid progress, how could the age-old ideas of the world pillar and the sky's support strut have endured?

Peoples along the Baltic Sea had become one experience richer. The sun had collapsed to earth before their very eyes. They needed their own pillar; their own sky-support.

Occasionally when I begin to doubt my explanation, I pick up a nail and lift my gaze to Polaris – in Estonian, *Põhjanael*: 'the north nail'. What

SILVERWHITE

might they have in common? What could connect them? What, apart from poetic fancy?

* * *

Recently, extensive research has been done into the global prevalence of myths, and efforts have been made to find Sampo's roots in faraway India. The fundamental importance of cultural contacts is an axis of this travelogue. So why not India? Because reducing historical cultural phenomena to a global archetype is reminiscent of the Book of Genesis and just as productive as distilling various wines into chemically pure water. In doing so, we sacrifice flavours, colours, and aromas to an experience that lacks any nutrient salt worth naming, all the while casting doubt over humans' ability to shape the world in their own image wherever they may be. This is not a critique of one method, of course, but preference for another. Extremes are undoubtedly necessary, as the most fertile soil collects on the front line of conflicts, and soil even more fertile than that gathers on the fields where they mingle.

'Every aboriginal treats the history of their totem ancestor like a telling of their own deeds since the beginning of time, and even since the very dawn when the world as it is known today was still in the hands of the all-powerful beings that gave it shape and form.' It's possible that the (paraphrased) quote by Theodor George Henry Strehlow is out of place here. Even so, my mind doesn't wander to the lands of the Mahabharata when I happen to read of Sampo's six-fathom-long roots, but rather to the Khanty of the Kislor River where I saw 'nine-fathom' spruce roots clutching the frozen ground of the taiga so tightly that they could offer shelter and protection to an entire people and nourish its worldview for countless centuries. Therefore, may the future be the one to decide whether and in what form the pillar found its way to the doorstep of Tallinn's Town Hall, and how it drifted into a wordless slumber as the post of an Estonian threshing barn that doubled as a dwelling.

III

WHISPERS REBORN AS WORDS

ANOTHER WORLD AND ANOTHER ISLAND

Humans die, disintegrate to dust, grow as grass, are reborn as birds in the sky, and slowly step back into the endless cycle of nature. Humans' manuscripts are copied, crumble into quotations, transform into folklore, and amplify or mute hopes and expectations alike. Still, words do not turn into grass, ants, or birds, but remain words. A word may wear down, and a word may find new meaning. Even so, its life continues after the speaker's death. In a way, every word that has ever been spoken and every thought that has ever been considered lives alongside and within us, reaching all the way back to the dawn of human civilisation in the buzzing cradle of Africa.

So does Pytheas's manuscript live on in the history of Baltic exploration? Are we able to recognise his words?

The Baltic comes up again a few centuries later in Pliny the Elder's *Naturalis Historia* (*Natural History*), a work based partly on new information and first-hand accounts. Eighteen years before the scholar was born, the Roman fleet under Tiberius first penetrated our waters:

> But the larger part of the Northern Ocean was explored under the patronage of the deified Augustus, when a fleet sailed round Germany to the Promontory of the Cimbri [Skagen] thence seeing a vast sea in front of them or learning of it by report, reached the region of Scythia and localities numb with excessive moisture. On this account it is extremely improbable that there is no sea in those parts, as there is a superabundance of the moist element there. [...] it is named the Gulf of Codanus [the Kattegat and straits], and is studded with islands. The most famous of these is Scatinavia; its size has not been ascertained.

SILVERWHITE

Near Ankara in present-day Turkey, Emperor Augustus ordered the course of his fleet to be inscribed: 'My fleet sailed from the mouth of the Rhine eastward as far as the lands of the Cimbri ...'

Even at the latitude of the Rhine, the summer sun rises more from the northeast than due east. The Romans' course would have led the fleet through the 'Scandinavian islands' to the Gulf of Finland. Was the reason behind this direction Pytheas's 'Britannia triangle', the southern edge of which, as we remember, shot northeast towards the rising sun? The Romans provided no further descriptions of the Baltic Sea. The empire's economic and military expansion was reaching its apogee. The riches of furs and amber were just as tempting as those of tin, to which southern England fell victim. In Kattegat, the Romans received reports of the 'region of Scythia'. The toponym is vague and had, since the time of Herodotus, been used for any eastern 'barbarian' land. However, the fleet could only have been interested in a 'Scythian' region that was accessible by ship. This may have been the eastern shore of the Baltic Sea, and perhaps even the Gulf of Finland, the key to the Silverwhite trade route. Pliny defined the 'European Scythians' as follows: 'The name of Scythians has spread in every direction, as far as the Sarmatae and the Germans, but this old designation has not continued for any except the most outlying sections of these races, living almost unknown to the rest of mankind'. By this, he meant the Germanic and Western Slavic tribes. For Pytheas, the 'most outlying' were the Baltic-Finnic peoples who dwelled on the most outlying Thule.

In this chapter, we will discuss the Baltic Sea and one of its characteristics as it was described by various authors who followed Pytheas. We must also quickly acknowledge that the sparse sources surviving to our day do not allow us to tie that characteristic feature to the mythical Baltic-Finnic Virokannas.

Pliny's military career was spent on the lower Rhine. The information he provides about northwest Europe is credible. He believes the Baltic Sea to be a channel that joins the long, narrow Caspian Sea far to the north via the Gulf of Finland and the Amalchian Sea. The notion of a circular route around Europe was based on rumours of a Baltic–Volga waterway and, as we will soon see, Pytheas's own work. The encounters were true, the conclusions false. Pliny was able to relay fascinating details about the Baltic 'channel':

> There is still living a Roman knight who was commissioned to procure amber by Julianus when the latter was in charge of a display

WHISPERS REBORN AS WORDS

of gladiators given by the Emperor Nero. This knight traversed both the trade route and the coasts, and brought back so plentiful a supply that the nets used for keeping the beasts away from the parapet of the amphitheatre were knotted with pieces of amber.

[…] forms an enormous bay reaching to the Cimbrian promontory [Skagen]; it is named the Gulf of Codanus, and is studded with islands. The most famous of these is Scatinavia; its size has not been ascertained, and so far as is known, only part of it is inhabited, its natives being the *Hilleviones, who dwell in 500 villages, and call their island a second world. Aeningia is thought to be equally big.

In addition to Pytheas's Thule (Land of Fire), another Baltic-Finnic word that we have failed to notice springs up: *Hilleviones, Livonians. We will discuss the people later, along with the Estonians. I have translated the original quingentis pagis, '500 villages', into '500 parishes' in Estonian (kihelkonnad), and for good reason: Pliny's choice of words came from the ancient Scandinavian word hund, which matches in sound the number hundra, 'one hundred' (Pekkanen 1980a). However, hund is an administrative term that corresponds to the Baltic-Finnic kond/kand – a stem that can be found in the Estonian words perekond ('family'), külakond ('village community'), kihelkond ('parish'), and perhaps even Kunda. The unexpectedly fertile translation error conveys valuable information, the most telling of which can be found in the words 'and call their island a second world'. Is this Pliny's ethnic-etymological explanation for the 500, and actually just 5, Baltic-Finnic tribes or parishes? Or can we trace it to Pytheas's On the Ocean? If you recall, Pytheas mentions 'a second sea' beyond Thule; the Gulf of Finland. Pliny speaks of 'a second world'. The ancient Scandinavian word vaerild, vaerald, vaeruld, meaning 'world', was borrowed into Baltic-Finnic as viro, viru, and evolved into a toponym that was used for the region of Estonia in general, and particularly for the northern Estonian coast. This was also the time in which the most archaic figure of Baltic-Finnic folklore, Virokannas, appeared. Perhaps the pillar of the 'second world'? Did the Kaali meteorite event push Saaremaa to the foreground of written accounts and other Baltic coastal regions to the back? Words, language, and literature live by their own rules which have little in common with economic relationships. Text conveys information, information is exceptional by definition, and everything exceptional sweeps the ordinary aside. The following texts also contain questions.

Aeningia is Denmark's Bornholm, which is half the size of Hiiumaa and twice the size of Muhu. Equating it with Scandinavia therefore reveals a

SILVERWHITE

trove of data. The phenomenon is quite familiar: frequent contact reduces distances and increases the prominence of trade hubs. Bornholm was an intermediate island where shipping routes to places both near and far intersected. Rudimentary Baltic rowing ships set off from its northern point at Hammer Odde (55° 24′), their bows aimed towards Sweden's southern point at Öland. Pliny emphasises Bornholm's importance, the consequences of which reaches back in time and might even extend beyond Pytheas's voyage. Traditions are not born overnight.

Yet even more fascinating are the Roman nobleman's adventures on the eastern shore of the Baltic: specifically, the fact that there were none. His voyage beginning from Carnuntum, near Vienna, is not described as anything out of the ordinary. He returned with a mythical treasure that suggests the safety of the trade route, shore rights, and the rules of safe harbour to which both parties abided. What were these parties? Pliny speaks of several marketplaces (*commercia*) and several shores, meaning they must have been located independent of one another. Diodorus Siculus writes of amber being transported from the island of Basilia, 'brought by the natives to the opposite continent, and that it is conveyed through the continent to the regions known to us' – a description a century older than that of Pliny. The tendrils of Rome's economy had had sufficient time to wrap around the eastern shore of the Baltic Sea. But where exactly? Market hubs could have extended from the mouth of the Vistula River to Cape Kolka, across from Saaremaa. Could they also have been located on Saaremaa itself, where silver and gold Roman coins have been found?

There is evidence of this possibility in Ptolemy's *Geography*, which he wrote less than a century after Pliny's work and without leaving the banks of the Nile once. His wanderings were limited to the Library of Alexandria, and Hipparchus was his paragon. In the *Geography*, the coordinates of several estuaries on the eastern shore of the Baltic are given with great accuracy. That of the Daugava is off by 1.5–2.5 degrees, true, but it is there. And Saaremaa? As Ptolemy wrote:

> Toward the east of the Cimbrian peninsula [Denmark] are four islands which are called Scandia, three of which are small, the middle of which is in 40° 31′ [east longitude], 58° [north latitude], the larger one is further east and near the mouth of the Vistula river; the extreme parts of this are, on the west, 43°, 58°; on the east, 46°, 58°; on the north, 44° 30′, 58° 30′; on the south, 46°, 57° 40′. This one is properly called Scandia.

140

WHISPERS REBORN AS WORDS

The first half of the numerical pairs are Ptolemy's meridians and should not discourage us, as his 0 meridian crossed the Canary Islands instead of Greenwich. Without a chronometer, the meridians are much less precise than the latitudes. Out of the 8,000 place names he lists, the vast majority were taken from older maps or writing, and only a small portion are based on genuinely measured coordinates. Ptolemy's geographical regions should mainly be determined in their relation to others. Most erroneous is the island's meridian; its latitude is slightly more exact. Most credible is the 'larger' island's position in relation to the three smaller ones, lying furthest east. No matter how we interpret Ptolemy's other information, let us proceed from the latter point.

The 'larger' island's easternmost location justifies asking if it might be Saaremaa. Ptolemy's island lies 2 degrees to the north of the Vistula estuary, Saaremaa 4 degrees. Ptolemy's island lies partly to the east of the estuary, Saaremaa entirely. The island's shape should be disregarded as being more guesswork than fact: a rhombus in either case, and Ptolemy's area is nearly three times greater than that of Saaremaa. However, it is harder to write off the 'island's' occupants as mere chance: *leuōnoi, levōnoi*, i.e. the Livonians mentioned earlier by Pliny. Their neighbours to the north are the Finns/Sámi (*phīnnoi*) and to the east is a people named the *phiraisoi, pheiraisoi, phiresoi*, whom, for reasons we have already discussed and will confirm again later, we may dare to identify as the Viru and the land called Virumaa. It may seem as if Virokannas is becoming too obtrusive. Nevertheless, unwritten literature nourished written literature with its juices long before Lönnroth and Kreutzwald. It will take millennia more before our Stone Age amber boar amulets find their way into writing. Roman coins have been found in at least seventeen sites across Estonia. This doesn't tell us much, either, so long as we don't see the coins being flipped over fingers and passed along.

Perhaps Ptolemy himself provides the greatest clue when he describes his large and easternmost island by saying: 'This one is properly called Scandia.' A second sea and a second world, a second world and the proper, true world? It is extremely difficult to tear ourselves away from the modern-day maps that bored into our subconscious when we were just children; to see the world through the eyes of the first maritime explorers and imagine the Atlantic surf rolling across a sandy Estonian shore or crashing into the limestone cliffs of Saaremaa. Why 'properly called'? As a working hypothesis, Pytheas's 'British triangle' had managed to acquire the realistic outline of England in the west, but in the east

141

SILVERWHITE

it crumbled into a chain of islands trimming the northwest shore of Eurasia. The Scandinavian peninsula did not exist in Ptolemy's day, nor did it exist ten centuries later. There were only smaller or larger islands. The distance from Gotland to the Livonians' Cape Kolka is almost exactly the distance that Ptolemy believed separated the smaller western islands from the eastern one 'properly called Scandia'. Still, the precision of his distance is more likely a fluke and certainly not fact. On the other hand, the easternmost location of Scandia is not accidental. What is accidental is naming that island 'Scandia'. Its similarity to Pliny's 'second world' is great, as both are inhabited by Livonians.

TACITUS READS PYTHEAS

Pliny the Elder, a former commander of a Roman cavalry battalion, died as a plump 56-year-old man of science while he was describing the eruption of Mount Vesuvius. At that time, Cornelius Tacitus was already around 20 years old. Like Pliny, he used Pytheas's *On the Ocean* as his primary source when describing eastern regions of the Baltic Sea. An account of Germanic tribes' clothing leads us to seafaring:

> They likewise wear the skins of savage beasts, a dress which those bordering upon the Rhine use without any fondness or delicacy, but about which such who live further in the country are more curious, as void of all apparel introduced by commerce. They choose certain wild beasts, and, having flayed them, diversify their hides with many spots, as also with the skins of monsters from the deep, such as are engendered in the distant ocean and in seas unknown. (17)

Tacitus uses awkward phrasing in 'diversify their hides with many spots'. Why? Is he referring to a mosaic of hides? We filmed the making of one during the shooting of *The Waterfowl People*. Two strips of hide, one light and one dark, are laid next to each other and a pattern is cut out with a knife. The light cut-out and the dark outline are used together, matched, and then attached to a hem, belt, pouch, rug, or other item using isinglass. The mosaic's patterns are angular because of the straight blade, and the geometrical precision of the intertwining positive and negative may look like an inexplicable work of mastery to one unfamiliar with the technology used. It is Nordic technology. Relying upon Tacitus's writing, we can assume the importation of skins from the Far North across the Baltic Sea.

Let's proceed to our latitudes.

142

WHISPERS REBORN AS WORDS

Tacitus calls the Baltic the Suevian Sea. 'Suevi' is a term that encompasses almost all the very ethnically diverse peoples east of the Elbe, primarily in coastal Baltic areas. We will discuss Tacitus's tribes in the next chapter. For now, let us concentrate on a mysterious quality of the Baltic Sea that should already be familiar. After describing certain tribes living on certain islands, Tacitus remarks that beyond them

> is another sea, one very heavy and almost void of agitation; and by it the whole globe is thought to be bounded and environed, for that the reflection of the sun, after his setting, continues till his rising, so bright as to darken the stars. To this, popular opinion has added, that the tumult also of his emerging from the sea [variation: falls into the sea] is heard, that forms divine are then seen, as likewise the rays about his head. Only thus far extend the limits of nature, if what fame says be true. (45)

The other sea, 'heavy and almost void of agitation', is Pytheas's frozen ocean; his 'sea lung'. There are unusual dawn and dusk phenomena in the region and, what's more, legends that the sun rises or sets from or into the sea with a 'tumult'. Is this, then, Pytheas's Thule – the distant glow of our cosmic catastrophe?

On the other hand, Ultima Thule had long become legend thanks to generations of Greek and Roman poets. Perhaps Tacitus believed it was appropriate to reprise the myth? Could his description be interpreted as literary cliché?

Fortunately, convincing evidence refutes this supposition. As it comes from Tacitus himself, it is credible. In his earlier work the *Agricola*, he describes the voyage of a Roman fleet around Britain (though here it would be more exact to call it England) and confirms: 'Thule too was descried in the distance, which as yet had been hidden by the snows of winter.' So Tacitus wasn't completely lacking in literary cliché?

The explanation is more straightforward. Unlike Eratosthenes, Hipparchus, and geographical mathematicians, practical geographers narrowly interpreted Pytheas's geodetic 'British triangle' as English islands. Tacitus adopted Polybius's and Strabo's false assumption that Pytheas circled around England and, as one who honoured tradition, had the Romans spot Thule in the given location. Who would return from Paris and admit to not seeing the Eiffel Tower? Over the four centuries that had passed since Pytheas's voyage, fanciful notions of Thule had not dimmed but flared. Thule had become a symbol, man's furthest goal, a glittering crown of masculinity. A fistful of pebbles from Thule's shore would have turned to diamonds in Rome and lifted the seafarer to the hero's pedestal, and Julius Agricola, the Roman governor of Britannia, was Tacitus's father-in-law. Thule is thus his polite bow or, rather, a belated tribute to his recently deceased relative. Tacitus was correct about one thing: no Roman had ever ventured so far north. He couldn't have known that no Roman would ever go that far again, either: the Pentland Firth at the uppermost tip of Britain corresponds to Estonia's Soela Channel, and the Orkney Islands are on the same latitude as the Saaremaa town of Osmussaar.

Tacitus falsely interpreted the 'British triangle', but he did not interpret Pytheas's *On the Ocean* with any prejudice. The Greek manuscript was Tacitus's primary, and perhaps sole, source when describing lands around the Baltic Sea. In addition to corresponding fundamental data, the theory is supported by unusual Greek case endings in Tacitus's writing, which has been pointed out by Professor Tuomo Pekkanen. Most importantly, Tacitus could not have repeated Thule as a symbol or as a toponym in the *Germania*, which was written after the *Agricola*.

The first thing to take from this, as comical as it may be, is Tacitus's credibility. Tales of the sun emerging from or descending into the sea with a 'tumult' were clearly borrowed from Baltic folklore, where the legend was apparently still in circulation. Folklore foremost represents unwritten history in Tacitus's works: 'their old ballads [...] are the only sort of registers and history' (2). This is so well put that it could work as a motto for our travelogue. The divine features and emanating rays are generally regarded as the myth of Helios and the sun chariot. Taking the study of geography into account, it becomes questionable whether the use of such archaic concepts can be ascribed to Tacitus when describing natural phenomena. Equally meaningless is the polite assumption that the author employed literary metaphor, which is precisely what he did. It was repeated dozens of times throughout the *Germania*, though not once for formal considerations. When he names Castor and Pollux as Western Slavic idols and claims Isis as part of Suevi rituals, the basis for the comparison is the gods' essential similarity and a need to explain unfamiliar practices or ethnographic phenomena to readers by using widely recognised symbols. It would be naive to see the end of the mythical world in Tacitus's words 'only thus far extend the limits of nature'. 'We do undoubtedly speak of the earth's sphere, and admit that the globe is shut in between poles,' wrote Pliny the Elder, with whose family Tacitus corresponded. Eratosthenes

WHISPERS REBORN AS WORDS

was aware that the world is a sphere and that one can sail from Spain to India, a statement in which I sense a gentle hint of Pytheas's return from the Baltic Sea to the Black Sea along riverways. Tacitus's 'another sea, one very heavy and almost void of agitation' is a direct loan from Pytheas, corresponding exactly with the geographic limits of sea ice that arcs across the mouth of the Gulf of Finland at a distance of one sea day from Saaremaa. Therefore, I would like to highlight Tacitus's words 'only thus far extend the limits of nature', putting them into sharp focus and comparing them with the Baltic-Finnic worldview, the Scandinavian visions of apocalypse, Pliny the Elder's 'second world', and Ptolemy's 'proper' (true) island. Tacitus is an appreciator of folklore. So, are his words folklore? We must ask ourselves: where does folklore end? It ends on papyrus, on paper, in marble, in the memory of a microprocessor. When writing about solar phenomena, Tacitus says what the local people believe; ergo, folklore. Yet when describing 'another world', he affirms that the hearsay is true. This is no longer folklore, it is literature.

With this phrasing, Tacitus is closer to Pytheas than any other earlier or later author with the exception of Apollonius of Rhodes. The place where Tacitus's queerly flashing phenomena take place is simultaneously 'the limits of nature'. This is a characteristic of Thule, is it not? Tacitus could not have allowed himself more than one.

Pytheas?

Ultima Thule was meant to embellish the memory of his father-in-law. There is only one Thule in the world. Thus, Tacitus placed all the strange features of Thule in the Baltic Sea without using the island's name.

Or have we simply failed to recognise the word and island and language and land?

SIX HUNDRED FACTS

Aestiorum gentes, the tribes of Aesti, were first recorded by Tacitus in the year 98 CE. Thirty manuscript copies of the *Germania* have survived to this day. All are third-generation manuscripts copied from the now-lost Codex Hersfeldensis, which was written between 830 and 850 CE. Palaeographic and stylistic analyses done over the last decades have identified microscopic differences between the manuscripts. Some *i*'s have distorted into *l*'s, some *u*'s turned into *v*'s. Can the present hinge upon a lost period? Can a newly found period change our future? Bizarre questions.

145

SILVERWHITE

The manuscript itself isn't very lengthy, coming in at just about 6,300 words. Nevertheless, it contains more ethnographic information than any other work of Greek or Roman literature. And here the emphasis is on ethnography, not geography. There are an estimated 600 facts in the work, of which only 70 can be found in earlier literature (Pekkanen 1976: 11). Tacitus used maps as reference when writing the *Germania*. Even so, he uses tribe names as a starting point when filling the geographic space; rarely toponyms. He calls not a single region east of the Elbe by name, making it rather difficult to interpret the geography of his laconic, elegant, and clear writing.

Tacitus's Germans, Sarmatians, and Scythians are economic and political terms that lack ethnic content. The Germans are primarily characterised by fertile soil, the Sarmatians by herding. The 67 peoples named in the work can be divided into six groups in six geographic areas to the east and northeast of the Rhine, more or less up to the Neva River. The most significant national attributes for Tacitus were appearance and anthropological type, and only then their customs and lifestyle. On only one outstanding occasion does he mention their language.

The fifth and sixth group of nations, provisionally labelled 'along the Baltic' and 'even more distant', are tied to the history of Baltic exploration. Distances should be understood in relative terms. Lively interaction reduces distances, lack of knowledge increases them. In Tacitus's day, the Baltic nations' immediate neighbours to the north, northeast, and east were the Sámi. Lappeenranta, almost directly across the Gulf of Finland from Narva, reflects those long-ago times. In addition to an ethnic quality, the city's Swedish name has preserved a colonialist attitude: Villmanstrand, 'savages' shore'. The Sámi, Tacitus's Fennians, neighboured the Vepsians, Izhorians, and Votians from the banks of Äänisjärvi (Lake Onega) to the mouth of the Neva. Lake Ladoga's name is of Sámi origin.

So, Tacitus first recorded the *aestiorum gentes*, the Aesti tribes, in the year 98. Aesti, Eesti, Estonia. But were Tacitus's Aesti really Estonians? Names wander, peoples wander, and name-forms can be passed down, inherited, and borrowed.

> Beyond the Suiones is another sea, one very heavy and almost void of agitation. [...] Only thus far extend the limits of nature, if what fame says be true.
>
> Upon the right of the Suevian Sea the Aestian nations reside, who use the same customs and attire with the Suevians; their language more resembles that of Britain. They worship the Mother of the Gods. As

WHISPERS REBORN AS WORDS

the characteristic of their national superstition, they wear the images of wild boars. This alone serves them for arms, this is the safeguard of all, and by this every worshipper of the goddess is secured even amidst his foes.

Rare amongst them is the use of weapons of iron, but frequent that of clubs. In producing of grain and the other fruits of the earth, they labour with more assiduity and patience than is suitable to the usual laziness of Germans. (45)

This is followed by a section in which Tacitus describes how amber is gathered.

LIVONIANS LEAVE UNWRITTEN LITERATURE

A myriad of texts should be read on the basis of rules that derive from the text itself. In order to identify the Aesti, we must first determine their closest neighbours. This is how Tacitus arrived on the Baltic shore and, as we will see, made it to the mouth of the Neva:

Beyond the Lygians dwell the Gothones [...]. Immediately adjoining are the Rugians and Lemovians upon the coast of the ocean. (44)

Next occur the communities of the Suiones, situated in the ocean itself [= on islands]. (44)

Beyond the Suiones is another sea, one very heavy and almost void of agitation. [...] Only thus far extend the limits of nature, if what fame says be true. (45)

Upon the right of the Suevian Sea the Aestian nations reside [...]. (45)

Upon the Suiones, border the people Sitones [...]. (45)

Here end the territories of the Suevians. Whether amongst the Sarmatians or the Germans I ought to account the Peucinians, the Venedians, and the Fennians, is what I cannot determine. (46)

In wonderful savageness live the nation of the Fennians, and in beastly poverty, destitute of arms, of horses, and of homes; their food, the common herbs; their apparel, skins; their bed, the earth; their only hope in their arrows, which for want of iron they point with bones. Their common support they have from the chase, women as well as men. (46)

What further accounts we have are fabulous: as that the Hellusians and Oxiones have the countenances and aspect of men, with the

SILVERWHITE

bodies and limbs of savage beasts. This, as a thing about which I have no certain information, I shall leave untouched. (46)

The final sentence is also the *Germania*'s sober and pleasantly undramatic finale.

Tacitus's chain of ethnonyms runs from southwest to northeast: the Lygians, Gothones, Rugians, coastal Lemovians, Suione communities, coastal Aesti on the Suevian Sea, Sitones (neighbours of the Suiones), and the Suevian 'border', across which are the Fennians, Hellusians, and Oxiones. The chain's outermost links are relatively easy to determine. Fennians are, of course, the Sámi, who can be traced through later geographical literature without interruption. The Hellusians' name comes from the Greek *ellós*, 'elk', which was the totem animal in what is now Savo. The word 'Oxiones', which has hitherto lacked a satisfactory explanation, comes from the Baltic-Finnic word *oksi*, 'bear'. Thus, it is one of our oldest written memorials and represented Tacitus's easternmost tribes: either the Karelians or the Ugandis, whose totem animal was the bear (Pekkanen 1981). Several ancient toponyms still carry lost totem signs: Hirvesaar (Deer Island, Russian *Vasilyevsky*) in the Neva estuary; Ohto > Ohta River (Russian *Okhta*), which was first recorded in the year 1300 (Popov 1981); and Otepää, a town which was translated into both Latin and Church Slavonic as 'Bear Head'. We should, of course, ask ourselves how credible and lasting such totem signs really are. One of Estonia's most renowned feudal lords was Bartholomaeus Savijärve (Sawijerwe), bishop of Tartu from 1441 to 1459. A bear was featured on his family crest and a bear's paw on his coins. Vassal Heidenreich Savijärve's ashlar emblem was half a bear with a paw raised in attack (Frey 1905: 58). Coats of arms are extremely conservative marking systems. That of Kuressaare Castle features the outline of Saaremaa, a stork, and two crossed ash branches (Buxhövden 1838: 295). If we take the heraldry as a hieroglyph, then we can conclude that the castle's name was bilingual, but the imagery was Estonian and thus older. Saaremaa's outline does not include the Sõrve Peninsula, which, as geology confirms, was still a separate island from Saaremaa. Kuressaare's coat of arms is the oldest surviving document of Baltic-Finnic cartography and conveys our ancestors' highly accurate geographical knowledge. Unlike Central Europe, the area of Baltic-Finnic settlement was not buffeted by storms of migration, and therefore our symbols, toponyms, and ethnonyms are also more intact. We can apply to them methods that cannot be used with their Central European

148

WHISPERS REBORN AS WORDS

equivalents. Tacitus's elk tribe and bear tribe are in pure harmony with Estonia's folklore, traditions, toponyms, and oldest symbology.

The beginning of the chain – Lygians, Gothones, and Rugians – can be defined by means of archaeological discoveries and the writings of Tacitus, Ptolemy, and later authors: the Gothones, aka Goths, lived along the lower Vistula River, and the Rugians settled nearby in its estuary and to the north. Both emigrated from Scandinavia, and sources allow us to track their later migrations with a sufficient degree of accuracy.

The most important of these tribes from our standpoint are the Lemovians. Comparing manuscripts shows them to be identical to Pliny's *Hilleviones* and Ptolemy's *Levoni*. Palaeographic analysis has given us the original forms of *leviones* (Pliny the Elder, before 79 CE), *levonii* or *levionii* (Tacitus, 98 CE), and Ptolemy's (170 CE) flawless *leuōnoi, levōnoi*. Tacitus places them quite far to the north and east of the Rugians, corresponding markedly to the Baltic coastline. The Lemovians' area of settlement also coincides with that of the Livonians before Baltic tribes reached the sea (Pekkanen 1976: 28). As a curiosity of history, let it be noted that Jaan Jung, one of the founders of Estonian folklore studies, already equated the Lemovians with the Livonians in 1876, albeit with entirely different argumentation.

THE SOURCE OF THE HEAVY SEA

Next, Tacitus names the Suiones, aka Sveans, whom we know now as the Swedes. Our dilemma is familiar: Estonians, *eestlased*, developed from the name *Aesti*, but were Tacitus's Aesti truly Estonians? So it is here: the Suiones have become the Sveans, but were Tacitus's Suiones ancient Swedes? Tacitus writes of their 'communities'. Why communities? Some Suiones also live on islands: does that mean the rest are on the continent? It is surprising to see the Suiones mentioned again (45) as being beyond the Aesti, but still on this side of the Suevians. Let us take a closer look.

Birger Nerman, a former professor of archaeology at the University of Tartu, organised archaeological digs in Grobiņa (near Liepaja, Latvia) and Apuolė (Lithuania), where he discovered a previously unknown ancient Scandinavian settlement on the eastern shore of the Baltic. Its oldest layers predate Tacitus (Stenberger 1977: 415). Tacitus's 'communities of Suiones' could perfectly describe small groups of immigrants living in a foreign ethnic environment. His second mention of the Suiones leads us to an even more significant conclusion: Tacitus's notion of northern

Europe encompassed all of Baltoscandia including the Åland Islands, the Turku archipelago, and southwest Finnish islands, where ancient Swedish settlements were indigenous. Tacitus's description moves from west to east and from south to north. The repeat Suione/Svean wedge corresponds with the highly trafficked mouth of the Gulf of Finland and is in complete accordance with the layout of ancient settlements. The other 'sea, one very heavy and almost void of agitation' also acquires a precise location, which Tacitus says lies beyond the Suiones (45): from the author's perspective, this means to the north and the east, i.e. the Gulf of Finland, which is east of the Åland Islands with their Swedish population and freezes over in winter.

Tacitus's knowledge of northern Europe is noteworthy and occasionally more detailed than that of Central Europe. The primary source for Pliny and Tacitus alike was Pytheas's *On the Ocean*.

Yet, strangely enough, our conclusions do not yet carry enough weight.

As the northern Europe section of the *Germania*, written in the year 98, is based on another work that was written four centuries earlier, it is more the Baltic Sea of Pytheas's day reflected in the work than that of Tacitus's time; more the ethnic map of Pytheas than of Tacitus; more the beliefs, customs, and contacts of Pytheas than of Tacitus. Some details have been added and others lost, but Pytheas's ring of questions remains intact: amber is brought from across the Baltic Sea; active seafaring is done on the Baltic; strange solar phenomena occur on the Baltic Sea; and beyond it lies 'another sea' that freezes over in winter: Pytheas's *Pepeegyia thalatta*, Pliny's *Mare concretum* or *cronium*, other authors' *Pontus kronios* or *Mare cronium*, and finally Tacitus's *Mare pigrum*, which he placed close to nations that traded amber. Earlier, we also showed that Tacitus used the island of Thule to embellish his father-in-law's voyage around Britain; consequently, he could not have mentioned Thule a second time in the later *Germania*. And, yet, Tacitus mentions twice that there lies a frozen sea near Thule: first in connection with his father-in-law's voyage, that type of sea being in quite plain contradiction to the geographic reality of the stormy waters surrounding the Orkney Islands. On the second occasion, he refers to a frozen sea without naming Thule, though it is in total agreement with Pytheas's account, the location of the Gulf of Finland, and local winter ice conditions.

Here, we come back to the Aesti.

Pytheas travelled and described, Pliny and Tacitus read and wrote. Pytheas brought distant Thule back from the Baltic Sea. Pliny repeats it,

WHISPERS REBORN AS WORDS

Tacitus does not. On the other hand, Tacitus is aware of the nation of Aesti, *aestiorum gentes*, which was unknown by both Pytheas and Pliny. Where did Tacitus find them?

THRESHING HOUSES ON AESTIAN LANDSCAPE

Let us leave this central question burning for a short while and return to what Tacitus had to say about the fertility of the land of the Aesti: 'In producing of grain and the other fruits of the earth, they labour with more assiduity and patience than is suitable to the usual laziness of Germans' (45). Such generalising assessments are mostly unfounded and driven by prejudice; furthermore, the author never set foot on the shore of the Baltic Sea. The account came from Pytheas. Let us attempt to imagine what a Mediterranean Greek might feel when arriving at our northern latitude. Pytheas was amazed by the white nights; he was astonished by the Kaali crater and its mysterious abundance of iron. The sea began to freeze in early winter and transformed into its direct opposite. What about that landscape, where everything was alien and incomprehensible, could stupefy Pytheas more? The miracle of farming so far north. In Pytheas's time, Estonia's islands and coastal areas were the world's northernmost grain-growing region. He tells us of it. Pytheas joined the celestial bodies, the earth's rotation, and the world ocean into a grand whole, and in the midst of this cosmogonical generalisation, he suddenly speaks of grain threshing, wild honey gathering, and beer brewing. These are contradictory magnitudes, a macrocosm made equal to a microcosm.

Humans' ability to thrive even at such northern extremes impressed Pytheas just as much as the Nordic environment itself. How were they capable of surviving? The threshing house is his accurate response. As summer is so brief, Pytheas says through Strabo's grudging lips: 'As for the grain, he says – since they have no pure sunshine – they pound it out in large storehouses, after first gathering in the ears thither; for the threshing floors become useless because of this lack of sunshine and because of the rains.' Pytheas correctly assumes that the adoption of threshing houses is due to the Baltic-Finnic people's agricultural climate. The sum of active temperatures in Estonia is 1,600–1,800 degrees, while in the Rhineland it is 2,500–3,000 (*Agroklimatichesky atlas* 1972). Estonia's growth period is 1.5 times shorter. The compressed growing season inevitably resulted in an impression of intense labour, which would lead to Tacitus's inaccurate conclusion that the Aesti were more 'assiduous'. Grain does not achieve

SILVERWHITE

full maturation at our latitude, and drying it in a threshing house is a method of post-harvest maturation – an invention whose importance did not go unnoticed by Pytheas. Linguistics adds temporal depth. The Estonian word for a threshing barn or house is *rehi*, the roots of which extend into pre-agricultural, proto-Finnic-Permic, in which it originally had many more archaic meanings. The Karelian *riihi* is an open-fire-heated sauna for drying nets and smoking fish, and in most Finnish dialects *riihtyä* means 'to become overheated or char in fire'. The content has changed, but the word has remained and preserves a long history: first, a fireplace or safe shelter, then figuratively as threshing, and, finally, when generalised, bread grain itself: *meil on rei üleval*, 'we have grain hanging on the poles' (SKES 3, 783; Wiedemann 1893: 1045). *Rehi* was lent across the sea to become *rie* in ancient Sweden and *ria* in modern Swedish. Linguistically, archaeologically, and in terms of exploratory geography, the *rehi* is part of Pytheas's landscape.

SEAFARING AND AMBER

The Aesti resemble the Suevians in customs and appearance. This tells us very little. Suevians is Tacitus's general term for all peoples from the Elbe to the Baltic Sea, including the ancient Swedes, ancient Livonians, and the Aesti. What caused him to use a common name for nations that differed both economically and linguistically? Firstly, the Suevian Sea itself. Estonians also refer ambiguously to peoples along the Mediterranean, even though a Bedouin herder and a Barcelonan mechanic have little in common. Another reason was traditions, which Tacitus believed were shared by most though not all peoples along the Suevian Sea. The Aesti were among the majority: 'Some of the Suevians make likewise immolations to Isis. Concerning the cause and original of this foreign sacrifice I have found small light; unless the figure of her image, formed like a galley, shows that such devotion arrived from abroad' (9).

Tacitus uses the word *liburna* for the light vessel or galley: a term that denoted ancient Baltic vessels in the Early Middle Ages and small boats of 'barbarians' on the Adriatic Sea in Tacitus's day. The ancient boat hadn't the slightest thing in common with the Egyptians' slim and beautiful goddess, leaving aside what matters most: Isis, being ruler of the sky, carried the life-giving sun disc between elegant cow horns. The symbol of the sky god on the Nile was the sun disc; on the Baltic, it was the ancient vessel. If Tacitus placed an equals sign between such dissimilar symbols,

WHISPERS REBORN AS WORDS

then they must have had a common function that the author understood, as well as origins that, as he lamented, remained a mystery. Limestone lies exposed in Estonia: a 600-metre-deep layer of bedrock that formed from animal skeletons which settled on the bottom of a tropical sea in the late Proterozoic Aeon. The steep limestone cliffs towering on both the mainland and islands represent one of the most dramatic chapters in our landscape and global geological history, though they crumble when subjected to sun and water and are unsuited for preserving ancient drawings. What occupies limestone's place in Finland and Scandinavia is even-grained crystal granite that the last ice cap polished to a shine in some places. These granite surfaces constitute northern Europe's oldest picture archives and have preserved over ten thousand rock paintings. Bronze Age artists' primary motif was a seafaring vessel and its crew. Above the boat is a sun disc and a person somersaulting through the air, whom, for lack of a better term, researchers have called an acrobat. Is it a ship burial? Judging by Tacitus's text, the traditions are being performed by the distant Suevians, of whom his knowledge is lacking. The Aesti? The sun symbol has the shape of an ancient vessel: is it the sun's death, its rebirth, a boat sacrificed to it? The sun-vessels chiselled into granite would nicely support Tacitus while simultaneously showing how little we've managed to penetrate the mindset of ancient humans.

Most researchers have identified characteristics of the Baltic peoples in what Tacitus writes next:

> Nay, they even search the deep, and of all the rest are the only people who gather amber. They call it glasing, and find it amongst the shallows and upon the very shore. But, according to the ordinary incuriosity and ignorance of Barbarians, they have neither learnt, nor do they inquire, what is its nature, or from what cause it is produced. (45)

Currently the richest amber quarries are located on the Sambia Peninsula and the coast of Lithuania. In the Early Middle Ages, the region was occupied by Prussian tribes who spoke languages belonging to the Baltic family. Following a devastating crusade, the Teutonic Order conquered them in the year 1283, and the last traces of the brave nation were assimilated, though their language clung to life until the 17th century. We should now cease trying to identify the Aesti as ancient Prussians, the reasons being as follows: Based on archaeological data, ancient Prussians reached the lower Vistula and the Neman rivers by the beginning of the Common Era at the latest. This could agree with Tacitus's information,

153

according to which the settlements of the Aesti were directly on the coast, but conflicts with him inasmuch as he used Pytheas as a primary source, and Pytheas recorded the conditions on the eastern coast of the Baltic four centuries previously. Insurmountable complications in our understandings to date arise with the question of how the name 'Aesti' could drift from the ancient Prussian tribes across the areas of Livonian settlement and fasten onto the Estonians, from which it can be followed from Tacitus's day to the present. Ultimately, linguists concur that the Sitones to the east of the Aesti were a Baltic people, and that the Venedians mentioned to the south were partly of Baltic, partly of Western Slavic, origin.

The Estonian language itself, which is astonishingly rich in a myriad of living artefacts, provides us with a weighty argument for equating the Aesti with the Estonians, as Tacitus did. The word itself is plain: *venelane*, Russian. Not one Slavic nation calls itself by that name. Apart from the Baltic-Finnic peoples, no one remembers the ethnonym anymore. Anymore? Yes, because classical Roman authors repeatedly mentioned peoples named the *veneti*. However, Pliny's and Tacitus's Venedians were located between the Germans and the Sarmatians, and they were already ethnically Slavic, Western Slavs, forefathers of the Poles, Wends, and Czechs. From this point forward, coastal Slavs' settlements on the southern shore of the Baltic Sea can be traced without interruption. Ancient Scandinavians used the ethnonyms *vindrir*, *vinedas*, and *winida* to refer to them.

The word *venelane* dates back to the time before Pliny and Tacitus. It is a Roman-era monument that Estonians alone have kept alive in their language. Sea-based contact between ancient Estonians, ancient Finns, the Venedians, and the Slavic language developed early and securely. When Slavic peoples migrated to the east of the Estonians nearly a millennium later, the latter's classical word for the Venedians was carried over on the basis of the similar language.

I first presented this hypothesis in 1976. The Estonian historian Evald Blumfeldt (1902–81), who worked in Sweden after fleeing Soviet occupation, doubted these conclusions in an article titled 'On the Origins of the Name "Estonia"': 'I have more to say about a theory put forth in L. Meri's work *Silverwhite*, his most convincing proof of which is the Estonian word for "Russian". Meri believes it came from the Venedians, was subsequently applied to the Western Slavs, and ultimately the Wendts. [...] Meri believes that the boar talisman is conclusive evidence of this, as Estonians worshipped the animals. However, this can hardly be definitive, as the Germanic peoples also worshipped boars, and quite popularly

WHISPERS REBORN AS WORDS

so.' Professor Blumfeldt's arguments are summarised in the following sentence: 'According to his theory, Estonians would have been the most important element in northern Europe.'

Here is my belated expression of esteem for the late academic.

I do not agree with depicting cultural history as a stadium marathon where someone is the first to reach the finishing line and someone else must come in last. In culture, there is no first or last, big or small, important or insignificant. Culture, as I understand it, formed about two million years ago, almost immediately after humans evolved and simultaneously developed language. It gradually picked up momentum and developed explosively over the last fifty-thousand years precisely because all populations, without exception, participated in the dynamic process of cultural exchange. Culture is a magic mirror in which we see every generation on every continent as alive. Culture is sovereign: any word, skill, or thought that is born lives on independent of the person and only visits the human brain to reproduce. It is unfortunately also true that culture will disappear along with humankind. However, it would be senseless to calculate, define, or weigh any population's part in creating culture according to that nation's modern-day population or vehicle production. Even continents wander! The coral islands on Saaremaa's western shore date to an epoch when the equator sliced through what is now Tallinn, and humans, given their relationship between power and intellect, have been in a state of even more dynamic transformation. 'The most important element' in northern European history is naturally not one tribe or another, but photosynthesis. What I wish to say is that, coming from a different generation, I also inevitably have a different sense of the past from Professor Blumfeldt, and the last thing I intended to do was to offend the Swedish nation. I am grateful for the review, as the academic's suspicions forced me to recheck my material. It stands up to the test.

In addition to Estonian and Finnish, the ethnonym *venelane* is native to Vepsian: *vena*, a Russian village as opposed to a Vepsian one. In the disappearing Votian language, *venäde* is 'the Russian language or characteristic of Russian'. Especially telling are *vändlane* and *vindlane* in the Tartu County dialects (Wiedemann 1893: 1343–5). Previously, we saw that contacts across the sea involved trade in skins in addition to amber. Subarctic dog breeds developed early on in northern Eurasia, being used as sled dogs and to protect herds of reindeer. Early in the Common Era, breeds more suited for hunting were imported from the southern shore of the Baltic: *vinttikoira* (Venedian dog) in Finnish (Erkki Itkonen to the author, 12.5.1977; SKES 6, 1779). The derivation of the ethnonym *venelane* from the Western Slavic *venethi* has been convincingly proved by Harri Moora, Kustaa Vilkuna, Erkki Itkonen, Paul Ariste, Richard Indreko, and Paul Kiparsky on the basis of archaeological, linguistic, and ethnographic evidence. One important conclusion is that contact with the Venedians was made exclusively by sea and already during the days of the Roman Empire. The Baltic tribes (later Prussians, Lithuanians, and Latvians) did not yet have such maritime connections in Tacitus's day, which means their name for the Eastern Slavs also comes from a later period. The Krivichi tribe became neighbours to the Latvians, and the name later expanded to include the entire nation: *krievs* is the Latvian word for 'Russian'. I have made it my duty to transplant this noteworthy linguistic detail into the fertile soil of ancient trading posts in order to ask in the next chapter: How did the first fortress towns develop from these trading posts? How did the trade routes work? What shores did they connect? What types of vessels have sailed our sea? Why have we forgotten them?

It would be risky to rest such weighty conclusions on Tacitus's shoulders if *aestiorum gentes* were to flicker only once in recorded history. However, the Aesti can be followed without interruption, starting with Tacitus. Five centuries later, the Gothic historian Jordanes speaks of a nation called *Hestii* and once again ties them to amber. The Christian priest Paulus Orosius, who was active mainly in Hispania, finished in the year 418 his seven-volume *History against the Pagans*, which the king of the Anglo-Saxons Alfred the Great later had translated. The king's notes include the land of the Aesti, 'Eastland', the nearby land of the Venedians, 'Weonodland', and, also near Estonia, *Cwenland* (Woman Land) and the Sámi *Scridefinnas*.

Thus, the Aesti are farmers who use a threshing house, gather amber, and navigate the sea. In accordance with Suevian custom, they worship 'the Mother of the Gods' and wear boar amulets as a sign of their faith. Such amulets have been worn in Estonia since time immemorial. But why is it a safeguard for the wearer still in the times of Pytheas and Tacitus? Did ancient invulnerability turn into a mark of power? Was this mark of power preserved in Estonian folk wedding tradition in the form of the *seamees* or *kärsamees* – a 'pig man' who represented the groom's family and served as master of ceremonies? Is Tacitus's 'Mother of the Gods' the Baltic-Finnic 'golden maiden'? Was the golden maiden originally the myth of a tribe's taboo holy site?

Unanswered questions also have a right to exist.

156

WHISPERS REBORN AS WORDS

If Tacitus's Aesti are Estonians, then should they speak the Estonian language?

IN THE LIGHT OF AN ETERNAL FLAME

The language of the Aesti, according to Tacitus, was closer to 'the language of Britain'. How should we interpret this? Let us first appreciate how outstanding this characteristic really is.

What forced the author, for the first and final time, to reference a linguistic difference between barbarian tribes? We should be more specific: a difference in terms of what or whom? Of Germanic peoples. Tacitus used the generic term for Northern and Western Germans, Western Slavs, and Balts, all of which belong to the Indo-European linguistic family. The coastal Suevian cultural region, if we may use such a term, encompassed tribes that spoke several different languages within the Indo-European tree, altogether twenty-six. Linguistic divergences were minor in that time. To a foreign observer, the differences between Western Slavic and Baltic languages may have gone unnoticed. However, as Tacitus's quill travelled further north while describing the ethnography of Baltic coastal areas, he came to a linguistic environment that differed significantly from the ones before. Up to this point, customs and lifestyles were the author's sole focal interest when differentiating tribes. Now, the linguistic dissimilarity was so striking that it rendered the former criteria insufficient, and he deemed it worth noting.

The comparison he employs is crucial from our argument's perspective: the language of the Aesti 'more resembles that of Britain'. By this, he means the language spoken by the British Celts. Although it also belongs to the Indo-European family, ancient Celtic bore certain similarities to Baltic-Finnic. Paul Ariste explains this as the long-term effects of a common substrate language on both our languages. Consequently, Tacitus's comparison is not formal.

Still, the question itself and another conclusion lie elsewhere.

Why did Tacitus choose British Celtic as a language for comparison? Celtic was also spoken in Gaul, a Roman province located in present-day France, which Tacitus knew well. In an earlier chapter of the *Germania* (28), he correctly theorises kinship between the Gauls and the Helvetians. Britain, on the contrary, was still a quite unfamiliar land to the Romans, who had conquered the southern part of the island just fifty years before the *Germania* was finished, with Tacitus's father-in-law Julius Agricola expanding Roman territory slightly to the north. Through daily contact and rich literature, the Romans' knowledge of Gaul was incomparably greater than that of distant Britain beyond the sea. So, how can we explain Tacitus snatching a basis for comparison from beyond the boundaries of his work, comparing the unfamiliar tongue of the Aesti with 'the language of Britain', describing one unknown with another?

The comparison was borrowed from Pytheas's *On the Ocean* – the sole source Tacitus used for his descriptions of the eastern coast of the Baltic Sea. Earlier in this travelogue, we recreated Pytheas's research voyage along with its motivation: tin and Cornwall, amber and the Baltic Sea, the sun's grave (*helios koimatai*) and Saaremaa. Comparing the language of the Aesti to 'the language of Britain' is logical and natural in Pytheas's work. Massalia's stable trade ties with Cornwall would have required a knowledge of the Celtic language and allowed Pytheas to describe the unknown tongue of the Aesti by way of the British Celts' known one. It would also mark the birth of Baltic-Finnish comparative linguistics in so far as the hypothesis, no matter how logical and pleasing, is followed by evidence, no matter how dry and sober.

Here we have returned to the question we left smouldering. Where did Tacitus get his 'Aesti'? We're no longer talking about tribes, but an ethnonym. Not even an ethnonym, but a bare word. Pytheas brought 'most remote Thule' from the Baltic Sea to Mediterranean literature: the Baltic-Finnic word *tule*, 'fire', which Eurocentric research has hitherto been unable to connect to a single language. Pliny repeats this Ultima Thule; Tacitus, for reasons we can now understand, not. Instead, Tacitus refers to the Aesti nations, *aestiorum gentes*, which were not mentioned by Pytheas or Pliny. So, what is the origin of 'Aesti'?

Indo-European languages convey the concepts of warmth, heat, fire, and flame using words with a common root of *ai-dh* (Pokorny 1947: 11). The Old Persian *āed* is 'fire', Greek *aidos* 'burning, fire' and *aidós* 'burned [to completion]', Latin *aestus* 'heat, glow, blaze'. While describing those settled on the eastern coast of the Baltic Sea, Tacitus relied upon Pytheas's text, which contained enough information to translate the Baltic-Finnic word *tule* into clean Latin: *aestus*, which was used to form the ethnonym *aestiorum gentes* and placed, correctly, the Aesti nations to the east of the Goths and southwest of the Sámi.

Are we Estonians bearers of fire? Fire-tenders? *Eisa* is 'fire' in Old Icelandic and 'hearth' in Norwegian. The Old Norse *eistir* and Old Swedish *ester* are 'Estonians' (SKES 1, 34). Are we Estonians fire-tenders?

WHISPERS REBORN AS WORDS

The Gothic word *aistan / aistada* came from the same root and originally meant 'fiery, blazing, choleric', later 'dread, honour, reverence', and led to an ethnonym that, apparently still in Tacitus's day, was used to refer to the resident of an inviolable (taboo) place.

Did the eastern direction become sacred? Or did 'sacred' become the eastern direction? Or is coincidence still playing its eternal game? Suffice it to say, Estonians are the westernmost easterners. Easterners?

Tacitus entered ancient Estonian seafaring with scant lines but with substantial information. The Baltic Sea had been explored from within and without alike by his day.

A NAME FROM BIRTH

If you observe history from a far-enough vantage point, it moves and transforms according to its own inner rules, and the further you are while watching, the more general they appear. Yet one needs only take one step closer and connecting paths begin to reveal their influence; in other words, an exchange of information that is more favourable in one region and therefore less so in another. Flint, obsidian, and slate are not deposited evenly throughout the world, and as the Stone Age came to a close, areas where tin and copper for making bronze could be found were at an advantage. The Iron Age thrust new regions into the spotlight and older ones out of it in turn: every technological age has its favourite child. History, which studies the causes and effects of past changes, naturally also deals with the pedosphere, latitudes, and solar radiation. These things can also be calculated, determined, bent into formulas, and shaped into laws.

This is a lengthy introduction to a short statement: in this chapter we are not tracing regularity, but anomaly.

Nature's blind chance struck Saaremaa. It could just as easily have struck Saratoga half a million years earlier or Sarasota a few thousand years later. Chance is blind: it selects neither time nor space, and this is also the source of its charm and exceptional value: thanks to ideal anomaly, we are able to better appreciate the idealness of regularity. Ideal anomaly is an experiment in a clean flask: everything becomes visible and transparent. Nature has been stingy in giving us Estonians such opportunities for experimentation. They are as unique as giant diamonds that are known by name and lack a market price.

Saaremaa's nearest neighbour across the Baltic Sea is Gotland, and the closest anomaly of the Kaali event is a shift in meaning that took place

SILVERWHITE

in the island's Gothic language: the Old Gothic *aistan* acquired a taboo meaning. It had already occurred by Pytheas's day and seeped through him into a literary backdrop which, from that point forward, began to tinge our fiery Thule in many a poet's writing.

Anomaly has applied a specific shade to natural sciences and the description of lands and peoples. The names of some countries derive from a few tribal names. This is a rather recent phenomenon. One prerequisite is the merging of former tribal lands or their joining into a single economic and political whole that could and should be called by a common name. Suione/Svean, *svidai* in Old Icelandic, originally meant one who 'died a natural death', i.e. not 'unnatural' (Pokorny 1947: 882). May we sense in this a contrast to the 'other world' of the world pillar? A place 'properly called Scandia' where a cosmic catastrophe razed, destroyed, and killed? Over time, the Suiones/Sveans gave their name to the entire territory of Sweden – *Sverige* – similar to the Franks and France. From their viewpoint, the closest neighbours were the tribes of Alemanni: *Allemagne* is 'Germany' in French. The Estonian word for German, *saksa*, formed in the same way. Originally, it was used for the Saxon tribes in the area around present-day Schleswig-Holstein, which they controlled until their invasion of England in 448 (SKES 4, 952). The Saxons have long since become the Anglo-Saxons, aka English, though Estonians still refer to the residents of Germany in our rather surprising old way.

And what do the Germans call themselves? In the Early Middle Ages, and even the Late Middle Ages, there was no need for an overarching name, so we also lack the given national terms. One simply said Bavarian, Swabian, or Lübeckian, and that sufficed. The Old High German word *diutisk*, 'of the people', was turned into the Latin *theodisce*, *theotisci*, and *teutonicus* for official documents, but no one knew what it meant, exactly. The people of the Netherlands used the word *dietsch* for their own language until the 17th century, and English has maintained the word Dutch to this day. The need to give a name to one's neighbours formed early, the need to find a common name for one's own people and language formed later, and there is solid logic to this idea.

Estonians have applied this system to ourselves with incredible unity. Our educated neighbours learned to call us Estonians at a very early date. We, however, only began regarding ourselves as Estonians rather recently: at about the time the Suez Canal was opened or Verdi's *Aida* premiered (the latter wasn't finished in time for the former's celebrations). According to popular belief, Estonians tended to refer to themselves more as *maarahvas*,

WHISPERS REBORN AS WORDS

'people of the land', until the term gave way to Estonian (*eestlane*) and Estonia (*Eesti*, *Eestimaa*) in the fervour of the first Estonian Song Festival (1869). We were born in the fire of song and carried along by its winds. It is a fanciful notion, of course, though the aim of this national-romantic interpretation is not to explain ancient geography or ethnonyms, but to contrast the indigenous community with the politically privileged urban populace.

Ancient Estonians' geographical knowledge preceded their socio-political development to a remarkable degree. We can call this our second anomaly and recall Tacitus's words: the boar amulet, a common mark of the Aesti nations, 'serves them for arms, [...] is the safeguard of all, and by this every worshipper [...] is secured even amidst his foes' (45). Could the shared symbol have expanded the tribes' common consciousness, fate, and identity?

In the early 13th century, Estonians' largest administrative unit was the county, which could temporarily unite with neighbouring counties in the case of encroaching foreign threats. As with Central Europe, we could expect our oldest geographical documents to show the names of these counties and the demonyms derived from them. Oddly, the opposite is true. The oldest surviving descriptions of the territory that is now Estonia can be found in Henry of Livonia's Latin *Chronicon Livoniae* (1224–7). In it, the author faithfully uses the ethnonym *estones* to describe inhabitants of all Estonian counties, the toponym Estonia, and also Estlandia as a generic term for the entire area. Especially telling are the instances in which his general name sits side by side with a county name: 'the pagan Estonians of Saaremaa' (one would expect simply 'the pagan Saaremaans') or 'to Ugaunia and later to all the Estonians', though the analogous phrasing 'to Bavaria and later to all the Germans' would have lacked any geographic substance and would have been unthinkable in 13th-century Central Europe. Henry's use of the encompassing label does not derive from a poor knowledge of Estonia's different counties: borne by a storm, Henry orders his ship to land 'in the Estonian county of Virumaa' and, after arguing with the Danish soldiers, describes the environs of his missionary work with sarcasm but precision: not the backwaters of Estonia, but 'amidst Järvamaa, actually in Virumaa'.

Henry of Livonia is not an anomaly himself. The ethnonym 'Estonian' was used earlier by the encyclopedist Bartholomaeus Anglicus, and the demonym and toponym alike were used with the same meaning in Livonia's oldest rhyming chronicle, written in the late 13th century.

161

SILVERWHITE

Henry of Livonia is far more uncertain in his use of a general name for Latvians and Latvia. Statistically, for every mention of Estonia, Henry writes Livonia 2.06 times and Latvia just 0.18 times. On the other hand, the 'Estonian' ethnonym is absent from the Latvian language. Instead, Latvians classically call Estonians *igauņi*, Ugannians, after the bordering historical county. Does this allow us to conclude that seaways amplified the pace at which general names developed and the Estonian toponym above all took hold on the homogeneously populated coastal areas? May we use this to draw further conclusions about the ancient ethnic map? To the Finns, Estonians are *virolaiset*, which is understood to be an application of the northern Viru County to all of Estonian territory. But now, even this explanation becomes questionable. The Virokannas of our folklore appears to personify the entirety of Estonia. According to Emil Nestor Setälä, *viro* means 'the world' and *kannas* is a 'pillar'. Virokannas comes from the most ancient strata of prehistoric Estonian culture. It appears that the Kaali event elicited an unusually early need to apply a name to the Estonian homeland and, as an extraordinary wave train, the names 'Estonia' and 'Estonian' were adopted in the local tongue unusually early as well. If a common identity did indeed promise success for the Baltic-Finnic tribes, as Tacitus claims, then the ethnonym's early adoption among contacts across the sea, as well as the nation's self-identification as Estonians, becomes entirely foreseeable. It appears to be confirmed by much later writing on a page of Joan Blaeu's 1662 *Atlas Maior: Livonia, vulgo Evestland*. In other words: 'Livonia, Estonia in the peasants' tongue'. Temporally, the order of the names comes as follows: 330 BCE: Thule, *tuli*, 'fire'; 98 CE: *aestus, aestii*, 'those of the fire'; 524: *Aesti*; 550: *Aesti*; 816: *Aisti*; 880: *Estum*; 1075: *Aestland*; 1154: *Estlandia*; 1167: *Estonum*; 1172: *Estia*; 1212: *Estones Osilianes*; 1219: *Estland*; 1224: Estonia.

As one might expect, the ancient toponym's dissipation coincides with the fall of the Roman Empire and the ebb of seafaring on the Baltic Sea as a consequence. Wulfstan (880) had the Vistula flow into the 'Sea of the Aesti'. Starting with Adam of Bremen (1075), the toponym refocuses on the present territory of Estonia. In addition to the accounts of the Danish king Sweyn II, the Bremen clergyman's sources included the classical authors Pomponius Mela, Solinus, Orosius, and the previously mentioned Lucan, who thirty years prior to Tacitus mentioned the strange inviolability of a certain island people.

It turns out that almost all of this is correct. Estonians are the Nation of the Fire World, as Pytheas wrote. Our sun has been black and our

162

WHISPERS REBORN AS WORDS

land sank into the sea, as the *Poetic Edda* believed. Saaremaa has been in flames, as our memory whispers. Fire dragons soared across the sky, as a Novgorod chronicler recorded. Blacksmithing gave us faith, as Apollonius and Virokannas knew, nourishing fields, covering threshing houses, connecting seas, protecting languages. Even so, we know ridiculously little about it, and so I will conclude this chapter with a passage from Tacitus. The Roman described a strange island that has not been identified to any degree of satisfaction to this day. It may not be Saaremaa. However, as the Aesti had 'the same customs of the Suevians', we can, with great probability, presume that several details could also appropriately describe Saaremaa, especially the muddy and tranquil Lake Kaali, surrounded by a great ring of stones:

> In an island of the ocean stands the wood Castum: in it is a chariot dedicated to the Goddess covered over with a curtain, and permitted to be touched by none but the Priest. Whenever the Goddess enters this her holy vehicle, he perceives her; and with profound veneration attends the motion of the chariot, which is always drawn by yoked cows. Then it is that days of rejoicing always ensue, and in all places whatsoever which she descends to honour with a visit and her company, feasts and recreation abound. They go not to war; they touch no arms; fast laid up is every hostile weapon; peace and repose are then only known, then only beloved, till to the temple the same priest reconducts the Goddess when well tired with the conversation of mortal beings.
>
> Anon the chariot is washed and purified in a secret lake, as also the curtain; nay, the Deity herself too, if you choose to believe it. In this office it is slaves who minister, and they are forthwith doomed to be swallowed up in the same lake. Hence all men are possessed with mysterious terror; as well as with a holy ignorance of what that must be, which none see but such as are immediately to perish.

Animal sacrifices and violent human death are front and centre in lore connected to Lake Kaali.

But how credible is lore? Archaeological digs in Isokyrö, Finland, have confirmed the performance of ritual human sacrifices in Baltic-Finnic settlements.

What does an archaeological discovery signify apart from itself? How credible is the literature that has been written? And memory?

163

SILVERWHITE

In spite of everything, I believe that nature has given us an incredible test for tracing the entropy and negentropy of information. Taking part in this test isn't, of course, the only reason why I enjoy being an Estonian.

IV

THE CIRCLE IS COMPLETE

PYTHEAS ARRIVES HOME

Perhaps preconceptions are necessary. Perhaps they slow changes and force one to farm one's fields more intensely instead of slashing and burning a new forest. Preconceptions possess great force.

Much has been written about Pytheas. Every researcher has noted and highlighted the ironic remark made by Polybius, conveyed to us in Strabo's writing, about Pytheas allegedly exploring the entire European coastline. 'Now Polybius says that, in the first place, it is incredible that a private individual – and a poor man too – could have travelled such distances by sea and by land' (Strabo). Strabo excuses his own doubtful assessment by referencing another source. Today's researchers have no doubt. Pytheas undertook other voyages as well. German professor Hans Joachim Mette stressed, without any prejudice, that Pytheas's travels after returning from the Far North, hugging the European coast from Gadeira to Tanais, should not be doubted in the very least.

I strive to envision northern Europe at a time when seafaring vessels still lacked rowlocks. The sea days must have been onerous and brief. In truth, that was already Pytheas's third voyage, the first having been to Spain to study the tides.

Gradually, a modern travel bureau with chattering telexes starts to take shape. Gadeira, I forgot to mention, is now known as Cádiz and Tanais was located on the Don, which flows into the Sea of Azov. It is so incredible, so illogical for the utterly logical Pytheas, who saw the world as a whole and described it as a single universe with all the heavenly bodies. If he was on

Saaremaa, on the frozen Gulf of Finland, perhaps even on the threshold of Silverwhite's waterway at the mouth of the Neva, then why turn back to Cádiz in order to reach the Black Sea? It would make perfect sense if he'd limited the expedition to Britain. In that case, travelling from Cádiz to the Black and Azov Seas via the Mediterranean would have been completely understandable. But given that he did not attempt to circumnavigate England, a third voyage to the Don is now extremely difficult to believe.

I wrote to Greek philologist Jaan Unt with my suspicion and was pointed to his exact Estonian translation of Strabo's text, which removed the final barrier on Pytheas's journey: 'That, then, is the narrative of Pytheas, and to it he adds that on his return from those regions [up or away from the Mediterranean, i.e. north] he visited [travelled] the whole coast-line of Europe from Gades to the Tanais' (Strabo).

Preconceptions hold terrible force. Every historian whose works I have read present those lines as 'Having returned from Thule, he travelled …' More accurate authors have added in brackets that Thule does not appear in the original. Yet given that Pytheas could not have leapt from England to the riverways of eastern Europe, Strabo's correction is seen as almost ethical. Instead of remaking the past, we truly could do a little more to improve the present. Jaan Unt comments: 'The verb *epanerhomai* may not merely mean to return. To me, it seems entirely impossible that the verb might simply mean Pytheas's travels "northward", and highly possible that he returned via the *Silverwhite* ring road' (Jaan Unt to the author, 14.7.1982).

With this, we have exhausted Pytheas's texts. We could, of course, ask ourselves if Pytheas's likely voyage from the Baltic to the Black Sea left any trace in literature.

Discussion of the Caspian Sea heated up during the time of Alexander the Great and Pytheas: was it an inland sea as Herodotus wrote? Or a bay that opened into the Arctic? The latter belief triumphed for a long time after Pytheas's voyage. The Gothic historian Jordanes added an unexpected but accurate toponym to the classical authors: the Caspian Sea is connected to the northern seas via the Vaga River, which also borders the 'island of Scandinavia' to the east. Jordanes alone mentions the Vaga in older texts. Tuomo Pekkanen found a simple and therefore outstanding answer: *Vagi fluvius* is indeed the Vaga River, now a 500-kilometre-long tributary of the northern Dvina, which the Komi people call the Výnva. The entire northern Dvina was earlier called by the name Vaga, its headwaters meeting those of the Volga and the Kama via a narrow

THE CIRCLE IS COMPLETE

watershed. This clarifies why Mercator's map of the Arctic (1569) turned out to be a greater sensation than the publishing of his atlas. The Gulf of Finland and Narva promised easy access to India, one key to which was the Vaga River. The conviction that the Caspian Sea was connected to the northern seas, which persisted for many centuries, appears to prove that Pytheas travelled down a long-established waterway from the Gulf of Finland, over the White Sea, and down into the Volga's headwaters.

Ptolemy, the greatest geographer of the Classical Era, was responsible for a false understanding of the Sea of Azov, which was extended incredibly far north – almost to the latitude of the Neman River (on a 1490 Roman map) or the Gulf of Finland (on Leonardo da Vinci's 1514 world map) (Nordenskjöld 1889: map IX and 77). This is also a deformation based on source. Earlier, we showed that Adam of Bremen possessed a remarkably accurate understanding of the locations of Estonia, Courland, and Semigallia on the eastern Baltic shore. However, he placed in northern Estonia's Viru County tribes that mythically or genuinely populated the shore of the Azov Sea (Pekkanen 1980b: 85). This likewise shows signs of Pytheas's influence and allows us to conclude that the final leg of his journey home ran from the Volga to the Don, the Don to the Sea of Azov, and finally to Olbia at the mouth of the Borysthenes. Pytheas is a strangely solitary figure in the history of exploration. He and no one else ventured to the eastern Baltic Sea and the Gulf of Finland. Notions of the Caspian Sea stretching towards the Gulf of Finland, and later the Sea of Azov reaching towards it as well, probably originated from his writings. Through Ptolemy, it had a lasting impact on people's understanding of the world and set the foundations for the early medieval T–O maps: a circle depicting the world ocean and a T within it to represent the Mediterranean. East was at the top of the circle, west at the bottom, south on the right, and north on the left. The horizontal top of the T runs from south to north. One side, the south to the intersecting line, represented the Nile flowing into the Mediterranean, and the other, running from the north to the intersection, was a waterway like the Volga, the Don, or the Dnieper. Where the top of the T touched the left side of the circle, the waterway flowed into the northern seas – the Gulf of Finland. If we proceed from the assumption that the world was understood differently in the past, then we must accept the logical conclusion that it was depicted differently as well. You always find little islands where the waterway crosses the circle of the world ocean, just a little to the left of the estuary, and next to that

SILVERWHITE

island there is always a familiar name: Thule. Few are the maps that are missing our home.

THE GREAT WATERWAY

The notion of the Baltic Sea being part of Oceanos persisted for an unbelievably long duration. Over time, the shipping route grew more exact, hugging the Swedish shore to the Mälaren trading post and from there onward to southwest Sweden and northwest Estonia, from which it split in the directions of Saaremaa and Virumaa. Nevertheless, the result remained unchanged. Oceanos's tiny 'Scandinavian islands' melted together into one great island, the coastline became continuous, and the Baltic Sea turned into a channel that still joined the world ocean and the Great Waterway at the rear of the Gulf of Finland.

Equally persistent was classical authors' belief in a direct connection between the Caspian–Volga waterway and Oceanos. True, a watershed gradually took shape between the southern and northern seas: the Riphean Mountains – one of northern Europe's oldest toponyms, the home of the northeast wind, where ice drifts and snow falls endlessly. The Ripheans survived into recorded history and in 1828 were identified as the northern Ural Mountains. Specifically, the Ripheans were originally more a folkloric symbol than a geographical one (if those concepts are distinct in the first place): an impassable barrier, a wall in the most literal sense of the word, complete with copper gates and a sentry, behind which are revealed shimmering fairy-tale treasures and terrifying perils. The Ripheans were nudged this way and that, and when it was positioned as running from west to east, then the barrier could only be crossed through the 'Caspian channel'. This conception, spiced with merchants' tales of horror, persisted long after the beginning of the Common Era while never coming into contradiction with the belief that there existed a connection between the southern and northern seas, located more or less around where the 'bay' and 'channel' of the Baltic opens up.

Was the reason behind such dogged conservatism the exceptional nature of Pytheas's journey? His travels that took place a millennium before their time? A lack of more precise sources? Only partially.

The experience, no matter how distorted it was by the time it reached classical geographers, appeared to confirm the existence of a Baltic–Volga waterway, ties with the North, and connections to exotic Sápmi and the White Sea region. Although sparsely populated, the western Estonian

168

THE CIRCLE IS COMPLETE

county of Läänemaa was one of the most prosperous in Pytheas's day. As the archaeologist Vello Lõugas wrote: 'Many items likely imported from the areas of the Dnieper, Volga, and upper Oka rivers have been found scattered among Saaremaa's Roman Iron Age artefacts, speaking to the wide-reaching cultural ties of the peoples of western Estonia via waterways. At the same time, the development of such ties requires that people to have a developed economy offering opportunities for trade' (1972: 168). 'Cultural ties', i.e. contacts, is a broader notion than trade. We will allow the specialised researchers to determine where contacts end and trade begins. For now, let's stick to a plainer truth: waterways were the corridors for contacts over great distances, and vessels were the vehicles. There were no other routes.

Whereas Läänemaa's prosperity vanished mysteriously and almost overnight, the populous settlement on the island of Hiiumaa's Kõpu Peninsula died out entirely in the 2nd century CE. Virumaa, Ptolemy's (170 CE) land of the *phiraisoi*, then rose to become the territory of Estonia's most prosperous county (if we may refer to the area using a concept that didn't yet exist at the time). This was a result of agricultural development. Why? The slash-and-burn method bore better results in areas where there was no mighty forest, but rather brush and shallow limestone bedrock. Along with this prosperity, Virumaa became the focal point of ancient shipping. The era also produced the name of the settlement at Toolse, which was then called *Toolisen*: a loanword that in the Gothic language meant 'throne', but as a place name was used for a site of rule or population centre (Vilkuna 1963: 215). The archaeology and toponomy are in fine harmony with Ptolemy's *Geography*.

At the beginning of the Common Era, the limits of the Roman Empire ran along the Rhine, for the most part. It was a political boundary. The tendrils of the Roman economy, however, stretched much further, choking the Baltic-Finnic peoples' sea ties with the Scandinavians and powerfully pushing forward new, Rome-oriented axes. Scandinavian iron production began only after the dawn of the Common Era. Prior to and during the Roman Iron Age, Scandinavia imported its iron (Calissendorff 1971). From Saaremaa? Interests and shipping routes swung to the southern coast of the Baltic Sea during the age of the Roman Empire. From there, Roman gold sprinkled as far as Saaremaa and Naissaar, and its mighty radiance reached even further north. Pelt-bearing animals disappear from Estonia's forests but remain in the territories of Latvia and Lithuania into the period of recorded history. The reason? That was the era when the

SILVERWHITE

Gothic loanword *skraha* entered the Estonian language no longer in the sense of a pelt, but money: *skraha > raha*.

What does this have to do with geographic explorations? What did gold have to do with the discovery of America? With the settlement of Alaska?

Pelts were the gold of the taiga from prehistory to the Late Middle Ages, and in the Komi-Zyrian language, the word *ur* is still used for both squirrel pelts and smaller monetary denominations. Pelts' trade value remained high throughout the ages in ancient Greece, Egypt, Rome, Harun al-Rashid's glittering Baghdad, and even faraway India, where the cost of an ermine pelt was a jaw-dropping 400 dinars. Amber came and went out of fashion. A powerful cowrie-shell fashion swept across the Finno-Ugric peoples, who could never have imagined that the 'snake heads' came from the Maldives in the distant Indian Ocean, the creatures' habitat. Fashions were temporary, but the demands of giant cultural hubs for pelts remained unchanged throughout ages and the passage of fads.

Let us add here a trade secret known by every trapper and tanner: the value of a pelt is determined by the density and length of the fur. The more northern the forest, the better the pelt. Archaeological digs in the Sámi area of Västerbotten, northern Sweden, have produced 4th-century earrings and jewellery that were crafted by Estonian blacksmiths (Stenberger 1977: 300). A quality hide only lasts for so long. Speed of travel was not secondary.

The waterways persisted and Thule paled, but the Aesti undertook dizzying journeys in the writings of learned men.

The silly name of the people's migration overshadows the complex societal and ethnic storms that brewed in Inner Asia and expanded with increasing intensity over the entire gigantic continent. The Eternal City fell in the year 410, but the Empire itself limped on a while longer, transforming and also changing the barbarians whose chiefs had become acquainted with Plato, sampled the pleasures of the imperial territories, and met the fires of Christianity.

Aestian ambassadors arrived in Ravenna to call upon the king Theodoric of the Ostrogoths in the year 524. The reasons for the visit were economic. The king's response to the Aesti has been preserved: it is about 270 words and explains that amber is formed when coniferous sap hardens in the sun and is 'purified' in sea water: 'We have thought it better to point this out to you, lest you should imagine that your supposed secrets have escaped our knowledge.' The king is illiterate, but that is the

THE CIRCLE IS COMPLETE

purpose of secretaries. More interesting in terms of travel stories are the following lines:

> It is gratifying to us to know that you have heard of our fame, and have sent ambassadors who have pressed through so many strange nations to seek our friendship. We have received the amber which you have sent us. [...] We send you some presents by our ambassadors, and shall be glad to receive further visits from you by the road which you have thus opened up, and to show you future favours. (Cassiodorus)

The Byzantine historian Procopius, born in Caesarea, was just finishing his greatest, eight-volume work at the time. Its chapters that touch upon Thule were written on the front line of the Italian civil war just three years after the mysterious visit of the Aesti to Ravenna. According to him, the Heruli tribe decided to return to Scandinavia after their defeat on the shores of the Danube. They 'traversed all the nations of the Sclaveni' (Slavs), then 'crossed a large tract of barren country', and finally 'passed by the nations of the Dani' (Danes) to come to the ocean, from which they sailed to Thule where they remained.

Thule stands apart from all other legends in that time did not cause it to dim but to brighten. Procopius's Thule is, of course, not Saaremaa. However, the grain of truth lies in its tie to the Baltic Sea:

> Now Thule is exceedingly large; for it is more than ten times greater than Britain. And it lies far distant from it toward the north. On this island the land is for the most part barren, but in the inhabited country thirteen very numerous nations are settled; and there are kings over each nation. In that place a very wonderful thing takes place each year. For the sun at the time of the summer solstice never sets for forty days, but appears constantly during this whole time above the earth. (6.15)

Procopius's description of the polar night forces us to wonder if he might have possibly been referring to Iceland, Greenland, or another mythical place. Yet another cliché? But no, our suspicion is unjust. He also reveals new and credible details: how can one determine time in the North when there is no boundary between day and night?

> For they said that the sun during those forty days does not indeed set just as has been stated, but is visible to the people there at one time toward the east, and again toward the west. Whenever, therefore, on its return, it reaches the same place on the horizon where they had

SILVERWHITE

previously been accustomed to see it rise, they reckon in this way that one day and night have passed. (6.15)

This is something far greater than Tacitus's compliment to his dear father-in-law. Behind it lies genuine first-hand experience in the Far North. And if we still had any right to give a doubtful shrug, then the following sentence dispels it. 'But among the barbarians who are settled in Thule, one nation only, who are called the Scrithiphini, live a kind of life akin to that of the beasts' (6.15). The *Scrithiphini* are the skiing Sámi. It is a Finnic ethnonym that was known already to Tacitus and may have been conveyed by the Aesti: Tacitus placed his Fenns directly to the east of the Aesti. Ptolemy knew them as the *Phinni* a century later, Paulus Orosius (early 5th century) as the *scridefinne*, Paul the Deacon (8th century) as the *scritobini*, and Adam of Bremen (1075) as the *scritefingi*. The Sámi gave Finland its name in every language apart from Estonian. The frequency with which the small and distant Sámi are mentioned is extraordinary. Cultural contact, as we usually call it, is such an ambiguous concept that it can encompass both the howling of the wind and the hum of telephone lines, and, lastly, unfortunately, humans with their fates, aims, journeys, and ships. In just a few dozen lines, Procopius paints a masterful picture of the Sámi hunting practices, pelts, skins sewn with sinews, and even the way they feed their infants – all balanced, precise, and factual. Commenting on their knee-length skin garments a few centuries later, Paul the Deacon writes: 'I have seen these.'

As usual, we have digressed. Our great rings around the Baltic Sea demonstrate the lack of local authors and sources. Procopius has the right to the final word:

But all the other inhabitants of Thule, practically speaking, do not differ very much from the rest of men, but they reverence in great numbers gods and demons both of the heavens and of the air, of the earth and of the sea, and sundry other demons which are said to be in the waters of springs and rivers. And they incessantly offer up all kinds of sacrifices, and make oblations to the dead, but the noblest of sacrifices, in their eyes, is the first human being whom they have taken captive in war; for they sacrifice him to Ares, whom they regard as the greatest god. And the manner in which they offer up the captive is not by sacrificing him on an altar only, but also by hanging him to a tree, or throwing him among thorns, or killing him by some of the other most cruel forms of death. Thus, then, do the inhabitants of Thule live. (6.15)

THE CIRCLE IS COMPLETE

Thirteen tribes reside on Procopius's Thule. The polar day lasts forty days, and the polar night is just as long. Such a day places Thule's northern limit at 68° 30′ and an equally long polar night pushes the boundary to the latitude of Murmansk.

The Scritfinns make Sápmi the northern limit of Procopius's Thule. Sápmi is the most distant and exotic part of Procopius's Thule. Contrasting the Sámis' customs with those of 'the rest of men' allows us to see the latter as living beneath the southern latitudes: 'ordinary' could, in Procopius's eyes, mean agriculture, not hunting. Setting sail from the Danes places Thule in the Baltic Sea. 'Thirteen tribes' appears to indicate that all Baltic coastal areas, the entirety of Baltoscandia, fits beneath Thule. A stone fell into water, and these are the last concentric circles on the nearly tranquil reflective surface; the largest, barely perceptible, wearily rebounding from the shores.

THE FIRST MAPS

The toponym 'Finland' was first recorded on skins in Estonia. Henry of Livonia describes the siege of the Somelinde (Soomelinna) fortress in the year 1212. However, the same chronicler mentions a Finnish missionary named Petrus Kaikivalda or Kaugovalda *de Vinlandia* working in Estonia from 1215 to 1220: Henry did speak Estonian to some degree, but it's unlikely that he realised our *Soome* (Finnish *Suomi*) could be equated to Finlandia. The latter came from the Sámi/Fenns and sprang from an awe-inspiring millennium-long tradition. It is an inherited moniker. Linguists find it likely that the name, which derived from the verb *suoda*, 'to wish, to promise', originally transformed into a tribal label and later expanded to the tribe's area of settlement, county, and ultimately all of southwest Finland, which even today is still called *Varsinaissuomi*, 'True Finland' (SKES 4, 1114). Contradictorily, the onomastician Viljo Nissilä believes that *Suomi* is a loanword from the Swedish *som*, meaning (a small migratory?) 'gang'. In this connection, it is fit to point out that both Finns and Estonian islanders are called by a common name in Latvian: *sāms*. Saaremaa itself is referred to as *Sāmu sala* ('Finnish island') or even *Sāmu zeme* ('Finnish land'). We may hardly take this as evidence of an ancient migration. Rather, the common name points to the extremely minor divergence of Baltic-Finnic languages, which the Baltic-speaking ear may not have perceived.

SILVERWHITE

Now, let us turn to a new source of information and momentarily leave Thule aside. The first Estonian toponyms to be recorded on a world map, dating to the years 1306–21, were Saaremaa (*Osillia*) and Estonia itself (*Estonia*). From this point onward, Saaremaa can be tracked even on maps where Estonia is missing. It is a nexus of sailing routes that were previously drawn straight across the map. Of particular significance here is the 1375 Catalan Atlas, on which we find both Tallinn and Vuarlant < Virumaa: each of the 24 loxodromes drawn onto the Baltic Sea meet at the island's south, pointing to Saaremaa's central role in the development and shaping of shipping routes. The latter hypothesis fits neatly with a medieval record that calls Saaremaa as a whole a 'great harbour'. That being said, it should be stressed that a map and seafaring may not have a bearing upon each other. Maps are born of measuring tools that allow us to measure the convexity of the earth and convey it on a flat surface. When such tools were lacking, regular maps were replaced by coordinateless pictorial maps, which will be the focal point of this chapter. Used alongside pictorial maps were sailing instructions – initially oral, later written. The oldest recorded sailing instructions in northern Europe date to the year 880 and describe routes around Nordkapp into the White Sea, through Norway into the Danish waters of Schleswig–Hedeby, and from Hedeby to the mouth of the Vistula River via Bornholm. A few centuries later, they were complemented with sailing instructions from western Norway to the western tip of Greenland, with eyes on Shetland, Faroe, and Iceland. One 13th-century Baltic sailing instruction outshines its earlier brethren in accuracy, describing a sailing route through the Kalmar Strait in south Sweden, winding between the skerries near Stockholm, to the southern shore of Finland, and from there on to Tallinn. These sailing instructions also convey, for the very first time, distances in units native to Baltoscandic seafaring: a *Weeke sees* measured at close to 8.3 kilometres (Lang 1968).

Pictorial maps, which I will simply call maps from this point forward, came into being before writing. Navigating in time and space is a fundamental human necessity. One did not need to describe an experience with phonetic markings in order to pass it down. Essentially, every map is a pictorial map; an ideogram with its own reading rules. For instance, maps of the Southern Sea are tied together with palm stalks. It is a device of mnemonic technology, but so are our books, highway signs, hard drives, and microprocessors. The further you delve into the past, the more the border between map and folklore blurs before disappearing entirely. As space is inseparable from time (and the measurement of it),

174

THE CIRCLE IS COMPLETE

we must judge the age of the first sketched maps as equal to the first attempts at developing a calendar, which is to say 35,000 years. The oldest surviving map is nearly 6,000 years old and depicts the Euphrates river valley with mountains ranges along its banks. It is recorded on a clay tablet but could easily have been a rock painting, in which case it would have been handled by an art historian instead, no doubt with different results. We are only barely capable of recognising humans' most ancient methods for navigating time and space. I'll give an example that, I must admit, has always moved me. It is the active state of mind that in Finnish is called *huomata*, in German *achtgeben*, in English *to pay attention to*, in Russian *zametchay*, and in Estonian *tähele panna*: an everyday grouping of words that has been worn dull and contains no question. Yet in literal translation, it reveals a distant horizon: *panna tähdelle*, *auf den Stern legen*, *to put down on the star*, *polozhit na avezdu*. Stars have guided our ships, a vessel's course was set by a star, the helmsman pointed to a particular celestial body, and reaching that star meant arriving home. Rarely, if ever, do we consider such simple things.

Just as rarely have we considered whether and how Pytheas's great circle and the path of our Silverwhite are reflected on old maps. They cannot be too ancient, and are basically new when compared with the Baltic Finns' seafaring tradition. The thing is that such old maps simply do not exist. There are a few clay tablets, too old; or a map painted on an Egyptian sarcophagus 4,000 years ago, which was meant to guide the departed to the 'Kingdom of the Dead' but which actually directs them quite accurately to the upper Nile. But from that point forward, a gap of a few millennia yawns. Many books have published the maps of Tacitus's *Germania* or Adam of Bremen's Baltoscandia or Ptolemy's Baltic Sea, whose strange contours and almost shakily written descriptions create the illusion of crumbling parchment. Nevertheless, the first was reconstructed by the geographer Konrad Miller in the year 1895, the second by the Dane Axel Anthon Bjørnbo in 1910, and only the third was copied in the 13th century: a good thousand years after Ptolemy's day and still merely a copy. In addition to Ptolemy, the last also inevitably reflects the 13th century or at least 13th-century source critique. The sole surviving map of Roman highways was copied in about the same period.

Luckily, we still have the geographical prose of several authors. Maps copied from various sources complement one another and enable us to draw retrospective conclusions that stretch much deeper into time than the era in which the copies were produced. We know from Roman

175

SILVERWHITE

literature that Augustus Caesar's advisor Marcus Vipsanius Agrippa commissioned a large world map that was largely based upon Ptolemy's *Geography*. The goal of this, and the majority of Early Middle Ages maps, was more ideological and symbolic than practical. Following Agrippa's death, the emperor ordered the work to be finished and a special arch to be built, from which the map was hung for all to admire. A table of more significant distances was added. 'All roads lead to Rome' could have been, and perhaps was, the work's motto. It was a powerful illustration of the global empire's might.

In about the year 776, a Spanish Benedictine monk named Beatus drew his famous T-O map, at the centre of which he placed Rome instead of Jerusalem and at the top, i.e. east, Eden. Although the map's ideological purpose was to exalt Christianity's triumph in both word and image, it is exceptionally free of mythological distortion: 60 per cent of its toponyms match the sole preserved copy of a Roman-era highway map with which Beatus could not have been familiar (Tooley and Bricker 1976: 22). The Beatus Map is the godfather of all European maps made in the Early Middle Ages, albeit primarily in terms of its ideology, not content. Its most famed heirs are a 3.5-metre altar map in Ebstorf Abbey, the original of which was destroyed in the Second World War, and a 1.5-metre circular map from 1280 in Hereford Cathedral on western England's border with Wales. The latter actually depicts the Roman Empire in the time of Diocletian (late 3rd century), but its text references the *mappa mundi* of 'King' Agrippa. One lesser-known descendant of the Beatus Map is the Sawley Map in Cambridge University, drawn around the year 1110.

Now we are coming to an important question. The Beatus Map shows a 'Fenno-Caspian passage' running horizontally, i.e. a waterway coming from the northern part of the world sea and flowing into the Mediterranean (including into the Caspian and Black seas), continuing down into the heart of Africa as the Nile. It is the roof of the T in a T-O map. In our area of interest there is also a waterway that traverses northwest Europe and flows into the 'Fenno-Caspian passage' near Constantinople. Is this an echo of Pytheas's route home? And, most importantly, was the waterway across northwest Europe ever depicted on the ancient Greek maps that were used to create Roman maps? Was our Silverwhite recorded on Agrippa's Roman map, which was a source for the maps of Beatus and Hereford? It appears we may now answer with a cautious yes. Lying at the mouth of the waterway crossing northwest Europe is, on the Beatus Map, the island of Thule (Tooley and Bricker 1976: 23). On the Hereford

THE CIRCLE IS COMPLETE

Map, the border of northwest Europe breaks apart into three bays north of the Vistula river: we can recognise them as the Curonian Lagoon, the Gulf of Riga, and Pärnu Bay. Mainland Europe ends at the coast of north Estonia, extending towards the 'Fenno-Caspian' passage, and to the east of Ultima Thule stands a Sámi on skis, very much in agreement with Tacitus's understanding (Spekke 1957: table III).

As the Hereford Map refers to the map of 'King' Agrippa, we may conclude that the classical geographical tradition had placed Ultima Thule in the area of the Baltic Sea and was not bothered by Tacitus, who made it so that his father-in-law could spot Thule in the waters of Britain. Thule is situated even with Estonia's northern shore, the latter simultaneously constituting the northernmost part of continental Europe and in the vicinity of the 'Fenno-Caspian' passage. Here, it becomes clear why Eratosthenes (3rd century BCE) drew the world's (and cultural history's) original meridian from the city of Alexandria, across Rhodes, and far to the north until it touched Ultima Thule, and why Ptolemy (2nd century BCE) drew the settled world's furthest line of latitude through Ultima Thule as well: the classical world's gaze reached no further, though it did touch Saaremaa. Once again, we are at a crossroads. Estonia's island became a symbol in literature and an object of debate in science. In cartography, however, it has been anchored in place for many centuries: here in the Baltic Sea, the mouth of Silverwhite.

IRISH EYES ON THE BALTIC SEA

If geography is prose, then world maps are iconography. Unlike geographical texts, cartography managed to maintain some degree of continuity despite the storms sweeping across the ruins of Rome and its libraries reduced to ash. New information was built upon classical tradition, perhaps because of the fact that new horizons were revealed in the gales of migration.

The Cambridge (1110) and Ebstorf (1284) *mappae mundi* added data and toponyms of northern Europe that were taken from so-called Aethicus Ister's *Cosmography* (Wuttke 1853; Hillkowitz 1934, 1973). The work, which has been preserved in twenty-two copies, was published around the mid-19th century with highly naive commentary. The wave of excitement was quickly followed by deep disappointment, which Konrad Miller, an authoritative historian of cartography, labelled a damning judgement. The work is fantastical, sloppily contrived, and written in

SILVERWHITE

such poor Latin that it is impossible to read. But even then, Aethicus's toponym 'Rifarrica' was equated to the ancient Estonian county of Revala and the name of Tallinn in the Early Middle Ages. Newer studies (1963) have determined that no geographer named Aethicus Ister ever existed. Behind the pseudonym was an Irish clergyman named Feirgil, later the bishop of Salzburg and commonly known as Virgil of Salzburg, who died in the year 784 (*Tusculum-Lexicon* 1963: 9). Irish culture is an unusual phenomenon in Europe and, according to some sources, Irish missionaries sailed to Iceland even before the Vikings. An unusual amount of space in the *Cosmography* is indeed dedicated to descriptions of seafaring vessels and the nature of sailing. Nevertheless, it is truly a hard nut to crack as a source. Feirgil disputes traditional German geography, all without holding back on tools and colours, mixing fact and fantasy. Without knowing the motives behind the polemics or Feirgil himself as an outsider, it would be both careless and fruitless to try to perform a more serious analysis and separate truth from poetry. The author shares classical geography's notion of a great waterway connecting the Baltic and the Caspian Seas. Here, we should interpret the Baltic quite conditionally as the part of the world ocean that rakes the shores of northern Europe. In Feirgil's words: 'Still, in the most distant North there exists an immense arcing route from the Caspian Sea to the gateway of the northern winds ...' It is, as we can calculate from the next part of the difficult-to-translate sentence, nearly 2,220 kilometres long: 'its length from the northernmost part of the world ocean in the west flowing and winding south to the Caspian Sea is 12,000 stadia, surrounded by mountains [inland?]' (40: 31–5).

Feirgil claimed that north of Rifarrica/Revala was an island named Munitia: settled by dog-headed people who used poles to make conical lodges, the land was infertile but supported plentiful livestock. Through the imaginative descriptions, we can discern realistic details that allow us to see his Munitia as Sámi-settled Finland. Apparently, the author relied on works that have since been lost.

Known to Feirgil was the ancient salt-trade route that crossed Jutland and ran along the Swedish coast. Rifarrica was described as a marketplace where weapons were also traded. Other forms of the so-called island's moniker are Rifargica, Rifartica, and, on the Sawley *mappa mundi*, Rapharrica. According to the map, the island occurs in a place we could simultaneously identify as the Gulf of Finland and the mouth of the 'Caspian channel'. In Feirgil's day, the ancient county of Revala and the county's harbour were pronounced 'Rebala'. Germanic languages

THE CIRCLE IS COMPLETE

lacked the Baltic-Finnic *b* phoneme, as a result of which it was usually replaced with *f*. Rifarrica was equated to Revala early in the 20th century for primarily linguistic-historical reasons (Busch 1921). Utilising maps created in the Early Middle Ages, we can now somewhat broaden the circle of conclusions.

Feirgil mentions the toponym *Olchis* as lying to the east of Rifarrica/Revala and the island of *Taraconta* to the west. The former is expressed as *urbs Choolisma in Olchis jugis*, the city of Choolisma in the Olchis mountain range (on a spring), while the Ebstorf Abbey map reads *Olchis fl. qui et Wolkans* – on the Olchis River, meaning the Wolkans. The Baltic-Finnic *Olhava*, or *Volhovo* in the *Chronicle of Novgorod*, is now known as the Volkhov: a turbulent, difficult, and important link connecting Lake Ladoga to Lake Ilmen. Linguistically interesting here is *Choolisma*, whose unusually long front vowel and *-ma* ending likewise point to Baltic-Finnic origin (Paul Ariste and Paul Alvre to the author, 18.11.1982). In the 19th century, the Russian researcher Vasiliy Peredolsky discovered the ancient Kolomtsy burial site on the shore of Lake Ilmen, along with an immense number of skulls (Popov 1981: 67). Finnic Votian burial sites are called *kolomishche* in a 1534 Russian chronicle. The suffix emphasises size and extent, but the name itself, just like the toponym Kolomtsy, derives from the Baltic-Finnic word *koolma*, which itself reaches back to Proto-Uralic. For some reason, places bearing that name are abundant along the upper Volkhov. The Olchis and Choolisma, with their unusually transparent origins, allow western-lying Rifarrica to be identified with Revala to a high degree of certainty, and also let us recognise a third toponym within the framework of the common waterway.

West of Rifarrica/Revala is the island of Taraconta, also written as *Tharaconta*, *Tareconta*, and, on the Sawley map, *Tarracanta*. The westernmost of Saaremaa's five ancient parishes (*kihelkond*) was and is still to this day called simply Kihelkonna, recorded in 1228 as *Kiligunde*. Kihelkonna's harbour was also favourably situated. The first part of the word, *kihel*, is borrowed from Germanic languages and originally meant 'deposit, joint guarantee, and joint obligations' – also, 'defence'. The second part of the compound word, reaching back into Proto-Finno-Ugric, once had the meaning of 'friend, connection, clan, army'. The ancient Baltic-Finnic languages lent *kund-*, *kond-*, and *kand-* into Old Norse. As we may remember, Pliny the Elder came into contact with the word. The name 'Kihelkonna' is unusual on Saaremaa and in Estonia as a whole, as an ancient unit of administrative governance turning into a toponym. Such a development can only seem

SILVERWHITE

possible if Saaremaa had no parishes, *kihelkonnad*, apart from Kihelkonna. The word must have developed before the formation of other parishes for it to take shape as a toponym, which means it is Saaremaa's oldest and, for a time, sole parish. Its prominent location on the island's west coast, directly in the trajectory of lucrative and also dangerous shipping routes, speaks to such a formation.

Colloquial language often shapes the toponym of the most dominant population centre in an area. Among southwest Estonian peasants, the word 'city' was primarily used in reference to Tallinn. Among the people of Saaremaa, it was Kuressaare, as there are no other cities on the island. Only the north Estonian shore possesses any analogy: Toolse, which originally meant 'throne, administrative centre, place of governance'. It was borrowed from the Gothic language and is one of Estonians' few memorials to ancient Gothic migration, meaning it can also be dated by the event. A few kilometres to the east of Toolse is the town of Kunda, which in 1241 was recorded as *Gundas*. The toponyms' reliance upon each other and Kunda's relationship to the word *kund-*, *kond-*, *kand-* are unclear, but they do confirm the age of such toponyms. Linguistically, it is therefore possible to tie Feirgil's island to Saaremaa. This also corresponds nicely to the Sawley map and makes Taraconta the westernmost link in the chain, Rifarrica/Revala the centre, and Olchis/Volkhov the easternmost.

With an author like Feirgil, is tempting but risky to move from overarching generalisations to specific details. Like Tacitus, he is ambiguous when describing the borders of Germania, though this does not necessarily mean he was familiar with the former. He moves from south to north, naming the Frisians, Danes, and clearly the Sámi (*vinnosii*). He mentions 'brilliant, very beautiful amber that hardens like stone' and then begins describing the neighbouring tribes, who, 'bounded by the sources of the northern wind', also occupy the island of Taraconta. The island's inhabitants use tar, skins, and leather straps to build their vessels, and also weave shields from the straps – they dry raw skin in the sun and on hot coals until hardened. The inhabitants of Rifarrica/Revala are said to be

> a keen and clever people who are skilled in destroying fortresses and fortified settlements, and are mentally prepared for skilled blacksmithing [...]. They have no king, but only leaders. The clans seal peace treaties amongst themselves and do not turn on one another. Even so, they honour a god and make sacrifices of promise to

THE CIRCLE IS COMPLETE

it, thereby demonstrating the enormity of their false ways. (Feirgil, 37–40)

Ships that carried salt? It sounds simple but it was difficult to execute. Salt was traded between tribes and required the observance of shore rights, shipping routes, and the legal definition of safe harbour. The oldest appearance of 'shore rights' (Estonian *rannaõigus*, constituting the right of coastal peoples to take possession of flotsam and jettisoned items) was made in an Anglo-Saxion document written during that same period, c. 666. Legal standards do not precede events, but come afterwards.

Lake Mälaren's unusual island of Helgö could look back upon five unbroken centuries and await just as many ahead. Tempests swept away Rome's five-storey barracks, aqueducts, sewers, and military roads; they threw tribes into such disarray that neighbour did not know neighbour; and, yet, Helgö thrived. 'One is almost tempted to call it an industrial centre,' admits Swedish historian Wilhelm Holmqvist (1970: 52). The painting doesn't fit the frame, but the frame itself is worth something, too! Seven thousand crucibles; the head of an 8th-century bishop's staff (Feirgil's time); a 6th-century Buddha statue (in the framework of cultural contacts) from Asia.

Now, Helgö acquires a sister. Along Pytheas's Bronze Age waterway, on the coast of Jutland near the mouth of the Schlei, the Baltic Sea's first fortress town takes form: *Hithabu, Haithabu, Haethum* to the Scandinavians and Anglo-Saxons; *Sliesthorp* to the Germans; *Sliaswich* to the Western Slavs; Schleswig, open to many languages and vessels, with roots extending back to the 7th century.

How far back do the roots of the Rifarrica/Revala fortress town reach?

Taraconta's inhabitants celebrate the summer and winter solstices. Here, I'll allow the reader to enjoy, and the researcher to evaluate, the incredible blending of spirited poetry and foggy truth that is so characteristic of Celtic folklore, and especially of Feirgil's prose:

They worship Saturn because in the days of Octavian Augustus [27 BCE–14 CE] they paid tax in the form of gold found on their shores, and then of their own accord, [but they did so] in the times of no previous or subsequent kings or emperors. Seeing that neighbouring regions also paid tax, they thought that a new god of days [*deus dierum*] had been born, and in that month of August they gathered the whole body of their people together, on Taraconta, quite a large island in the sea of Oceanus, and they made a great heap held together with stones

181

SILVERWHITE

and bitumen, setting up enormous structures of remarkable size, with sewers underneath built of marble [...], and they called it Morcholom [Marcholom, Marcholon, Marcholum] in their language, that is the star of the gods, which they call Saturn through [etymological] derivation of the name. And they built a huge and well-defended city there called Taraconta. This people will work much destruction in the days of Antichrist and they will call him the god of days. (Feirgil, 32)

Feirgil was one of Adam of Bremen's main sources when describing the Baltic Sea.

FATEFUL HARMONY

Earlier, we showed that Saaremaa was the first Estonian toponym to appear on the world map. Now, we should correct this with Arab geography. In January 1154, Muhammad al-Idrisi completed his *Opus Geographicum* on the island of Sicily. In it, he mentions Tallinn (Kaleweny), Turku, Pärnu, and several other sites in Estonia and Finland. The toponyms travelled from our land to al-Idrisi's work in Sicily in their Baltic-Finnic form. This is extraordinary. We were 'barbarians' then and are still nameless and faceless in the eyes of western Europe, aren't we?

The information could have found its way to the Mediterranean island in two ways: across the Atlantic Ocean from the west or along the Silverwhite waterway from the east. I believe the latter is more likely, despite the fact that al-Idrisi's knowledge of Russian geography was very much lacking.

While being so consequential to us, the first-ever mentions of the names Tallinn, Turku, and Pärnu are barely a speck of dust from the standpoint of European cultural history. Al-Idrisi's worth does not lie in the sum of the facts, however he may have found them. Let us see him as a synthesiser; as the uniter of two extremely dissimilar cultures. On the one hand, he proceeded from early medieval barbarian European geography, from the rudimentary T-O arcs that retained the distant echo of Roman geographical tradition. However, al-Idrisi divided the world into parallel climactic zones, and those in turn into vertical sections. This reveals the influence of Greek geography, and primarily Ptolemy's model of latitudes and longitudes. The fate of Greek cultural heritage was in a somewhat happier state than that of Rome, as it found temporary shelter in Byzantium, where Greek manuscripts were seamlessly passed along to Arab scholars. Ptolemy's geography found its way into western

182

THE CIRCLE IS COMPLETE

Europe only three centuries after al-Idrisi's death. There is no such gap in Arab culture. Consequently, we should ask ourselves whether the Arabs may have conveyed other Greek information about the Baltic Sea that we haven't yet recognised.

Our touchstone is, of course, Ultima Thule – the Arabs' Sūlī, Tulia, Tule.

Let us begin with the al-Khwarizmi school, which adopted Ptolemy's tradition and composed its own geographical tables. Ibn Khordadbeh (820–912), one of Ptolemy's translators and annotators, wrote (Nansen 1911: 2, 144): 'As for the sea that lies beyond the Slavs and on the shore of which the city of Thule stands ...' It is impossible to imagine Ibn Khordadbeh's Thule on the Norwegian coastline, Iceland, Greenland, or, as Tacitus and Winston Churchill would have it, Orkney. The statement is unambiguous. Al-Battani (852–922), a student of the same school, wrote:

> Next, the south–north direction was addressed and it was found that the habitable portion stretches from the equator up north to the island of Thule [...] where the length of the longest day can reach 20 hours [...]. The British islands lie in the same ocean waters in the northern direction, and they are twelve.

These are the length of days given to us by Geminus and Pytheas's account of the 'British triangle'. Ibn al-Faquih continues in about the year 900 (Nansen 1911: 2, 145): 'Thule is an island in the sixth sea between Rome and Khwarazm.'

I could never have written such a fine sentence myself!

Our next author inhaled the salty air of the Atlantic, Indian, and Caspian seas: al-Mas'udi (900–56). Having claimed that the Baltic Sea was connected to the Sea of Azov, he adds (Nansen 1911: 2, 146): 'On the northern edge of the settled world is a great sea [...] and close to it is a land called Thule, beyond which there are no other settled places.'

Famed Khwarazmian scholar al-Biruni (973–1048) named an ocean stretching from Gibraltar to the settlements of the northern peoples. From there, as he writes (Krachkovsky 1957: 157, 248, 329–42): 'a great channel or bay splits from the sea and extends quite near to the Islamic Bulgar state [= the Volga]. This [channel, bay] is known as the Varangian Sea. The latter is in turn a people who live upon that shore. Then beyond that, the ocean turns to the east ...' Thule is missing, but there is a notion of a waterway that stretches from the Baltic Sea to the Bulgars on the Volga River. A channel? A bay? Al-Biruni's hesitant choice of words, calling the

SILVERWHITE

Baltic Sea and the Gulf of Finland a channel, points to its intensive use as a waterway but not to insufficient knowledge of geography. The scholar was familiar with northern Europe. He was the first to mention the coastal people of the Baltic Sea, the Varangians, as well as their closest neighbours, the Vepsians. Al-Idrisi completed his work a mere century after al-Biruni. The dissemination of information from the Baltic Sea, across the Volga Bulgarian state, and into the Arab (and Byzantine) cultural hubs is proved through al-Biruni and may be provable through al-Idrisi as well.

Viewed from the distant Arab world, the northern limits of Volga Bulgaria could easily blend with coastal lands on the Baltic Sea. This is demonstrated in turn by the well-travelled polymath al-Ghazali's (1080–1169) writing (Mongait 1959: 169–81): 'The limits of Bulgar power extend to the areas that pay it tribute. They lie one month's distance from Bulgar. One is called Vepsia [Vepsians, Visu]. There, the people trap beaver, stoat, and squirrel ...' He was complemented by al-Quazwini prior to the year 1283 (Nansen 1911: 2, 145):

> Burgan [Bulgar?] is a land far in the north. The day there shortens to up to four hours and the night lengthens to twenty, the opposite in summer. It is populated by pagans and idol-worshipers. They war with the Saqaliba [Slavs]. In most respects, they are similar to the Franks [Central Europeans]. They are skilled at many types of crafts and shipbuilding.

The word 'Thule' is missing from this passage, but one feature of Thule is present: a summer day extending up to twenty hours. This leads us only to the latitude, but the longitude we seek appears to lie in al-Quazwini's comment about 'pagans and idol-worshippers'. Arabs were remarkably informed in terms of people's faiths; and out of all Europe's coastal lands, Estonia was the last to be subjected to the Roman Church by fire and sword.

Our last Arab text here is a passage from a work written by Saadi Shirzai in about the year 1300 (Nansen 1911: 2, 160): 'The longest day lasts up to twenty hours [...]. There is an island that is named Thule. It is said that the islanders live in heated saunas because the climate is so cold. It is generally regarded as the northern end of the settled world.'

And, as a final chord that harmonises so beautifully with our folkloric notions of Sampo and the spinning heavens, al-Dimashqi's (1256–1327) concise sketch of the lands beyond the Arctic Circle (Nansen 1911: 2, 161): 'There, the dome of the sky spins around like a mill.'

THE CIRCLE IS COMPLETE

Thule, our Saaremaa, was a touchstone.

Important common themes emerge from the excerpts presented here: the very earliest authors unanimously place Thule in the Baltic Sea, 'on the other side' of the Slavs, and halfway from Rome to eastern Khwarazm. It is connected to the Volga waterway and is, from the Arab perspective, a crucial link between the east and the west. Yet the closer we get to the present, the blurrier Thule's landscape becomes. In the end, only the island's astronomical properties remain.

The Arabs' orientation agrees with that of western Europe but is more specific in its details and temporally much closer to our day. Based on this, we can conclude that Saaremaa/Thule was an active link on the Silverwhite waterway during the 1st century of the Common Era and that Pytheas's geography persisted, allowing Arabs to place Thule in the Baltic Sea. It simultaneously means that Arab authors relied upon works that have not survived to our day, but that tied, with sufficient clarity, the tradition of Thule's moniker with the Baltic Sea, Saaremaa, threshing houses, blacksmithing, shipbuilding, and ancient religion. They all call the Thule islanders fire-worshippers, a term al-Idrisi applied to the inhabitants of an island in Estonia's coastal waters as well.

The classical world's literary heritage lived on in the careful hands of Byzantium, and especially there. Anna Comnena, an Anatolian princess with an insatiable thirst for knowledge, described in about the year 1100 Constantinople's imperial guard: barbarians, 'Varangians from Thule', armed with battleaxes. Seventy years earlier, al-Biruni called the Baltic Sea the Varangian Sea. The origins of the word are uncertain, but a Baltic-Finnic root is probable: the Votian *varo* and Estonian *varu* originally meant 'vow' (SKES 5, 1652), a connection that bound the crew of a vessel or a military unit with an oath of 'spirit and blood', which on the Baltic was called *vaering*. However, this is of secondary importance from the point of the question that interests us.

The problem itself is important, but it flees from us like a kingfisher. So, let us grasp it in our palms as gently as its wings will allow: even as late as the 14th century, Arab literature *still* places Thule in the Baltic Sea; and European maps *already* place Saaremaa in the Baltic Sea during the 14th century. Saaremaa was simultaneously Thule to the Arabs and Osilia to European cartographers. Al-Idrisi mastered both cultures and was unlikely the only one of his kind. Consequently, our island could have split into two and appeared on maps either as Thule or Osilia. Or else yet another, even more ancient word pair. On the Hereford *mappa mundi* mentioned earlier, Ultima Thule is drawn near a skiing Sámi figure and next to the island lies another named Ysland.

Can we find a standard that might allow us to read old maps more accurately?

Ysland is likely Iceland, though perhaps not always. The Old Norse *Eysysla*, Saaremaa, is a compound word: *ey* ('island') and *sysla* ('county'). It was used in the Icelandic sagas and was later Latinised as *Osilia*. However, translated names are very common among toponyms. In this case, Saaremaa would be Eyland, Yland, and, in certain folk-etymological cases, also Ysland. This digression has no argumentative value. Let us turn our attention to words only under certain circumstances. And these are simple: if Vinland, a land discovered by the Viking Leif Erikson on North America's northeast shore, is located near the island, then it is no doubt Iceland. And if the name is indeed conveyed differently, then let it be corrected to Iceland on the map.

Yet, as we may recall, Henry of Livonia mentions a Finnish missionary named Peter de *Vinlandia* working in Estonia in the years 1215–20. Parisian chronicler Bartholomaeus Anglicus de Glanville calls Finland *Winlandia* in the year 1230 (Juva 1968: 32). It isn't until 1532 that 'Finlandia' first appears on the world map (Gordin 1967: fig. 2). The 'Vinland' form of Finland's dual name must have been quite a headache for cartographers starting from the year 1000. Vinland is located in the western Atlantic Ocean and Ultima Thule lies near Vinland? That means Ultima Thule must also be in the western Atlantic Ocean! And another island, perhaps Osilia or Yland, or one yet entirely nameless? Was that corrected to Ysland?

The criterion by which we should review old maps with a new perspective is as follows. If Ultima Thule, Iceland, and Vinland lie east of the Scandinavian 'archipelago', then they should be interpreted as Saaremaa and Finland. Scandinavia, as unbelievable as it may seem, might also be missing entirely. In this case, Ultima Thule's position in relation to the Silverwhite waterway would be crucial. If it is located at the mouth of the bay, strait, or channel that extends to the Caspian Sea, then it is Saaremaa and is simultaneously used to identify the neighbouring areas, no matter what names they may bear. But if Ultima Thule, Vinland, and Iceland lie in a clump west of the Scandinavian 'archipelago', then the Vikings' Vinland has already left its mark on the map. The cartographer placed Ultima Thule in the Atlantic Ocean because he erroneously applied the entirely correct understanding that Ultima Thule and Vinland form an indivisible tandem.

THE CIRCLE IS COMPLETE

Thus, harmony may prove to be decisive. The foggy horizon swallows the historical island along with the memory of memories. 'Ethnological misrepresentations, once they are born, never die,' wrote the University College Dublin professor Proinsias MacCana in defeat (1973: 11). Indeed: who still holds any interest in Ultima Thule today? Apart from Thule itself?

REBU GIVES ME GOLD

A celestial body emerged from the cosmos and pressed onward – cold, dark, and lifeless. The blind laws of nature directed it and kept it intact. Then its straight path crossed a certain solar system. The third planet from the sun caused it to list ever so slightly.

Is this the point where astronomy becomes history?

Not much was needed. Contact was made. Lifeless material transformed into energy, light, fire, thunder. What remained gradually cooled and turned into axes and spear tips, words and poems, worldviews and world maps. History, in short.

Those lines could have been more precise. The Earth drew it towards itself. But it also drew the Earth closer: impacts are always mutual. No one can interfere in another life without changing one's own.

We happened to witness the event, and that's history.

A limestone outcrop had risen from the Baltic Sea – a handsome green island, slightly smaller than present-day Saaremaa. That is where it fell, boring a deep hole into the bedrock. That hole filled with water. We named it Kali. Now, we write Kaali. Why Kali?

There are phenomena that raise no questions, and others that oblige us to ask. It would be pointless to ask: why Earth? However, it is quite sensible to ask: why Kali? Whether we are able to, or wish to, answer the question is another matter. Back when we witnessed the event, we were asked the same question by Nature, which delights in rolling questions onto our path every day. That, if you may, is a game that we have played with Nature since childhood, always ending in a draw. We may not beat it, just as it may not beat us.

At the time, we gave Nature a plethora of entirely satisfactory answers. We said: that is fire. So: Kali. And many more things that we may perhaps even remember, but are unable to tie to the event. Like children who remember more than we do and know less. Our own Estonian children still sing that long-forgotten song:

SILVERWHITE

Mina kakku Kalevile	I give bread to Kalev
Kalev mulle rauda	Kalev gives me iron

Why Kalev? Why an exchange of bread for iron? What iron? Is Kalev *kalev* ('a giant') aka *kali* ('fire') aka *Kali*? And why does the song continue in this way:

Mina rauda Reole,	I give iron to Rebu
Rebu mulle kulda	Rebu gives me gold

The fallen sun became Tuli ('fire'), Tuli became Ultima Thule, and that in turn became *aestiorum gentes*, the people of Fire Land or what have you. Rebu transformed into Rebala, the Irishman's Rifarrica, Revala, a harbour on a great waterway, and the Tallinn fortress town that served as a nexus for shipping routes both near and far. The sea was our life's axis. Does a simple children's song remember the days that a pietist legend of dark slavery has so doggedly attempted to wipe from our memory? Perhaps we remember more than we know.

I would like to circle back to the starting point, which will, of course, be the end point of this section. A ring around home, a home circle, our own land and sky (*maa* and *ilm*), the world (*maailm*). Poetry, like technology, knows no bounds, though there are certain laws to it and I do not believe this statement carries any contradiction. Poetry offers iron to Rebu? 'Rebu' is also the Estonian word for 'yolk'. So, iron to yolk? Yes and no. In Finnic folklore, the sun was born from a yolk:

mi munass on ruskiaista	My egg has a yolk
ne on päiväks taivoselle,	it is the sun in the sky,

and Tallinn's oldest name emerged from the same word. There are no catastrophes in nature – nature knows only nature. 'Catastrophe' is a human judgement, the anguish of the defeated. In scientific literature, and also at points in this text, the inaccurate word pair 'Kaali catastrophe' is used. The Estonian language and folklore have retained a distant event, a question, and also the answer to an extraordinary event. It is a calm answer, and here it is.

* * *

Vessels still clung to the shoreline. Not too close, as the shores were treacherous, but also not pulling too far away from the limits of visibility,

THE CIRCLE IS COMPLETE

because the open sea posed an even greater danger. Nights were spent on barren islands or bare headlands, rarely in the ship. The Åland Islands were relied upon to cross the Gulf of Bothnia. Occasionally, the seafarers would leave something behind. So it was on this day as well. Was it a storm? An attack? A senseless fight around a night campfire? The 25 coins left in the Åland soil were stamped sometime between 591 and 628 in the farthest reaches of the Baltic–Volga waterway: Sasanian Persia, a mere dozen or so years before the Arab invasion (Lewis 1958: 145).

At that same time, the ancient Baltic ship acquired sails.

V

RED SAILS

WHERE CAN I FIND A SHIPSMITH?

Baltic vessels acquired sails between the years 600 and 750 and thus became stable, obedient vehicles for seafarers, suitable for both long trade voyages and rapid attacks.

Ninda panin pääle mastisida	So I raised the masts
kui neid pilliroogusida,	like those reeds,
ninda panin pääle plokkisida	so I raised the blocks
kui neid pideli pulkesida,	like those lace bobbins,
ninda panin pääle köüesida	so I raised the lines
kui neid kannel-keelesida,	like those zither strings,
ninda panin pääle purjesida	so I raised the sails
kui neid neiu põllesida,	like those maiden's aprons,
ninda panin pääle lippusida	so I raised the flags
kui neid neiu lintisida.	like those maiden's ribbons.

(Kolga coast)

Until 1962, no one had seen an ancient Baltic ship. True, two 9th-century vessels were discovered in the Oslo fjord in the late 19th century. They were called the Oseberg and Gokstad ships, after their respective discovery sites. Researchers presumed that ancient Baltic ships must have been similar in construction: at least 23 metres with a tall topside, a deep keel, and multiple rowlocks.

It was a naive presumption and caused a great deal of confusion. A ship's construction is determined, to the utmost degree, by the nature of its sea. And when Danish archaeologists discovered an entire fleet in Denmark's Roskilde Bay in 1962, enclosed it with a watertight dike, pumped the

SILVERWHITE

sea floor dry, carried every last fragment to shore, and reconstructed the ships in a village museum, then every previous conviction was turned upside down.

The keel was primarily responsible. Suddenly, it turned out that former conclusions regarding the ability to navigate Baltic harbour sites, trading posts, and rivers were no longer valid. Ancient Baltic vessels lacked the Gokstad ship's large, heavy keel with a T-shaped crosscut – that is their feature. The chain of conclusions stretches far. A vessel was able to land directly on the shore almost anywhere. It needed no deep harbours but did require stops sheltered from the winds. Calculating for Baltic waves, its optimal length would have been 15–25 metres. A slim, flexible ship with a low keel and shallow draught would be capable of 'riding' the waves. This is not a poetic metaphor but a technical concept. Reconstructions have proved the ancient vessel's excellent seafaring abilities. The small number of rowlocks on the Roskilde ships confirm that a mere 'handful of men' could power an ancient vessel.

Attached to the keel was the mast step, which could have been of spectacular size and up to 10 metres in length, distributing the force of the sail evenly across the entire hull. In older ships, stays were drawn from the mast's tip to the bow and stern; not to the topsides, as we are accustomed to seeing today. It was easy to lower such a mast if the need arose. However, the positioning of the stays also inherently limited the square rig's movement, thus reducing the manoeuvrability of the vessel. A new manner of rigging became prevalent around the year 790, as a result of which the ancient vessel also learned to sail into the wind.

In the old rigging, the ship's yard must have been attached at a 45-degree angle to the hull. The lower corner of the sail, attached to the sheet, was bunched up like a bag, giving the sail its characteristic ancient triangular shape. The yard came near to the ship's x-axis and was apparently drawn parallel to the mast when docked or anchored. So long as they were not sewn from sealskins, the sails were painted red.

Ma mies merime poiga,	I'm a man, a seaman's son,
purjetan punasta merda,	I sail the red sea,
oma merda oigejada,	our own sea the right sea,
Viru merda virgilista,	the Viru sea the furrowed one,
Rootsi merda ruugejada	the Swedish sea the rouge one,
sinisilla siilavalla,	beneath the blue jibs,

RED SAILS

punasilla purjeella –	beneath the red sails –
sinised on servarihmad,	bluc are the straps,
punased on purjeained.	red are the sails.
	(Kolga coast)

Estonian folk songs, not to mention colloquial language, have preserved the expression *laevasepp* – 'shipsmith'. It is anything but poetic metaphor. Shipbuilders were called smiths all across the Baltic Sea. Among other sources, this is confirmed by the song of Olaf Tryggvason and, even more vividly, a shipbuilding song from the Tõstamaa region:

Kust ma leian laevaseppa,	Where can I find a shipsmith,
laevaseppa, raudaseppa,	a shipsmith, an ironsmith,
targa tööde toimendajat,	a smartly skilled craftsman,
virgu tööde valitsejat,	a deft master of the crafts,
kes taub mulle rauast laeva,	who'll hammer me a ship of iron,
tammest laeva, puudest purju,	a ship of oak, sails of trees,
madarasta laevamasti?	a mast of bedstraw?

IS THE VIKING AGE IMAGINED?

As is always the case in history, a technological innovation that in this travelogue we may refer to as the shipbuilding revolution served as a summary of economic and social development up to that point. On the other hand, shipbuilding technology had an accelerating and intensifying effect on social and economic development in turn, shaping a unique maritime culture, tailoring a legal system to the coastal environment, and creating a cycle of folklore and beliefs, whose bold contours we can now follow in somewhat greater detail than in earlier eras.

Snorri Sturluson (1179–1241), a lawspeaker and poet in Iceland and Norway, conveyed to us a fascinating description of a long-ago event. A Norwegian named Harek wished to take his ship from the Baltic Sea to his home harbour on the North Sea. He sailed through the narrow Øresund/Öresund strait and feared a Danish attack. So what did he do?

First, he lowered his mast. Then, he spread a grey sail over the deck. Why? He needed to hide the numerous rowlocks, as his was a well-crewed warship. Then, he ordered the men to lie flat against the hull, leaving only a handful of rowers at the bow and stern. As a result, the Danes believed

SILVERWHITE

his vessel was a heavily laden cargo ship and allowed him to pass through the Sound unhindered.

The story allows us to draw several conclusions that the ships discovered in Roskilde reinforce wonderfully. Firstly, it tells us of the ruling system of maritime law at the time, which has little in common with our conception of the Viking Age that gained traction through literature. Let it be stated that in this travelogue we bid farewell to the Viking Age. For it did not exist, at least not in the way it has been depicted in recent times. The Viking Age is a fiction that was invented first by Christian priests who heaped their own sins onto the pagan Vikings, and even more by later Scandinavian romantic-nationalist literary tradition. Our most significant conclusion comes from the construction of Harek's ship: with its oarlocks concealed, the vessel bore no difference from an ordinary cargo ship. Is this true?

Recent archaeological discoveries prove that it is so. Three of the six Roskilde vessels were cargo ships. Wreck no. 3 is nearly intact, down to 'the very tip of the stem' (Olsen 1963: 28).

The ship was 13 metres long, 3.2 metres wide, and made of oak planks. The strakes were spaced at 94 centimetres: large for a cargo ship, which shipbuilders understood. Three timbers were installed from board to board to brace the hull. The second was in the centre of the vessel and supported the mast. Small decks were built at the bow and stern; cargo was stored in the centre. Two oarlocks were found on both the port and starboard sides at either end of the ship, making eight oars per vessel. The oarlocks were only lightly worn, meaning sailing was the predominant means of propulsion. Rowing was only done in case of emergency: apparently when entering a harbour or river where it was difficult or dangerous to master the wind. The stem was curved, its tip curling handsomely back. It was made from a 3-metre-long oak trunk and carved on the edges to imitate the clinker planking. The three uppermost planks were reinforced with rubbing strakes. It was a cargo ship. Wrecks no. 1 and 2 had similar proportions: lengths of 19 metres and widths of 4.2 metres. This makes the ratio of length to width 4.06 : 4.2 – a length the width of four ships or slightly more. Their strakes were spaced at slightly smaller intervals: from 90 centimetres (no. 1) to even 75 centimetres. Heavier traverses at the centre of the vessel were due to the weight of cargo, which put stress on the outer shell. Wreck no. 1 also had planks at the bow and the stern meant to block spray: bulwarks.

RED SAILS

In what ways did an ancient warship differ? As Snorri Sturluson's account reveals, its outer appearance wasn't unlike that of a cargo ship. Construction details were the same. The main difference was its proportions. Cargo vessels were wider and squatter, reinforced with thwarts. And yet, even with such insignificant variations between the two types of vessels, the sailing qualities of a warship were entirely different.

Warships were longer and narrower, lacking thwarts and with larger strake intervals. But why did Harek take down the mast while sailing through Øresund/Öresund? Let us try to be more specific in our phrasing: in what way might the sail, or even an empty mast, give him away?

We come back to the keel. A heavily loaded cargo ship sat deeper in the water. Proportional to the lightness of the keel, it must also have had a smaller sail area. This in turn made it slower than a warship. Masts were crafted using very high-quality wood, with shipbuilders showing exceptional care in their selection and hewing. No one would put an unnecessarily tall mast on a cargo ship. A warship could be betrayed by its large sail area, and especially by a tall mast. We can understand Harek's simple cunning. We can also understand that cargo ships were protected.

The greatest draught of a low-keeled Baltic cargo vessel was no greater than 1.2 metres; usually, 0.8–1 metre was sufficient. The draft of a warship was even less.

It all began with the keel.

WHO WERE THEY?

Notions of the Viking Age have been the most stubborn and galling of all Nordic literary clichés. Since the days of Adam of Bremen, it has been described as pagan tribes' biological lust for plundering. Nationalist-romantic historians later attempted to give it a noble patina by pointing to overpopulation in Scandinavia, which archaeological and economic studies have disproved with modern statistical methods (Lönnroth 1963: 12–19). In literature, the Viking embodied mythical Nordic strength. Their raids were seen as an explosion of almost natural intrepidness, and the Vikings' bellicose nature was amplified and falsified 'to the extent of downright foolishness'. However, the sagas' literary sources have turned out to be worthless or, in the best case, a mosaic of tiny scattered facts, whose causal relations are false. Baltic history, as one Scandinavian historian remarked, is not plagued by a lack of source material, but by a plethora of tawdry sources. They are taken as fact and have distorted

SILVERWHITE

reality beyond recognition. Even worse: far too often, 'Vikings are equated to Scandinavian seafarers' (Olsen 1963: 16).

The Swedish historian Erik Lönnroth asks who they truly were, these helmsmen who travelled the Baltic–Volga waterway. As whom and what should we regard them, now that the Christian missionaries' fearmongering depiction of the Viking Age has been shattered? The tranquil smoke of the Helgö iron-smelting ovens; the protected privilege of cargo vessels; the shipbuilding hub on Gotland; the Viking comb production in Peenemünde – yes, all form a truly stark contrast to the clashing of swords and shields and the primeval battle cry.

The word 'Viking' will no doubt stray onto the pages ahead, as it would be ridiculous to resist such terms. Words are not to blame. At fault is the ignorance that has given the word false substance. Let us try to shed it. An equals sign should be drawn between the Vikings – making jewellery and ploughshares, trading, and sailing the seas – and the seafarer, creator of the unique Baltic nautical culture. It's worth adding that the word 'Viking' comes from the Old Norse *wic*, which was used for a bay under the protection of trade peace – a man of the bay. It was first recorded in the year 680. Estonia's western county of Läänemaa was also called by the name *Wiek* in Old Norse, lasting until the 20th century.

THE SHIP KING

We are accustomed to calling the person in charge of a ship a captain in modern Estonian: *kapten*. However, the Estonian language formerly had no such title, nor did Old Norse. The division of labour on ancient vessels was done according to completely different principles that can only be reconstructed through indirect sources.

Archaeological ship discoveries tell us that keels were sometimes joined in two or even three places. We know from folklore that a new vessel could be sewn together from two old ones. A ship was a precious treasure that required the efforts of several craftsmen and, at least in the early Viking Age, was owned jointly by a community or village.

However, ships are also a rather odd catalyst of social relationships. We have long held to the notion that a single captain must govern a ship – otherwise, it cannot be navigated. The division of labour on board reflected the social relationships on shore, though only to a certain point. A captain's word was law on the sea. Raids and trade accelerated differences in wealth

RED SAILS

and material possession in turn, and most likely also the prestige of the captain as an inherited profession.

The captains, who over time also became the vessel's owner, were called *styrimadr* in Old Norse: 'steersman'. Men who took part in voyages for trade and warfare acted as partners. Revenues and expenses were shared equally, though the steersman had the right to a larger share. This system lasted until the Middle Ages.

Another word seems to have been commonly used to refer to a captain in Estonian: *kuningas*, 'king'. As a name for a leader or commander, this probably developed even before the time of Tacitus (Moora and Ligi 1970: 74–83). Other Nordic peoples also used 'king' in this meaning during the Viking Age. The extent to which the concept was prevalent in ancient times is demonstrated by its preservation in 63 place names, many medieval surnames, and, until the year 1805, it was the title given to the leader of almost every major Estonian rebellion.

A crew could be composed randomly for trade voyages, depending on willing hands. On warring expeditions, the relationship between captain and crew was more clearly defined. Sailors made up the leader's retinue and were connected by blood or oath. Later sources call these men *cognati et amici*, relatives and friends. A fee was to be paid if a friend fell. Thus, an *amicus* is a companion, friend in war, friend in trade, and fellow crew member. The word pair points to the ancient composition of a crew: blood relatives, mainly. As family ties loosened, outsiders were taken on board. However, to be an equal partner on trade voyages and raids, they were required to confirm their obligations by oath. We know how the Livonians gave such an oath: by stepping on a sword. The concept of a friend thus had a broader, more militaristic, and more nautical definition in those days. We can find quite fresh proof of this in ethnography: friendly trade between the shores of Viru (north Estonia) and Finland continued until the Second World War.

Laeva kipper, poisikene,	Skipper, boy,
ära lase laeva liiva joosta,	don't let the ship run a'sand,
puutuda punakivisse!	touch the red stone!
Lõunatuul tuli lõhkumaie,	Southern wind came crushing,
kagutuul tuli katsumaie,	Southwest wind came testing,
põhituul tuli põikamaie.	Northern wind came 'cross the bow
Viskas nied mehed meresse,	Tossed those men into the sea,

SILVERWHITE

soapaseared salme'esse,	Shins into the strait,
udusärgid umikusse,	Shirts into the waves,
linajuoksed liivakusse.	Blond hair into the shallows.
	(Jõelähtme)

Memories of an ancient ship-king live on in coastal dialects, which equate a king to a skipper: *noodakuningas* ('seine king') / *noodakippar* ('seine skipper'). The Estonian loanword for skipper, *kippar*, most likely developed in a later time when ships had become private property. The Old Norse *skipherr* > *kippar* > *kipper* originally meant the master or owner of a vessel, whose responsibility it also was to perform ancient sacrificial rituals.

* * *

The outer appearance of a cargo vessel did not differ much from that of a warship and, as we saw earlier, clever deception could render the two identical. All the more surprising, then, is an ancient Baltic custom that limited the ability to practise deceit. When a vessel set out on a hostile voyage, an animal head was affixed to the stem as a sign. On a peaceful trade voyage, it was removed. The minor advantages that came from deception apparently did not pay off, which shows the effectiveness of maritime law and adherence to it.

We encountered the figure of a boar earlier when discussing Tacitus. Here, let us delve into the more distant history of the unusual custom, both geographically and genetically. The boar was used by Estonians and Livonians as a symbol of strength, power, success, and fortune, but it was also common among the Volga-Finnic Erzyan and Moksha tribes. This was written about by Vasily Klyuchevsky (1849–1911), one of the greatest 19th-century historians, who was also ethnically Mordvinian (Klyuchevsky 1956: I, 305; Aleksejev 1972: 45). As the symbol migrated onto the stem of a ship, it appeared wherever ships did and, in a sense, can be used to map both superstitions and shipping routes. Another possibility can be outlined: the head of an animal on a ship stem could only perform its function in places where people knew how to read the symbol.

Let this remain a dotted line for now. Perhaps we will be able to add new dots someday and trace the line further. Our travelogue could indeed serve as a bridge between poetry and exact science.

RED SAILS

The captain gripped a tiller, which in Estonian was called a *saps* – a word also used for the forearm of an animal, usually a horse. It was his sceptre and club, but also the men's fate: home shore or a shipwreck. The ship king/helmsman sat on the starboard side of the vessel where today's ocean-going giants have captain's quarters and navigation rooms: sacred altars to the art of navigation along with their handsome, old-fashioned sextants, signal flags, long rulers, and well-worn logbooks. Home shore or a shipwreck.

Laeva kippar, poisikene,	Skipper, boy,
oia ohjad, keera köied:	hold the reins, turn the ropes,
meil onvad mered madalad,	our seas here are shallow,
meil onvad karid kavalad,	our reefs here are sly,
neemed neljanurgelised!	our capes four-cornered!
	(Haljala)

* * *

'When a boat grew old, it's remains weren't set alight, but rather it was pulled ashore and left there until the wind and the water finished it off. An old boat's remains were considered sacred [...] and using it for firewood would bring misfortune upon a house.' (Audru)

SHIPS BURN, SPIRITS ROAM

Sea culture created sea beliefs, at the centre of which was the ship. Its mute witnesses reside in our coastal villages to this day. The windmills of Koguva village were sawn down for fuel over time, but ancient hulls still crouch alongside its chest-high stone fences, tar-black and mysterious in the shade of age-old verdant maples and ash trees, within reach and yet off-limits to every axe.

Ships could be burned, but only on one day a year when the sun reached its zenith at midsummer. Nowadays, Estonians celebrate it on the eve of St John's Day (24 June).

The customs that we unscientifically called 'sea beliefs' are actually divided between two environments. One was the ship, the other the sea. The ship was home, a projection of the hearth upon the sea. Though

SILVERWHITE

mobile and wandering, it carried the protectors of hearth and the living spark of home everywhere.

* * *

'The shipwright would place the first slivers on the stones in a sauna and warm them there, then take them back to the ship. The ship's spirit came with the slivers. The spirit would leave the ship before she wrecked.' (Pärnu)

To the east of Tallinn towards Viru, a ship spirit was called a *haldja*, 'fairy' or 'sprite'. To the west, it was also known as a *kotermann*, 'Klabautermann'. It was born of the first splinter and was temperamental, easily offended, unsocial, and even cowardly. Nevertheless, there was no friend more loyal to the ship or its crew. When in peril, it helped to reef the sails, hold the tiller, and warn the crew of a storm or certain shipwreck in time to save them. Yet if sinking was inevitable, then the spirit would save itself. If it jumped overboard empty-handed, then the men would escape empty-handed as well – meaning at the cost of the ship. If it departed the vessel with a hatchet in its belt and a pack on its back, then there were hopes of salvaging at least part of the cargo. One legend from Jõelähtme tells of three ship spirits who appeared simultaneously on a three-masted ship and began bickering over who was first and would become its Klabautermann.

The mermaids of Nordic folklore entered Estonian sea beliefs back in shadowy ancient times (without reaching Latvian folklore in the process) and even taught us how to play the zither (Loorits 1949: II, 262). Comparing the dissemination of beliefs with the map of ancient settlements allows us to date certain aspects of the former with some level of accuracy and credibility, i.e. to estimate their age.

In Kuusalu, there was legend of a coastal spirit that chose a king from others of his kind (a reference to the captain/ship king); and on the island of Muhu, people spoke of a *lõkitseja* – a flame that ranged in size from that of a candle to an entire haystack – that coaxed seafarers further and further away from shore, towards hidden treasure. Relations with sea fairies in Jõelähtme were particularly amicable: sometimes, they would call out to a captain and tell him what direction to sail to rescue a crew in peril. They possessed sympathetic human characteristics, asked to borrow tobacco (Lüganuse), and complained about their difficult lives.

A coastal spirit is indeed a kind of link between the friendly ship spirit and the much more severe sea spirit. The ship was a hearth and the shore

RED SAILS

a community. The sea was an alien environment and relations with it tempestuous: benevolent one day and malevolent the next, depending on how well the captain performed his duties to the sea. Kindness was not guaranteed, and to achieve it, one had to take great pains, bear costs, acknowledge the supremacy of nature's power, and appease it.

Sacrifices were made to Old Man Kolju on Kuusalu's Kolju Promontory. The limestone cliff of Toompea, Tallinn's Upper Old Town, was likely a sacrificial site, though the feudal city erased all local heritage. Even so, according to one legend recorded 'at the last minute', swine heads – those strange symbols of nautical power – were sacrificed in Tallinn. Naturally, the custom could not have developed in the Christianised city during the Hanseatic Era; and its survival through the Middle Ages, all the way until M. J. Eisen recorded it in the late 19th century, is proportional to its age and roots stretching deep into the soil. A similar well-known sacrificial site was Mustjala Cliff on Saaremaa, where there was said to have even stood a 'holy pillar'. Nineteenth-century local historians wrote of the following legend:

> The people of Panga village say that long ago, one living being, a human or a pet, was flung from the cliff every year and seized by the deep grave where it was devoured by the sea god. To prevent such a tragedy, the people decided to win the sea god's favour with an annual voluntary sacrifice that was said to be beer and vodka. Even now [1861] on certain days, men row out to a place where the sea churns and froths and roars. There, they pour beer and vodka into the sea, and the sea god is placated ...

When relations with the sea were conducted correctly, seafarers' attitudes towards their environment became harmonious, even friendly. Musician sailors were often invited to play for a 'sea wedding' – the reason why the men preferred not to touch an instrument at sea. Although the musician was wrapped in an apron like a water sprite (Tõstamaa) when diving to the seabed, it was quite a great risk. There was work to be done at the bottom of the sea, but sailors were not allowed to disturb the Thursday evening peace, as one folk song confirms:

Meie katel keetanessa	Our pot is set to boil
Neljapäeval nõidadele	To the witches on Thursday

SILVERWHITE

Thursday is Tooru/Thor's day, the holiday of the Nordic 'lightning god', behind which we can still see the flash of our Fire Island. Friday (Estonian *reede*) was dedicated to Freyja, the Norse Venus, mother of Tooru/Thor and Lemminkäinen, whom Tacitus thought he recognised as Isis. Hopefully we are not mistaken by seeing Friday as the end of the ancient lunar calendar's four-day work week – a taboo day on which one was forbidden to do chores or pull up anchor, and transgressors were threatened with doom and destruction.

In ancient Estonia, Thursday night was a sacred time when all kinds of witchcraft were performed, and even in the early 20th century, it was still unsuitable for working. January was Tooru/Thor's month, *Thorrablot* in Icelandic, the latter half of which meant 'sacrifice'. Still, any conclusion would be premature. The need to count time evolved into the ability to count time so early on that we can regard it as one of humankind's major achievements. Already in this work, we have recognised that humans tend to equate technology with culture and the lack of technology with a lack of culture. Appreciating history is part of good form; actually, we still very naively and egotistically believe that 'proper' life and 'proper' culture only began with us. We are the world's navel and take offence when our children regard themselves, just as self-evidently, as culture's navel without leaving room for their mothers and fathers. My thoughts ran away from me here and I should condense them so as to leave only the question: why Thursday? And that in spite of the fact that the Baltic-Finnic numeral system was based on seven, both divisible by the phases of the moon and adjustable to measure time. The Estonian *kaheksa*, 'eight' (*kahdeksan* in Finnish), is a later invention that originally meant ten minus two. *Üheksa*, 'nine' (Finnish *yhdeksan*), meant ten minus one, and *-heksa* (Finnish *-deksan*) itself is believed to be an Indo-European loan (*dekn*, cf. Latin *decem*, 'ten'). Questioning is the obligation of all, answering the power of few. Why Thursday? We do not know, cannot remember.

* * *

Once on a beautiful day, a ship stood far from shore at anchor. The sun was hot and not a single breeze ruffled the surface of the water – that's why the ship couldn't sail, but had to stay at anchor awaiting the wind. Suddenly, the sailors hear banging on the seabed beneath their vessel. They go below deck to see if they can't maybe hear what's banging on the seabed better. There, they could hear quite clearly what sounded

RED SAILS

like a lot of carpenters busy chopping and hewing wood on the seabed right below the ship. Some were already afraid those carpenters might want to chop through the hull and send them to their doom that way. Suddenly, as clear as day, they heard a loud voice call out from below: 'Kai, bring a ladle of *taar*!'[1] – 'Oh-ho, what do you know!' the sailors marvelled, 'There's a whole family of people living down here on the seabed!', for the girl Kai was there with her ladle of *taar*. Neither the sailors nor the captain could wrap their minds around it. 'Wait just a second,' one clever sailor said. 'I want to see if this "Kai" might bring us our own ladle of *taar*.' And as loudly as he could manage, that sailor cried, 'Kai, bring me a ladle of *taar*, too!' All the men went up on deck to see if Kai would board their ship with her ladle of *taar*. After a little while, the surface of the sea rippled and a beautiful young woman with long hair surfaced, grabbed the gunwale, and hoisted herself aboard, where she walked up to the men and offered them *taar* from a wide ladle scrubbed white. The men accepted the *taar*, drank it, and said it was very fine, sour *taar*. After the ladle was drunk dry, the sailors asked, 'What's that banging down there on the seabed?' – 'They're making the wagon that'll be used to cart off to hell the possessions of those who keep working into the night on Saturday and Wednesday!' she replied, then leapt into the sea. Men say they haven't heard any banging since then. (Viljandi)

WHO WERE THE CHUDS?

Earlier, we smirked while demonstrating that the strange boar heads of Tacitus's Aesti endured in Estonian traditions, and there is no doubt that if scrutinised, half of the examples would turn out to be coincidence. However, the name for a chimney in an Estonian farmhouse does come from the same root that long ago was used for a swine snout: *turp* 'pfost bey den Rigen Ofen' (Göseken, 1660); *turv* (*turb*, *turp*) i.e. *turvapost*, 'support post' – a round or rectangular wooden post used for support on the outer corner of an oven; *patsas*, i.e. *patsasammas*, i.e. *patsatulp* – a wooden support post on the outer corner of a more modern country oven (Saareste). Now, we will see that the world pillar / sacrificial post / hearth post even found its way into geographical sources, albeit distorted. Of course, we can only speak of very old information.

How might we check the veracity of this decoded information?

We must, for a moment, stray into a more abstract discussion. It involves our memory device. Linguists use Morris Swadesh's mathematical method which allows one to track the correlation between time and a common lexicon: in other words, the durability of the societal memory device. Ethnography and folklore studies lack such a method. Still, it isn't impossible that we haven't grasped all the conclusions of time's accelerating current. Perhaps the following comparison might help explain the chain of logic. We have several close, far, and extremely distant islands from which we receive a communication. The further away, the weaker the signal. However, if the sender's strengths double over certain distances, then the communications can reach us at an equal intensity. And so it is: the durations of the Iron Age, the Neolithic, the Mesolithic and so forth double in relation to one another. To put it another way, historical periods continually lengthen as one moves towards the past, as if balancing the distances that separate us from them, and information from last week exists in our present side by side with information from the last millennium. The size of the latter is smaller, of course, but not its intensity. Small-sized information in turn increases the signal's ability to penetrate. It might be time that is rediscovered; time that is reborn. In short, the strength of any given signal or field is inversely proportional to the square of the distance, and this can likely also be applied to ethnography and traditions if we interpret 'distance' as Nature's fourth barrier: time. In another connection (and another book), I hypothesised that breaking the time barrier is one of humankind's most important qualities – one that raises it from the level of biological organism to that of societal being.

Testing the Sampo hypothesis leads us back to geographical discoveries.

Though we previously noted that the word 'Peipus' is four times older than Egypt's oldest pyramids and comes from the Kunda people, who disappeared without a trace (or else disappeared while leaving the type of traces that we cannot yet decode or connect to any other living nation), now is the time to add that Lake Peipus holds yet another secret. The Russian name for Europe's fifth-largest body of water is *Chudy*, lake of the Chuds.

Who were the Chuds? By tracing the fine mechanism of phonemic mutations, linguists have determined its Finno-Ugric origin beyond any doubt. So what did 'chud' mean? Linguists based their conclusions on the name used for Votians, which long ago meant 'keel'. This is not merely a hypothesis. The Votians occupied the area between Narva and the Neva, with pockets also found near Alatskivi, Jõgeva, and Avinurme. Russian historians translate the 13th-century north Tartu county of Vaiga into Russian as *klin*, 'keel'. Votian territory is indeed keel-shaped – assuming that the chronicler was able to accurately track the ethnic map of the time from the height of a satellite.

However, the Russian word *klin* isn't 'chud' yet!

Sámi is used to bridge the gap. Ancient Sámi legends use the word ćudde for an enemy, raider, or enemy force. Historians have found that the peoples of the Gulf of Finland's southeast shore – the Virus and Votians – would travel to Sápmi, land of the Sámi, to collect tribute in furs. And Sápmi, as we may recall, began from the northern shore of Lake Ladoga. Using this information, one of Finland's most renowned linguists has restored the oldest meaning and pronunciation of the Sámi ćudde: šūta, which meant 'keel'. The mysterious word's evolution would have been as follows: Votians collect tribute in furs from the Sámi. Finding out that the hostile tribute-collectors' name meant 'keel', the Sámi translated it into their own language as šūta. Over time, Slavic populations pressed further northward until finally coming into contact with the Sámi. Hearing the Votians called by that name, the Slavs adopted the word and used it to name Lake Peipus in their own language.

The Achilles heel of this linguistically flawless leap is the logic of landscape. Slavic settlement in the taiga expanded from south to north along the Dnieper waterway and not in the opposite direction. First, they met the Balts, then the peoples living along the eastern shore of Lake Peipus and the lake itself, and lastly with the Sámi. It would be beyond strange to presume that they corrected a long-used name for Lake Peipus according to the Sámi.

Thus, there arises a temptation to point to simpler possible interpretations that bring us back to Sampo. The Russian chroniclers sometimes used the word *chudy* as a broad term and othertimes to refer to a specific tribe. Its original Baltic-Finnic form may have been *suu*: 'mouth', i.e. the mouth of a river, the mouth of the Silverwhite waterway, transforming in dialect to *suue*, *suudme* (Wiedemann). The word for 'mouth' stretches back to the common Finno-Ugric lexicon and provides, in metaphorical use, an important geographic term: one that was inevitable and solely conceivable at the neck of the Gulf of Finland. Thus, in a broader sense, 'Chuds' as those who used the estuary, and 'Chud' aka Votian as one living at the mouth of the gulf in a narrower sense?

Nevertheless, we should emphasise that the laws of phonemic mutation do not permit such an interpretation if we presume a direct loan from the Baltic-Finnic languages to Russian. Those who communicated the tribe's

SILVERWHITE

name had to have been the Balts, ancestors of the Latvians, Lithuanians, and Prussians.

Yet we have now lost sight of the Russian word *klin*, which was used to translate the ancient county of *Waigra/Wagia/Wagya*, and not only that: in the old chronicles, *votsky klin* was used to refer to Votia itself. Isn't that strange? The thing is that in literal translation it would mean 'keel's keel', i.e. a pleonasm akin to the Sahara Desert or Lake Võrtsjärvzee. Still, it would be premature to accuse the chronicler of an insufficient knowledge of the Baltic-Finnic languages. He may not have realised the overlapping meanings of keel and Votia/Vaia. Upon closer inspection, the Church Slavonic word *klin* (Vasmer 1953) was used to mean both 'keel' and *vai*, 'stake'.

Vai is a loanword from the Baltic languages. The Lithuanian *vāgis* and Latvian *vadzis* mean 'keel', though shifts in meaning are completely commonplace with loanwords. Much more significant here is the possibility that Slavic tribes encountered the name for Votians through the Baltic tribes. Thus, 'Votian' could have been the Balts' ethnonym for the peoples settled on the southeast coast of the Gulf of Finland, bringing into focus the image of a stake, a *pole*.

The possibility finds unexpected support from the ancient Russian word for Narva, *Rugodiv*. It is also the name of the Baltic-Finnic god of good harvests, the oldest recorded form of which (1551) is *Rongoteus*, in folklore also the *Runko Pole* and *Rõugutaja* (?), probably originally *Rukotivo*. The first half of the compound word means 'rye' and the second half 'stake' (*teivas*), which, for some unclear reason, is in semantic kinship (again!) with the extremely ancient loanword *taevas*, 'sky'. Leaving the lending problem aside, let us acknowledge that this is yet another instance of the pole symbol and is documented by an entirely credible toponym from a Russian chronicle. We cannot completely cast overboard the pole's part in shaping the Votian ethnonym or in translating Vaiga as *klin*.

A NAIL IN THE NORTH

Movement, not regarded from the perspective of modern physics, required space–time that was bounded by a familiar, home-like beginning and an equally tangible but misty end. Seafaring presented incessant cosmogonic questions and simultaneously also answers – poetic moments that did not need to be preceded by analyses. The Baltic Finns' cosmos comprised two contrasting parts: the living environment, called *maa* ('land'), and the sky above it, called *ilm* ('world, weather'). The universe,

RED SAILS

i.e. the Baltic Finns' *maailm*, attached to the North Star. In this sense Stella Polaris is not a direction but a foundation. From it sprang the cardinal directions, which could be imagined as the poles of a summer lodge. There are four directions, and draped over them is the sky, which seems to have been called *juma*. The Estonian word for 'sky', *taevas*, was borrowed from the mysterious Kunda language and is related to the Sanskrit *daivash* – originally 'shimmering, brilliant'. The expanse of the *juma* contains countless tiny holes: stars. *Juma* revolves together with all the constellations, the sun, and the moon. As meteorological warmth and cold are determined by the direction of the winds, *ilm* acquired its second meaning of 'weather', *ilmastik*. Lake Ilmjärv (*järv* = lake), was, by all logic, named after the reflection on its surface.

The world's end was not a temporal but a geographical notion. Time was infinite, space finite. People's perception of time's endlessness was perhaps even more acute than it is now, because ties to long-gone generations were alive: every decision and step was made with consideration for one's ancestors, and the illusion of their proximity gave even greater depth to the physical world. Objects were passed down from generation to generation, preserving the warmth of many touches and the whispers of many events. The finiteness of space stood in sharp contrast with this. At the world's end was a slushy sea, in which we can recognise characteristics of the Arctic Ocean, or else a steep cliff, even a stone obstacle, beyond which dog-headed monsters (South Estonians' *peninuki*) prowled vigilantly for their prey.

Experiences polished poetry, and the poetic moment did not conflict with the mundanity of the temporal world. Among the Baltic historical sources of the Early Middle Ages, three navigational guides have been preserved, the oldest recorded around the year 1073 (Adam of Bremen, IV; cf. LUB III, no. 244 and CCXVI; Johansen 1933: 162). They are not a part of this time frame and will come up later. For now, let us just note the most important detail: out of the three, not one has Central European origins. All three were shaped by Nordic navigational experiences. The most precise of the three describes a route that is totally consistent with Pytheas's route from southern Sweden, through Birka and southwest Finland, and into the Bay of Revala.

The navigational guide contained a certain number of toponyms and the distances between them. The latter are given in day journeys, which is the guide's primary unit. However, the length of such a journey varies between 110 and 165 nautical miles. Thus, a day's journey at sea came

207

not from theory but practice. The primary unit reflects the complexity of navigation, the probability of winds, the number of reefs, and likely also the opportunities for anchoring overnight – in other words, an experience with a degree of accuracy that depended on two factors: the length of recording the experience and the manner in which the experience was conveyed.

If we reach back from the era of the 'red sails' to Saaremaa's ship burial sites, we get an awe-inspiring 1,700-year tradition, though it might only have been passed down orally. Written navigational guides are also laconic in style and wording, just like texts meant for memorisation. And no doubt they were. Adam of Bremen, who recorded the oldest surviving guide, was thus acting as a collector and de-poeticiser of nautical folklore. Prior to the advent of literacy, the rational core of navigational guides had to be recorded just like the rational information contained in any other folklore. Parallelisms, alliterations, and poetic metaphors were mnemonic techniques that eased the preservation of knowledge-based experience. Forgetting one element did not yet render the entire information useless. To use contemporary scientific jargon, parallelisms could be called repeat information. Repeat information has a limit after which it becomes noise.

Consequently, folk songs were the Baltic Sea's oldest form of pilotage. Let us call the period the Folkloric Era of navigation.

However, day journeys had to be checked. That meant assessing speed. Speed is the ratio between length and time. Determining length requires measuring the distance between two points. Determining time requires dividing a day into definite units in turn. This was a tough nut to crack. The lengths of nights and days changed rapidly in the Nordic region at the peak of the navigational season, offering no support. In this sense, a Baltic seafarer was in a much trickier situation than one on the Mediterranean Sea or in the Bay of Bengal, for whom the length of a day remained practically constant. For a moment, let us leave the dawn–dusk phenomenon aside and focus on the sun, which we are accustomed to regarding as a definite point for orientation. This isn't a particularly precise line of reasoning. The thing about the sun is that it can only be used to determine direction if we know the precise time, and vice versa: we can only use it to determine time if we're aware of the direction. Catch-22! The exception is astronomical midday, which can be determined by the shortest shadow – the point at which it points straight down. That was the only means to *precisely* measure direction in the days before the compass and the clock. Over time, it developed into the sundial, the oldest navigational instrument in

RED SAILS

the Nordic region. I once came across a sundial on the mossy stone wall at a family home in Koguva, Muhu Island. It was a handsomely carved square limestone slab about 10 centimetres thick, 35 x 35 centimetres in size, and with a circular face that resembled that of a compass. The family told me it was a piece of local craftsmanship from the early 18th century. The Estonian language appears to remember as well: a sundial was called a *päevakivi*, 'day stone', and a compass a *põhjakivi*, 'north stone'. As the tools do not share the slightest similarity in appearance, the similarity of names must have come from their comparable functions.

The day journey is a unit derived from local navigational peculiarities. It is more a notion than a unit, and its limit points – supposed overnight anchorages – could be of interest to archaeologists. In the North, there was another well-defined nautical distance used alongside the day journey: *ukaesiö/ugeso*, which on a chart would be equal to 4.5 nautical miles. This is equal to 8.3 kilometres, roughly comparable to the archaic Estonian unit of land measurement called the *penikoorem*. We must remember that the *ukaesiö* is interpreted as the furthest visible point on the horizon still observable from a vessel. Thus, the value of an *ukaesiö* depended on the position of the observer – the height of the mast. The distance of the horizon in miles is calculated by the equation

$$2{,}075\,(\sqrt{a} + \sqrt{b})$$

in which a is the height of the observer from the sea and b is the height of the observed object from the sea, both in metres. Sailing on open sea, the value of b is zero. To a layman, it seems incredible that one's gaze could lock upon an empty expanse of sea. We're used to identifying the sea as a place of incessant movement, the incessant play of swelling and cresting waves; poetically,

> emerald, wave-soft extends the bed,
> all afloat, here and far, dreamlike in its spread.

This is both true and false. The morning sea and placid evening sea can maintain traces of frothy bands – the frozen crowns of waves lost to the bottomless deep – for an unbelievably long time. To the professional eye, a calm sea is full of points for orientation. As many are sown there as there are marks on a steppe, so monotonous to a Northerner. In any case, returning to the equation, what we have is a special situation where 'the height of the observed object from the sea' is equal to the surface, and the value of b is consequently 0 metres. Simple calculation shows that a

SILVERWHITE

4.5-mile *ukaesiö* thereby gives *a* the value of 4.8 metres. The result is fully credible, as the height of an ancient Baltic vessel's yard from the sea was, depending on the length of the ship, between 4 and 5 metres. Naturally, the observer was located at the intersection of the mast and the yard. The crow's nest was a much later invention.

Thus, folkloric navigation relied upon constant observation, memory, and organic tradition. Perhaps we should explain what is meant here by navigation. It isn't reefing the sails of a luxury yacht, elegantly manoeuvring a cruise ship, or squeezing the maximum velocity out of a speedboat. Navigation's primary quality is more boring, important, and universal, applying just as rigorously to the most modern submarine as it did to the most rudimentary ancient Baltic ship: navigation means the ability to determine a vessel's location with maximum precision at any given moment. In the wheelhouse of a modern ship, a pencil steadily traces its course, and the point where the navigator pulls its pin-sharp tip from the paper with an abrupt jerk signifies the vessel's central point at that given moment. Losing the continuity of a course is a crime that costs one's degree. Even with modern-day technological equipment, restoring continuity is much more difficult than maintaining it. In the time before compasses and clocks, the continuity of a course was the sole thread guiding men and their fates. It ran before one's eyes as an unbroken stream of images. Morning began with a familiar promontory, evening arrived with a familiar islet. Scrolling between them were the day's stretches of open sea: no landmarks or seamarks, not even the sun on overcast days, and certainly not the North Star on white summer nights. There was the *ukaesiö*, there was wind that could shift in strength and direction, and there were unequal day journeys. This inequality stemmed from the fact that in addition to every other factor, the length of a day journey was determined by local shore rights. We'll return to those later.

The navigation from one anchorage to another along a course that was refined to its most optimal route over centuries of experience became so deeply established on the Baltic Sea that Hanseatic cogs later adopted these same routes. For a moment, we must reach beyond the framework of time to explain the reasons for such exceptional conservatism in nautical routes. We tend to think that the sea is free and limitless, forgetting that this is a poet's truth that does not coincide with that of a seafarer. Deep-draught Hanseatic cogs were unsuitable for navigating ancient sea routes. And yet they clung to traditional sailing courses as if the Baltic were not an open sea, but a poorly marked rural road. It could have been no other way.

RED SAILS

Was Scandinavia an island? Was Estonia an island? No one knew, or if they did, then they didn't give up the knowledge. Navigation was grounded in oral heritage. It was local experience; a mental creation that could not be coaxed out, taken by sword, pocketed, or shipped away.

THE DAWN–DUSK PHENOMENON

We should determine the longitudinal extent of Baltic Finns' nautical routes or, to put it more simply, see how far to the south and the north our voyages may have reached.

Even before Tacitus, the notion of Western Slavs and Russians had taken root in our language. Consequently, the southern shore of the Baltic can be proved to be the southern limit of Baltic-Finnic seafaring.

A leading Finnish archaeologist has studied the northward dissemination of jewellery produced in northern Estonia. The silver Iron Age pendants called *rinnalehed* are unique to Estonia, and the blacksmiths who made them lived in the vicinities of Revala, Rakvere, and Tartu. In addition to the Revala area, they have been found in Ingria, the Vaga river system (a tributary of the northern Dvina/Viena), in northeast Finland, and, most of all, west of Vardø (Sámi Várggát) on the northern Norwegian coast. Vardø lies about 175 kilometres west of Murmansk. The amulets, about 7–8 centimetres in diameter and decorated with magical braided patterns, were good luck charms. They have recognisable depictions of snakes and drinking vessels similar to those found on old runic stones in Ridala, Karuse, Hanila, Muhu, Pöide, Valjala, Käärma, and Kihelkonna. Warriors, drinking vessels, and sun wheels carved into the trapezoidal headstones are so unusual that they are believed to be monuments to ancient Estonian sacred sites, allowing them to be dated to the pre-crusades period (Lumiste and Kangro-Pool 1969: 30–6). The silver amulets also sometimes feature depictions of vessels and – let us add as a hint – tools that resemble anchors. One amulet found in Vardø, on the eastern Varanger Peninsula, includes an Estonian eight-pointed star pin, as a result of which the find has even been called 'outright Estonian'. Nevertheless, with the frankness typical of an archaeologist, I will add that the geography of trade ties between Estonia and Sápmi has not yet been researched and therefore any broader conclusions would be premature (Kivikoski 1970: 88–94, cf. illus. 4, Botnhamn discovery).

Whereas archaeology merely points to possible trade between Estonians and Sámi, folklore, language, and traditions appear to confirm

SILVERWHITE

it with remarkable consensus. Sápmi is the most recurring geographical notion in Estonian folklore. It was associated with witchcraft, illnesses, hunting fortune (which means hunting!), the disappearance of migratory birds (rookeries?), animal diseases, prophesying, and, above all, the image of a shrewd and dangerous enemy. Sápmi was not an abstract symbol of distance; rather, the memory of an actual landscape shimmers through it.

What does language have to say about it? The oldest meaning of the Baltic-Finnic word *lape* was 'a remote or distant place' (Wiedemann); *lappele minema* meant 'to go around something'; and in a *lappepaik* ('*lappe* place') one could encounter a negotiator or even a witch. In *Lapu*, 'Lapland', there lived the 'Lapus', who in ancient times pillaged the land together with the 'dog-headed people'. Let us keep in mind that language connects the Lapu and cynocephaly. This, as we will see, leads to several interesting possibilities.

Colds were said to come from the North, i.e. Sápmi, as a 'Lapu witch' (Räpina, 1863). A Lapu witch appeared during a bird hunt on Saaremaa, and when the hunter accidentally shot her, the bullet flew back and struck him. On Thursday evenings, our own home-grown Estonian witches flew to Sápmi to celebrate after their spinning was finished – similar to the Northern Lights (Võnnu). Total strangers who showed up at a wedding party were sometimes called Lapu, dog-paws, or bone-legs: 'Dog-paws must be given beer and vodka, otherwise their curses will bring misfortune upon the young couple' (Vigala).

Let us lengthen the string of synonyms: Lapp aka Lapu aka dog-head aka dog-paw aka bone-leg. A parallel explanation for the last-mentioned is the skeleton, which is to say a corpse, though older explorers also used the word 'skis', which in the taiga were covered with skins on top and bottom, giving reason to compare them to legs of bone. Total strangers were to be treated with the utmost caution at celebrations: nettle sprouted wherever a Lapu urinated. All in all, whirlwinds were nothing more than an invisible Lapu witch trying to steal the year's hay crop from the field. The Estonian language knew Lapland better than Latvia!

Nevertheless, let us separate the concrete from the abstract. Around the globe, every nation's folklore had witches and wise men live on the world's peripheries. So it was in Estonian folklore, with the slight difference that we also had a specific geographical region in mind. Let us point out another interesting detail. In the Late Middle Ages when Estonian seafarers' geographical erudition narrowed, the folkloric witches and wise men were moved to Viru, Sõrve, Hiiu-Kõpu, and the island of Ruhnu:

212

RED SAILS

in other words, to places with the common characteristic of seafaring, contacts both near and far, and contrasts. At the same time, there is no notion of 'Seto[2] witches' in Estonian folklore. The web of phenomena that we could connect to the list of common Lapu/dog-head names was not ambiguously set beyond distant forests, but tied above all to sea routes. It is the most direct and remarkable instance of Sámi cultural heritage that can be found in ancient Estonian beliefs (Loorits 1949: II, 475).

We should wonder: what on earth are we looking for here? And we do. Our subject remains the dawn–dusk phenomenon. There is no shame in wondering.

First, we must explain the reason. In the Early Middle Ages, the ivory jewellery worn by Europe's royal courts and clergy – not to mention bishops' crosiers, thousands of crucifixes, tens of thousands of rosary beads, reliquaries, altar diptychs and so on – was not actually made of elephant ivory, but that of walruses. Walrus tusks were the source of Europe's ivory in the Early Middle Ages.

The walrus is, of course, a mammal similar to the seal, only about ten times heavier and 4–5 metres long. Tusks can grow to up to three-quarters of a metre and weigh up to 4 kilograms.

We have come to the reason and are forced to pause on it for a moment, all without letting the dawn–dusk phenomenon out of our sight. The Estonian word for walrus, *morsk*, is borrowed from Baltic-Finnic; more specifically, Sámi. In Estonian-language literature, it is also referred to as a *merihobu*, or 'sea horse', which comes from the Nordic word *walross*, walrus. Behind it is a funny story.

Wal, you may recall, was borrowed in turn from us: Baltic-Finnic *kala* (fish) > English *whale*. In Norse and in Baltic-Finnic, the second half of the compound word – *ross* or *rus* – meant both 'horse' and 'ship'. Out of ideological or taboo considerations, Skaldic poetry avoided the word 'ship' and instead used the poetic metaphor 'horse' (Haavio 1965: 125). The ship was a sailing horse, a wave horse, even a bronze stallion. The last symbol becomes clear as soon as we imagine the rows of round shields fastened to the gunwales of a warship. Now, we can also figure out the rules of Estonian seafarers' folk songs and translate the strict system of symbols from lines already familiar to us

> Skipper, boy,
> hold the reins, turn the ropes ...

or the lyrics of a song recorded in Järva-Jaani:

SILVERWHITE

> I sing the song into a horse,
> The words into a stallion [...],
> Then I ride to Finland,
> To crush the Finnish bridge,
> To snap the supporting beams ...

which, contrary to earlier beliefs, is undoubtedly a war chant despite the possibility that the bard who sang it two centuries ago may not have been aware of the metaphor's original meaning. Thus, the Estonian *merihobu*, the German *Walross*, and the English 'walrus' would actually mean 'fish horse' or, more precisely, a fishing boat, and altogether a great and valuable catch. Compared, the English 'walrus' also has a direct relationship with the original Sámi loanword *morse*.

Walruses populate the Arctic Sea's narrowly defined coastal zone. Like elephants, they formed massive herds in ancient times. As late as August 1778, Captain James Cook noted that walruses covered entire ice floes in 'terrifying numbers'. On his voyage to the White Sea in the year 875, the Norwegian Ohthere of Hålogaland killed sixty of the animals in two days. Walrus tusks were treasure, a gift fit for a king in the literal sense: Ohthere gave them to the Anglo-Saxon king Alfred.

Now, the 'walrus-tusk coast' in the distant North has stepped up alongside the fur forests and 'amber island'. This is just what we were looking for, to return to the reason for the voyages. Our task is still to determine the extent of north–south cultural contacts.

The southern limit was the south- and southeast Baltic coastline. Determining the northern limit is more difficult. The furthest corner of the Gulf of Bothnia appears to be beyond doubt. A network of inland waterways to the White Sea via Äänisjärvi (Lake Onega) is probable. It was an old trade route by which greenish eastern Karelian slate found its way to the Pärnu and Emajõgi river regions as well as to Saaremaa during the ancient era of hunters and fishers (Clark 1953: 246 and map no. 134). However, during the period in focus, the southern limits of Sápmi, ancient Finnmark, the land of Tacitus's *Fenns* and the Icelandic sagas' *Finnride*, ran through central and southern Finland, northern Sweden, and northern Norway. Sápmi was not far from Estonia. As late as the 17th century, a trial was held in Haapsalu over a 'Lapp witch' named Mart Johan Canutipoeg, who was born in the little village of Töjby, 56 kilometres south of Vaasa. On the other hand, walrus have not populated the Baltic Sea, at least not in recorded history, and the Norwegian Sámi have also appeared elsewhere in Estonian folklore.

RED SAILS

Thus, we could, within the bounds of reasonable probability, draw the northern longitudinal limit of seafaring at the coast of Norway's present-day county of Finnmark, including the country's easternmost town of Vardø – a place where that 'downright Estonian' amulet credited to Revala silversmiths was discovered. *Something* had to be paid for whale tusks. Again, within the bounds of reasonable probability. All secondary proof.

Now, let us add direct proof. Shouldn't frequent references to the Sámi in Estonian folklore have had a mirror image of equally frequent references to the Estonians in Sámi myths? It did. As late as 1673, ancient Sámi tradition preserved the memory of a 'gold woman' called *Wiru akka*, 'the Viru (Estonia) spirit' (Scheffer 1674). The Viru spirit was anything but incidental. According to Sámi belief, she was the wife of supreme deity Tooni. Let us try to ignore the complex family relationships between mythological figures and focus our attention on the geographical ties: the dispersal of Sámi in Estonia corresponds to the dispersal of Viru in Sápmi. Had we not similar evidence from an entirely different collection of national folklore – the White Sea Karelians – it would be nothing but an isolated detail, from which it would be risky to draw conclusions. According to ancient beliefs, an animal's life was on equal footing with that of a human: when killing a bear, a hunter was required to ask his victim for forgiveness and, if possible, pass the blame off onto another:

> Otso, thou my well beloved,
> Honey-eater of the woodlands,
> Let not anger swell thy bosom;
> I have not the force to slay thee, [...]
> Thou hast from the tree descended,
> Glided from the aspen branches,
> Slippery the trunks in autumn,
> In the fog-days, smooth the branches

fibbed 'wise old Väinämöinen' in the *Kalevala*'s 46th rune. The White Sea Karelian hunter employed the same cunning when he persuaded the bear to believe that the fateful hunting knife was not his (Kuusi 1963: 47):

Ei ole veitsi minnu tekemä	The blade is not my handiwork
likä toisen kumppalini;	nor that of my companion;
Virossa on veitsi tehty [...]	the blade is made in *Estonia* [...]

Consider that to both the bear and the hunter, Viru had to lie at a safe distance and exist in the quite familiar real world simultaneously.

SILVERWHITE

Otherwise, the trick would have been useless. But again, let us refrain from describing hunting customs in greater detail to make space for geography: the Viru toponym was documented in the White Sea region. Given these facts, mutual contact between Estonia and the North should be seen as highly probable. It goes without saying that this could only be possible by sea.

So: walrus-tusk hunting lands in the north, a tusk market in the south, and a sea between them.

It is 1,300 kilometres from the southwest corner of the Baltic Sea to the rear of the Bay of Bothnia, and from there 1,800 kilometres to the shore of Finnmark. The coast of Revala is approximately halfway between the two.

Now, we hold the dawn–dusk phenomenon's diagnosis. Yet we must side-track yet again, this time into literary history. When 44-year-old Friedrich Robert Faehlmann read his rendition of the Estonian 'Dawn and Dusk' myth aloud at the Learned Estonian Society on 9 June 1843, it was met with astonishment.

To summarise the myth, Vanaisa (literally 'grandfather') employed the aid of Dawn and Dusk – a man and woman who enjoyed eternal youth – to light the sun in the morning and extinguish its flames in the evening. The two fell in love in summer when they stared deep into each other's eyes and shared a kiss. Vanaisa heard of this and, as he was content with their service, offered to let them marry. Yet the couple preferred to nourish eternally their young love, so vibrant in courting, and meet only at the time of midsummer.

Here, I will only highlight the final passage of Faehlmann's story where Dawn and Dusk suggestively come together in the north at summer's peak:

> Only once a year for four weeks do the two meet at midnight, and when Dusk hands the fading Sun to her beloved, it is followed by a hand-squeeze and a kiss, and Dusk's cheek blushes and glows rosy from the sky until Dawn ignites the giver of light once again and a yellowish radiance proclaims the Sun rising anew.

It was all so beautiful, so literary, and above all so unexpected, that the Dawn–Dusk myth was seen as a literary fake despite Faehlmann claiming that he wrote it down precisely as told first-hand. With some hesitation, modern literary studies have sided with its folkloric authenticity. It was only in the 1930s that folklorist Oskar Loorits came across genuine Dawn

216

RED SAILS

and Dusk myths in the Setomaa region (1949: I, 535–7). According to the legends, the length of a day depends on the sun being woken up on time; or on when the sun-bonfire is lit; or even by an old man who walks before the sun, showing it the way through the heavens, and increasingly needs more time to rest as he tires from the long journey.

The varying length of days, this length's dependence on the seasons, the changing of seasons, the sun's apex in relation to the horizon – in other words, a whole array of specific astronomical phenomena – serve as the mundane foundation for this poetic myth. Igniting the sun requires it to have been extinguished and points to a possible connection to the Thule/Tulemaa/Fire Land legend. For the moment, that is secondary: we are interested in the possible accounts of longitudinal nautical journeys in folklore.

First, let us conclude that the phenomenon of brief summer nights has found attentive interpretation in Estonian myths. We may add that the myths involving personified and anthropomorphised celestial bodies, celestial herds, and celestial shepherds differ significantly from the Latvian tradition. Why is this? The changing of seasons is as natural as gravity, is it not? A stone slips from your hand and falls before your feet. It needs no explanation. Mundane and 'self-evident' phenomena do not spark curiosity. They do not inspire one to explain the phenomena or in this case to personify gravity.

However, the Baltic Finns live at a quite unusual latitude. At midsummer, one must only travel a few days north or south to experience a remarkable change in the length of days. These changes are not evenly distributed over latitudes. If, for example, you travel from Naples to Venice on 21 June, then the day lengthens by only 32 minutes. The extension is practically unnoticeable although there is a 4° 45′ difference in latitude. Yet if you travel just as far north from the Revala coast, i.e. to the latitude of Finland's Kalajoki or Kajaani, then the day lengthens 4.3 times faster (Peep Kalv to the author, 28.2.1974). The phenomenon's astronomical aspect is actually more complicated and its psychological effect even more powerful. Here, we use sunrise and sunset as the limits of a day. It defines the length of a day at southern latitudes where the sun drops straight below the horizon like a stone into water. The sun disappears, and at the next moment, a southern night has spread its black wings, the constellations spark one after another like streetlights, and the land immediately falls into deep slumber. But not in the North! Here, the sun dips so low below the horizon for the night that dawn arrives before dusk fades away. In other

SILVERWHITE

words, the composite number 4.3 is incorrect. Even a short trip amplifies the astronomical effect exponentially.

Estonians are unable to realistically appreciate the impact the phenomenon has because we're a sedentary people. We are accustomed to the white nights, we write lyrical poetry about them, the electricity bills for streetlight networks fall to zero in late June, and none of it keeps us from enjoying a good night's sleep. It's different for our neighbours. In the summer of 1973, a group of Polish university students came to work the fields in Tartu County. A newspaper quoted one as saying, 'The unusually white nights don't let you sleep.' The indirect speech expressively insinuates the reporter's surprise and a fair dose of disbelief. But so it is. To notice the dawn and dusk effect, one must move in a longitudinal direction.

Now, let us turn the spear tip of these conclusions upon itself. Voyages to the southern shore of the Baltic Sea are proven. Voyages to the furthest corner of the Baltic Sea lie within the bounds of sound probability. Here, we're no longer talking about short distances, but a 1,300-kilometre-long journey by ship that must have amplified the dawn–dusk effect exponentially. The shift in its astronomical manifestation was sudden and dramatic; it had an unpleasant, downright physiological effect on people, forcing them to compare, debate, and seek answers. The Dawn and Dusk myth could be a fragment of longitudinal sea voyages left in Estonian folklore.

Should we wrap it up? Not yet! Let's have a read from the *Iliad*.

WHERE ON EARTH DO THOSE BIRDS GO? ...

For a change of pace, the *Iliad*: 'the Trojans advanced as a flight of wild fowl or cranes that scream overhead when rain and winter drive them over the flowing waters of Oceanus to bring death and destruction on the Pygmies, and they wrangle in the air as they fly' (*Iliad*, III).

It is like plummeting into a deep well, as the *Iliad* is believed to be nearly three thousand years old. Let us prune the passage down to cranes, winter, and death. Cranes are migratory birds. In autumn, they fly over Homer's homeland in endless Vs, and they winter on the Nile's sandy banks, lakes, and the fields of Egypt. Owing to the unusual structure of cranes' respiratory organ, their melancholy *cloo*-ing can be heard on land from even the greatest heights. The volume of the call has been compared to that of a trumpet, and the *Iliad* simply calls it a scream.

RED SAILS

Is it coincidence that both Greeks and Estonians believe cranes are birds of death? The Estonian word for crane is *toonekurg*. *Toonela* is the land of the dead in ancient Estonian belief, Tooni its distant and mysterious ruler.

There are many species of cranes. Some nest in Estonia, others pass through and build their nests in the marshes and swamps of the Far North. The Latin name for the latter is *Grus grus*, in Finnish *kurki*, in Sámi *guorga*, in Mordvinian *kargo*. In the now extinct language of the Finno-Ugric Kamasins, who lived in the Sayan Mountains, it was *kuro*. It is difficult for an Estonian to imagine the immense diversity of birds in the tundra. I can remember that at one collective farm in north Sakha (formerly Yakutia), the pay for a day's labour included a kilogram of goose or swan meat. That was before environmental protection laws. Moulting and flightless birds were netted on tundra lakes, clubbed, cleaned, and preserved in permafrost as in a refrigerator. Every spring, an Inuit boy named Kanilu went to Little Diomede Island (Imaqłiq in Inupiaq) in the Bering Strait to collect eggs. As we rode horseback through the late autumn tundra in North Kamchatka, our guide Äiteki asked just a single question over three days: 'Do you know where the geese fly to?' I explained how far California was from where we were. Äiteki thought and sighed. 'Oh. Then there are many hunters on the way who shoot our geese.'

In the year 875, Ohthere characterised the peoples along the White Sea as fishers, bird trappers, and hunters. The final detail is especially accurate and telling, as trapping birds and gathering eggs were more important than hunting. It is only comparable to fishing in Northern lifestyles: both provide reliable, guaranteed sources of food in certain seasons. There is an old Estonian saying: a hunter goes into the woods, flip over the kettle; a fisherman goes fishing, put the kettle on the fire.

The dense flocks of migratory birds were the bread of the North.

But the 'pygmy' people? Pliny, who has been berated for his naivety for two thousand years, can add that when spring arrives, the pygmies ride goats and rams to the sea to eat crane eggs. Of course that would drive the critics mad! How could they have known that Pliny's description matched, with ethnographic precision, the annual cycle practised by the peoples of the Arctic Sea? Pliny used the word 'ram' because a goat wasn't enough: one characteristic of rams is their mighty horns. How could he have been any more exact when not even the Estonian language has a unique word for reindeer, simply calling them 'north moose' (*põhjapõder*)? Reindeer herders made the spring migration to the shore of the Arctic Sea

219

not only in search of eggs: the tundra's most bloodthirsty predator, the mosquito, made the journey unavoidable. Winds blow along the shore and the clouds of mosquitoes are relatively harmless there. Deep within the tundra, humans, not to mention reindeer herds, are defenceless. Swarms of mosquitoes and gadflies force the animals to flee in terror, they can no longer be kept together, and the herds dissolve.

The only aim of this diversion into livestock practices is to tie the cranes of Toonela to the tribes along the Arctic Sea. We aren't interested in how the authors of the *Iliad* might have acquired their information, nor in that information's serpentine path to Pliny, who added a feature that betrayed reindeer.

Let us filter out a sole connection: the cranes of Toonela, birds of the land of the dead, fly to the Arctic Sea. Why? To at least partly grasp the perspective of an ancient seafarer, their conception of the physical and the spiritual worlds as well as humans' place within them, we must add a few words about an ideological earthquake. It took place around the year 500 BCE and manifested in funeral pyres, which, according to Henry of Livonia, were still practised in full force during the locals' ancient struggle against the crusaders. The change in religion has been connected to the spread of slash-and-burn farming practices, to which we will now add the system of notions and belief inspired by the Kaali event. Cremation burials, which were notably sparser in areas of Latvian and Lithuanian settlement, do appear to be condensed around our Thule/Tulemaa. The basis for the practice was the notion of the spirit setting off on a long journey to Toonela. In order for it to depart, the spirit had to be totally freed of the body.

Souls of the dead were thought to arrive in ships or as birds. In autumn, the fatty, well-fed birds flew south and returned hungry in the spring. Thus, the belief took root that migratory birds are spirit birds, communicators between the living and the dead, transporters of the souls of Toonela. Just as summer turned to winter and vice versa, so people believed that life harmoniously traded places with death, which is nothing but another form of life. This cycle had no beginning or end, no conflicts or contrasts, no creation or destruction, but was merely the eternal rotation of life's different manifestations. The Estonian understanding of the Milky Way — *Linnutee*, 'the bird path' — was a poetic outcome. It was the path walked by souls, the Sámi *birds' guide*, the Mari *forest geese path*, the Udmurt *geese path*, the Komi *bird path*, the Mansi *forest duck path*.

RED SAILS

Now, let us remember that the sky was spread out over the four directions above the land. Within those limits, everything is real. Where was Toonela situated? The answer is surprisingly modern: it turns out that where the land and the sky meet in both the north and the south, there is a low, little hole; a narrow strait; a crevice; almost a cave through which migratory birds enter the world's reflection. Everything there is different and opposite: a true Stone Age Alice in Wonderland, though it no longer concerns navigation. May the world of the living be the limit of our interests here.

But that boundary must be guarded, right? What if one of the living strays into the 'anti-world'? And so it is guarded. Naturally, the edge of the world where land and sky meet is very low. A tall person couldn't stand up straight. Therefore, the Ingrians believed that dwarves guarded the edge of the world. Finns believed that 'bird-house men' lived there – *Linducodon mies* (1637). Mari, in turn, believed there were dog-nosed people, *pi-nereške* (Toivonen 1924: 95). Estonian myths about the place are extremely detailed. As the edge is a boundary, there are guards on either side. Here, humans keep watch, and on the other side of the great mountain are dog-headed people (Järva). Cynocephali guard the end of the world to keep anyone from crossing (Läänemaa). Occasionally, the mountain takes the form of a high wall. The dog-headed people constantly battle the guards, trying to escape through the narrow gate and devour every human on Earth. They can swiftly hunt a man down on their single leg and use their acute sense of smell to find anyone hiding.

So, reality and fantasy melted together in the ancient seafarers' minds, and geography and mythology fused into a single worldview that allowed them to confidently set sail and leave their home shores behind. True: the sea gives, the sea takes, the sea buries men. But another belief was this: a man is a ship who carries home, a woman is a city who stays standing. Dangers and perils were placed at the outer limits of the known world, and those limits were far away. Many legends tell of men who build a strong ship and use it to sail in search of the 'end of the world'. However, they find themselves in a sea that clots like blood and ultimately becomes as thick as porridge. Sometimes, the sailors would reach a great wall and lower a scout down the other side. Alas, he would not bring reports of the 'other world':

> The sailors tied one man to a rope and lowered him. When they
> pulled it back up an hour later, the man was gone: a horse skull hung

SILVERWHITE

in his place. Another man was tied to the rope and lowered. When they pulled it back up after a while, they found only half of the man dangling.

The sailors wondered what to do. They tried one more time, putting a man in a bottle and lowering him. When they pulled it back up a while later, they found the man intact inside.

Right away, the sailors asked what it was like in the underworld. The man tried to speak but was suddenly tongueless. He tried to act out the underworld things but was suddenly blind.

So, the sailors were forced to turn around without finding the end of the world. (Pootsi)

SPARKS ARE TULEMAA'S CHILDREN

We have lost the code to nautical words and many symbols will probably remain a mystery forever. A 'wave horse' and a 'bronze stallion' denoted a ship. Rowers sat upon cargo chests and the Estonian word for chest – *kirst* – was long used as a symbol for a ship in our oldest rites. We also know that *laev*, 'ship', was a taboo word along with *uss* ('snake'), *huss*, and *uisk*, all tied to old beliefs. But what was the relationship between ancient ship worship and the village swing? Or the swing and the fire-snake?

> Long ago, an Estonian prince or giant fell in love with a princess who lived on a faraway island [variations: Gotland, the North, a deserted island]. Finally, the young man ventured out to ask for her hand. But he had a jealous adversary who was a witch. That man sent wind and storms to batter the ship, finally casting it upon a deserted island where enemies took the prince prisoner. However, the princess of the Island of Sparks also knew witchcraft and thus found out her suitor was imprisoned. She immediately began building a ship that was enchanted to withstand any storm. The princess sailed that ship to the prison island and freed her suitor, then carried him across the green and placid Turja Sea to the shore of Viru, where they set the ship alight and burned it down to ash so they could remain there forever. As it happened to be midsummer eve, bonfires have been burned at midsummer in our land ever since and swings are set swinging just like a ship must rock on the waves of the sea ... (Kohl 1841: II, 269–71)

In that legend recorded before 1840, the only credible detail is the custom of burning ships during the summer solstice. The tradition is spread

RED SAILS

throughout Estonia, Finland, and Livonia. But midsummer is the sun's sacred time! So, one more hazy memory of the connection between ship and sun persisted even in the 19th century: one of the sea beliefs from Tacitus's day, the sources of which vanish into the twilight of time.

A swing is the link between a ship and a midsummer bonfire. In Estonia, there are eight known variations of expressing a child rocking in a cradle. One was *lemmis lendab*, and the second word in the phrase – 'is flying' – was visually accurate: cradles were once hung from a long, flexible pole and did indeed fly up and down. But what does *lemmis* mean? Other Finnic languages come to the rescue. The Votian word *lemmüz* meant 'fireball, moonstone, meteorite'. It comes from the same root as the Finnish word *lempi*, which in modern parlance means 'love' or 'passion', though it originally had a much simpler meaning: 'spark' (Paul Ariste to the author, 1.7.1974). A ship tied the swing and the cradle to 'flying', and that in turn to a fireball, a fireball-like volcano shooting sparks, and that was not a symbol of fear but promised fortune and prosperity. We should think back to Tacitus again: the Aesti were protected by 'boar amulets', which Adam of Bremen called 'dragons' centuries later. The fire-breathing dragon hid and protected those who wore its depiction, sacrifices were made to it, the children of Tulemaa were called 'sparks', decoded and de-poeticised as shards of meteor.

We heard the splash of Pytheas's oar.

Is it too little? Is it too much?

And yet, each of these elements – ship-burning, the flare of witches and auroras, wondrous white nights, journeys to the walrus-tusk shore, a wall with dog-headed men behind it – every one of these shadows of memory survive in Arab culture in one way or another: in travelogues, literature, and even folklore. Can this be possible? And is there any tie between them? It is probably time to turn towards Mecca.

What does 'the Arab world' mean?

VI

THE SILVERWHITE WAY

THE SULTAN OF HANDSOME FEATURES

Here and henceforth, when referring to the Arab world we mean the Early Middle Ages cultural circle that in various eras encompassed different peoples in Central Asia, the Middle East, North Africa, and the Iberian Peninsula. The bearers of unity were Arabic and Islam, or at least literature written in Arabic script. During western Europe's dark centuries, the Arab world was a haven for classical culture. Warmongering Christianity destroyed more than just the Huns, Vandals, and Goths. Just as it destroyed Estonians' ancient culture, so it pillaged the cultural treasuries of Greece and Rome. The comparison comes from the man who recorded the legend of the Island of Sparks. Thanks to the Arabs, cultural continuity was not interrupted in spite of everything. The fertile seed of the Renaissance was stored in their barns.

According to the words of the Prophet, the house of Islam rests upon five pillars, the fifth being a holy pilgrimage to Mecca. Here, also, the glorification and poeticisation of travel that, hand in hand with extensive trade, resulted in beautiful and poetic travel writing, which still proceeded from practical considerations. It was the birth of a new genre. Alongside scientific geography, nonfiction began to take root and bear golden fruits. 'Seek knowledge, even as far as China' are words traditionally attributed to the Prophet Muhammad, and word became deed. Over an astonishingly short period, literacy became universal among Arabs for religious reasons. More than that: writing was high fashion in the Arab world, akin to 18th-century Germany, along with a deepening interest in genealogy, history, geography, and ethnography. In those areas, unlike in the case of philosophy and theology, the Koran did not limit the freedom of thought or research. Almost every merchant kept a diary that he later

SILVERWHITE

made available to writers of encyclopedic overviews. The translation of Ptolemy's *Geography* into Arabic set a firm foundation for the scientific method in Arab geography.

Just as with all moving peoples, just as with the Baltic Finns, the first page of Arab geography began with folklore. The following description, which contains allusions to the Ob-Ugrics that are difficult to decode, was written in the 10th century:

> The image of the world consists of five parts: the head, two wings, breast, and tail of a bird. The world's head is China. Behind China is a people called Wakwak. Behind this Wakwak are people whom no one except God can count. The right wing is Hind [India], and behind Hind is the sea; behind this sea there are no creatures at all. The left wing represents Khazar, and behind Khazar are two nations each of which is called Manshak and Mashak [Ob-Ugrians or Madjars?]. Behind Manshak and Mashak are Gog and Magog, both of which are nations whom only God knows. The breast of the world represents Mecca, Hijaz, Syria, Iraq, and Egypt. The tail begins west of Alexandria. The tail is the worst part of the bird. (Ibn al-Faqih)

The genesis stories of Finno-Ugric peoples also include a water bird that created the world by raising mud from the bottom of the primordial sea with its beak. Even so, the aim of that passage is not to point to the intersection of both cultural circles, much less to emphasise the accuracy of the Arabs' mythical geography. First, we require background; a more intimate acquaintance with that distant world so we may make judgements concerning the directions of interest, methods, and credibility of individual travellers. Around the year 891, al-Ya'qubi, the son of a postman, wrote the *Book of Countries*. In its foreword, he describes the explorer's method as follows:

> When I was in the prime of youth, possessed of an adventurous spirit and a sharp mind, I took an interest in reports about countries and about the distance from one country to another; for I had travelled since childhood, and my travels had continued uninterruptedly and had taken me to distant places. So whenever I met someone from those countries, I asked him about his homeland and its major city; if he told me about the place of his home and where he resided, I questioned him about that country concerning ... his birth ... what its crops and who its inhabitants were, whether Arabs or non-Arabs ... what its people drank. I even asked about their clothing ... their

THE SILVERWHITE WAY

religions and beliefs, and who held power [and leadership] there … how distant that country was and what countries were near it and … for riding camels. Then I verified everything he told me with someone I could trust […] I attached each report to its proper country, and everything I heard from trustworthy inhabitants of the major cities to what I already knew. I realized that no creature could encompass the entirety of it and that no human being could reach the end of it. […] Thus we have composed this book as a summary of information about the countries; therefore, if someone finds any information about a country we have mentioned not included in our book, this is because we have not intended to include everything. The philosopher once said, 'My quest for knowledge is not a desire to cover every detail, however remote, nor to command every last point, but rather to know what it would be wrong to ignore and what no intelligent person would contest.' (al-Ya'qubi 2018)

There is something moving about this. It is no longer the cold marble of Rome, but the nervous and skittish mind of our era, propelled by the awareness of knowledge's depthlessness. Of course, al-Ya'qubi was a child of his time. To him, the world's navel was Baghdad and its closest Islamic environs, and his descriptions of these are disproportionally lengthy. No doubt we are the children of our time as well, oftentimes even more childish. We tend to criticise medieval clergymen for writing chorales instead of 'La Marseillaise' and not delving into historical dialectic – something we believe we ourselves understand so fully. Al-Ya'qubi stands beside us; perhaps even ahead. He maintained a strong curiosity and was enticed by the distant North and its dim secrets:

Let us now give an account of the other countries and the distances between one country and another and one city and another, in four parts, according to the four regions of the world: east; west; south, the direction of the qibla, which is where Canopus, which the astronomers call al-Tayman, rises; and north, which is the abode of the Bear, which the astronomers call Polaris. We shall describe each country according to the quarter in which it is located and what is adjacent to it. May God grant success.

Arabs were, among many other things, outstanding merchants – a fact that may place them in an equivocal light from our perspective. Let us refrain from doing so. Christopher Columbus was also a merchant first and foremost, not to mention Lapérouse, James Cook, and even Adam

SILVERWHITE

Johann von Krusenstern, who transported one thousand pails of Viru Valge vodka from Tallinn to Kamchatka at the price of 48 roubles per pail, with a twelvefold profit. From the standpoint of cultural history, it is as insignificant as the buttons on Einstein's pants, which he absentmindedly left unfastened on at least one occasion. Al-Masudi continues with descriptions of the North – we will determine the geographical details later. At the moment, let us focus on his sketch of the national character in the *Book of Admonition and Revision* (956 CE). Why should a merchant root through the labyrinths of distant peoples' natures? The ethnographic scope is most definitely broad:

> As for the inhabitants of the northern quarter, from whose zenith the sun is distant, specifically the Saqaliba [a general term for inhabitants of the taiga in al-Masudi's day] and the Franks [the Carolingian dynasty, which extended from the southern shore of the Baltic to the mouth of the Oder River] and their neighbouring peoples who all settled in the northern reaches, the further they move away from the sun, the weaker the sun's strength diminishes there. Cold and dampness dominate their lands, snow and ice perpetually interchange, and hot tempers are rare. Their bodies grow tall, their characters dry, their manners stiff, their intelligence crude, their language difficult. Their complexions are pale to the extreme, blending from white to bluish. Their skin becomes frail and their bones burly; even their eyes turn light blue, matching their coloration. Their hair is straight and golden from the excess of dampness. The qualities of the cold and lack of warmth have caused a capriciousness in their beliefs. Those of them who have penetrated further north submit to foolishness, crudeness, and brutishness. The further north, the stronger these qualities grow.

Where do facts end and poetry begins?

Strangely, that question is always directed at the past and only rarely towards the present. Fairy tales about the Tunguska event, a laser beam created and dispatched from the Cygnus constellation, black holes, and God knows what else, all belong to our era. We must listen! However, our attitude – even the most sceptical – is humble and submissive towards scientific and technological poetry. The science world's secret jargon, betrayed to neither accountant nor finance minister, has long since raised a wall around laboratories, in the shadow of which fun and lively alchemic experimentation take place. These words are unfair, of course; even foolish. They merely contain bitter despair over the unfathomable

THE SILVERWHITE WAY

Weltanschauung of modern science. I can no longer even say how many elementary particles there are anymore – the building blocks of nature that we wish to call our own, just like all things that slip from our fingertips. Only the past still appears simple, clearly arranged, and whole. And so we dump our scepticism onto the past because it is mute and still, powerless to defend itself. It is so much more ignorant than we, the wise, so it may be tolerated, even loved, and we can climb onto its shoulders to make ourselves great once again.

We are still confusing technology and culture.

Poetry is also scientific fact, as it is fed by experience. This confession is unfair in al-Masudi's case, however: his century was that of the Arab caliphate's decline, but also of the greatest flourishing of Arab culture, which some researchers have somewhat mysteriously called the 'Islamic Renaissance'. An Islamic atlas was published in addition to fiction, regional geography, and philosophy, though individual accomplishments were not what mattered most: it was a time of synthetic personalities. Fact, style, and idea blended together, and narrow subjects merged into a single culture without losing their scientific depth.

Al-Masudi has, somewhat tritely, been called the predecessor of modern reporters, and parallels have been drawn between his style and 'contemporary newspaper jargon'. True, the latter accusation was made in the late 19th century. After finishing his studies in Baghdad, the direct descendant of Muhammad acquired the bulk of his knowledge from long journeys encompassing lands between the Atlantic and Indian oceans, the Caspian Sea, most likely also China, and even the Malay Archipelago. Out of al-Masudi's numerous works, the only one to have survived and gained widespread renown is *The Meadows of Gold*. Such was the title given to it by romantic English translators in the 19th century: the original sounded more like a dusty technical manual: *The Washing of Gold and Mining of Gems along with Gifts to Noble Rulers and Men of Letters*. Al-Masudi had an appreciation for style, elegance, and literary form. It is hard to surprise yours truly with the extent of geographical coverage. We do wish to demonstrate that today's little big world was a big little world in olden days. If al-Masudi surprises us with anything, then it is the objectivity of a scientist in questions of nationality and religion. He dedicates remarkable space to the history of Christianity, from which the Occident was forced to look away in shame for many centuries. Al-Masudi concludes the book's foreword with poetic verses that have the feel of being written more by a traveller than a scientist, more an author than a geographer:

229

SILVERWHITE

> I am a follower of Hijrah: if anyone is in their homeland
> in any country,
> then the camels' humps are my homeland.
> My tribe lives in Syria, my beloved in Baghdad,
> but I am in Raqqa, though my brothers are in Fustat.

And, yet, how is this possible? How did all these experiences pile up? Ones that later nourished the Occident for several centuries, giving Vasco da Gama the Arab carrack, prompting Columbus to sail to India? How did all these experiences collect in Arab hands? How did they travel?

Travelling (we may confess in a travelogue) is a dangerous illness, and is furthermore infectious. The following lines were written about twenty-four years after al-Masudi's death in answer to that question. They come from al-Maqdisi, the grandson of an architect of Jerusalem, praised by a plethora of researchers whose worth is reduced in turn. However, the description itself, despite its vain rhythm and self-assertive pathos, provides a realistic depiction of travel comforts around the year 980 (Krachkovsky 1957: 215):

> I have owned slaves, but have had to carry baskets on my own head. A number of times I was close to drowning; our caravans were waylaid on the highway. I have served the judges and the great ones, have spoken with rulers and ministers. I have accompanied the licentious on the way, sold goods in the markets, been confined in jail, been accused of being a spy. I have witnessed war with the Romans in battleships, and the ringing of church bells at night. I have been a bookbinder to earn money, have bought water at a high price. I have ridden in sedans and on horseback, have walked in the sandstorms and snows. I have been in the courtyards of the kings, standing among the nobles; I have lived among the ignorant in the workshops of weavers. [...] I have worn the robes of honour of the kings, and they have ordered gifts to be given to me. Many times I have been naked and destitute. [...] heresy had been imputed to me; I have been accused of greed; [...] Experiences of this kind are many, but I have mentioned such a number of them that the reader of our book may know that we did not compile it haphazardly, nor arrange it in any random fashion. The reader will thus be able to distinguish between it and the others; for, after all, what a difference there is between one who has undergone all these experiences, and one who has compiled his book in perfect ease, basing it on the reports of others.[1]

THE SILVERWHITE WAY

Our goal is not characters, but the world; not the micro-structure of natures, but the macro-structure of the era – but who can stay true to their goal till the very end! I would undoubtedly prefer the son of a postman to the grandson of an architect as a travel companion, and yet the latter was more valuable as a scientist. He was pedantic, consistent, and critical, at least of his predecessors, who did not number few. Al-Maqdisi walks at the end of a long procession – he is the anchor of the classical 10th-century Arab school.

Let not the term 'anchor' disorient the reader. Many more followed al-Maqdisi; travellers and men of letters even greater than him. Here, the 'end of the procession' only denotes the end of one period of literary history; something like Voltaire in France. Having exhausted all earlier possibilities of form and substance, he fecundated a new way of observing the world without even realising it. Practical experience was the basis for the Arab worldview. There is no longer any doubt. Yet in what way does their experience differ from that of a man from Ardu, a potter from Kõmsi, or a Baltic Finn sailing in the shadow of a protective boar's head?

The answer is given by Ibn Yunus, the senior researcher at an observatory near Cairo. The year was around 980 CE (Krachkovsky 1957: 106–7):

> The study of celestial bodies is not unknown to religion. Their study alone allows one to determine the times of prayer: the moment of dawn, when a man beginning his fast must refrain from food and drink; the end of evening twilight; the length of oaths and religious vows; the periods of eclipse, which must be known in advance to prepare the traditional prayer for the occasion. Their study is essential for a man to face the Kaaba during prayer; for determining the beginning of the month; for recognizing various unusual days and the time to sow, the growing of trees, the harvesting of wheat, the position of one place from another, and finding one's direction of travel without becoming lost.

That is precisely the world's enchantment: not one flower is like another; not even a single birch has two completely identical leaves. The Arab world can only be compared to the Arab world. From neither before nor since then can we name such an explosive development that so rapidly transformed nomadic tribes of shepherds wandering the semi-deserts into the inheritors of classical culture, receiving it and evolving it further. It is to them that we owe algebra, chemistry, stereometry, the most accurate achievements in pre-optical astronomy, and, to return to our main topic, even the word 'admiral'.

SILVERWHITE

It is easy not to know that distant world, and even easier to over-idealise when getting to know it. They were humans above all. Wise, ignorant, generous, jealous, precise, avid readers of poetry. True, reading their texts, one sometimes suspects that a fear of distances was foreign to them. Perhaps it can be explained by their outstanding knowledge of astronomy. The movement of constellations had to hold an equally important place in the folklore of a nomadic people endlessly on the move as it did in that of seafaring peoples, and by blending it with knowledge of mathematics, the result was a more-or-less accurate idea of a wanderer's coordinates. Thus, we could understand the fatalism that guided the wanderer's steps, as well as the calm, unprejudiced gaze with which they observed changing landscapes. 'All roads lead to Rome' is something that was said in Roman times, but glinting through that saying is Rome, not a Roman's greatness: a road had to come before anything else. The stars sufficed for Arabs. And when those were no longer enough, they started composing astronomical tables. People visited the great libraries to copy them. One such library was in Merv – where today's Samarkand–Ashgabat railroad crosses the Marghab River – and that is where the wanderer hastened to save, at the very last minute, all that could be saved; to read and to copy into his future encyclopedia the texts written by earlier geographers before the enemy approaching the library could burn it to the ground. Muses and cannons, blockades and Symphony no. 7 are a topic equal to the age of human culture. The continuity of Arab culture was great, and information circulated quickly. Much was saved as a result, as, in addition to much else, the board-level steppes and expansive deserts of the Middle East and Central Asia have played the part of artillery range over the last few centuries.

The Arab geographical tradition thus passed from hand to hand even more so than for classical authors, branching out between multiple simultaneous scribes, collecting in comprehensive works, transforming or gaining detail depending on the characters and possibilities, and *still* it has outstanding surprises in store. One such surprise came to light in a dusty provincial town in Iran in the early 1920s; another, even more important, in an archive in Madrid half a century later: an aristocratic diary, thought long since lost and to survive only in lone fragments that were almost impossible to piece together, detailing a fifteen-year journey. The Word awakens, multiplies, dissolves, becomes flesh and blood, and from beyond the abyss of the ages, we suddenly find ourselves staring into a pair of attentive, raven-black eyes.

THE SILVERWHITE WAY

The classical world and the Renaissance.
A broken anchor chain, a man holding the ends together.

WHY THE THULE MERIDIAN DID NOT BECOME GREENWICH

One of the bearers of the Central Asian tradition of astronomical tables presents, in nonfiction form, an overview of the North. In terms of facts, accuracy, expertise, and even inexpressive generalisations, it is third-rate writing. The author was an astronomer and a geodesist; he was not a geographer. And that is his merit: the excerpt quite objectively details the average level of Arab geography in the years 852–922, giving us a sense of its connection to classical geography, degree of accuracy, and unprejudiced application of logical derivation (Krachkovsky 1957: 101–4):

> The world is round [...] and air surrounds it on all sides. [...]
>
> Starting from the islands in the Western [Atlantic] Ocean that are called the 'eternal' islands and number six in total, all six inhabited [the Canary Islands; Ptolemy's and al-Battani's starting meridian was the Canary Islands' longitude; see Ptolemy], the earth's inhabited area extends to the furthest reaches of China; it is estimated at twelve hours [we would say: 180 degrees]. They determined that when the sun sets in China's furthest reaches, it begins to rise above the first inhabited islands that, as mentioned, are located in the Western Ocean; and when it sets there, then sunrise begins in China's furthest reaches, and that is half of the earth's circumference, which is to say the length of the zone of human settlement, with which we are now acquainted. It is 13,500 miles, if we calculate in the miles used for measuring land [the author contrasts with nautical miles].
>
> Next, width [south to north] was measured and it was found that the habitable area extends from the equator northward to the island of Thule [original: Sūlī] [...] where the longest day lasts up to 20 hours. [...] The width from the equator to the island of Thule is about 60 degrees. That is the sixth part of the earth's circumference.
>
> As for the Western Ocean which is called the Encircling, nothing is known of it but the region west and north of the farthest point of the land of al-Habash [Abyssia; i.e. Africa] towards Bartdniyah [Britain]. It is a sea in which ships do not sail. The six islands in it opposite al-Habash are known as the Fortunate Islands. [...] In (the Western Ocean) also to the northward are the islands of Bartdniyah, twelve in number. Then it extends far away from the inhabited land, and no

SILVERWHITE

one knows its character, nor what is in it. [...] As for the part that may be settled or uninhabited, but we do not know, it is an entire eleven-twelfths. [...] Deductions and conclusions enrich a man, however, and this is not denied by one who has partaken in the sciences and walked the path of analogies. And so, the sun, moon, and stars move above us, and dependent upon their movement, proximity, and distance are summer and winter, plants, animals, inhabitation, and everything familiar to everyone. If the sun and the moon rise above each individual place on the rest of the earth the same way as they do for us, then it is possible that there are also plants and animals, seas and mountains, like ours. And so it must be.

And that wasn't even the Middle Ages yet!

Where is the dogma? The slavish prostration to authorities? Ptolemy, that demigod of geographers? No, no – read what he has to say about Ptolemy, balanced and polite: 'In this book, there are errors in longitudes and latitudes. We will speak of them where necessary.' That suffices.

Classical cultural heritage – the part of it that interests us – was in good hands. No, even that is a half-truth. It had been planted in fertile soil and was bearing unprecedented fruit. But from the Baltic-Finnic perspective, what matters most still lies ahead.

From Pytheas to Tacitus, all geographers entered the Baltic Sea through our back door – the Jutland portage or across the Danish channels.
Arab geographers' information passed through the Baltic Sea's main door, through which we ourselves once arrived at its shores: along the great Baltic–Volga waterway. Front door and back?

Geographically, Europe is a tiny peninsula; a little burr on the tail of a giant continent. And that's exactly how it was seen, given the right perspective. Estonia itself was regarded as a peninsula as late as the 19th century. So, if we look truth in the eye and see past what we're used to, the back door.

True, information's path through the main door wasn't shorter. And yet ...

Even from the passage we just read, it's clear that information became more *precise*. It started moving faster. The ground before the main door was smoother. Al-Battani – and he it is whose words we quoted, a man who only gained renown in the Occident five centuries later under the name Albategnius; not quite Einstein, but certainly Nils Bohr from the year 2440! – as I was saying, al-Battani does not deny his fondness for Ptolemy, much less his critical approach. Indeed, he corrects Ptolemy's

THE SILVERWHITE WAY

coordinates. On what grounds? We don't know. Like Roman geographers, he places Thule in Britain. Do you recall Tacitus's compliment to his dear father-in-law? 'Thule was also seen,' he reports, and that crackling bouquet is extended from century to century with a courteous bend of the knee. Al-Battani had no reason to drop it. Nevertheless, he brings Thule's latitude 3–4 degrees to the south of Tacitus's and Ptolemy's, adding that even that is approximate.

Can we use this knowledge to deduce that another tradition was still valid in the Arab world? Information more accurate than Ptolemy's regarding Thule / Tulemaa's actual location?

Al-Battani (852–922) remains silent on the subject. However, Anna Comnena (1083–1150), the educated Byzantine countess and chronicler, was able to say that sailors who entered the emperor's service came to Constantinople from Thule. Tulemaa, which in western Europe had been diluted into a poetic symbol of great distance, retained the definition of a specific region in the Middle East.

Al-Battani was a mathematician and student of the famous al-Khwarizmi, whose Latinised name gave us the word 'algorithm'. Following their introductory texts, let us look closer.

HOW MONTESQUIEU'S PERSIANS AND THE SULTAN'S AMBASSADOR WOULD DESCRIBE WILD STRAWBERRIES

Symbolism is the building block of poetry and the landmine of exact sciences. Generally, we are only able to describe what is already familiar to the author and the reader from prior experiences, and what is consequently unworthy of description. A mere mention, allusion, or symbol suffices. It is much harder to describe a genuinely new phenomenon that neither the author nor the reader has encountered before. The causal associations could break. A description of gunfire could sound like this: 'Every time he closed one eye, a shot rang out.' There is nothing false about the statement apart from a deformed connection of which the describer is unaware. The next author who, let us assume, obtained his information from the first, would subsequently arrive at a certain generalisation: 'Far beyond the mountains lives a wondrous people. When they close one eye, a shot rings out, but when they close both eyes when going to bed at night, there is no shot at all.'

Thus, the literary genre of *mirabilia* was born. Renaissance Europe relished it as the unrestrained fireworks of Eastern fantasy, and only with

SILVERWHITE

our century's methods of investigative thought and textual criticism did we begin to discern true events, landscapes and fates within it. 'One day when we were with the king we made camp,' tells an Arab who journeyed to the Volga, and to whom the following chapters are dedicated.

> I went in among the trees with my companions, Tikin, Sawsan and Bars, and one of the king's followers, who showed us the stem of a plant. It was small and green, like a spindle in thickness, but longer and at the base a large leaf spread out on the ground which was carpeted with new shoots which bore a berry. If you tasted them you would think they were seedless pomegranates. We tasted them and found them incredibly delicious, and spent our time hunting for them and eating them. (Ibn Fadlan 2012)

The sultan's ambassador ate wild strawberries. Now let us imagine him attempting to describe the traditional brooches and pendants of Finnic jewellery.

The same year that Alexander Theodor von Middendorff embarked on a round-trip thousands of kilometres long to borrow a geological drill from Barnaul in Russia, two learned men debated a problem in a cold, polite, and vicious way: whether a traditional Estonian women's pin was the size of a large saucer, a small cake plate, or even a soup plate. Now, we know that a conical south Estonian brooch can be up to 26 centimetres in diameter, 11.5 centimetres thick, and weigh up to 560 grams in silver. That isn't exactly insignificant, we must admit, though we do not ask: why? That wasn't all. Multiple necklaces with silver coins connected one after another in rows were also worn. 'Especially in Põltsamaa and Viljandi Parish, they go too far with it,' claimed August Wilhelm Hupel in his soft and balanced manner: some young women were said to wear more than forty silver roubles around their necks and on their chests! Still, Estonians have no difficulty describing the phenomenon because we know what matters most: they were jewellery.

But if we were unaware? Our traveller could have written, 'On their chest, they wear a silver box that is round,' and it would have been just as approximate as our description of a gunshot. But how was it to be understood by the Persian encyclopedia author al-Hamadani two centuries later? 'Every woman has a box attached to her chest [...] and these boxes are round ...' All it took was using plural instead of singular, and the brooch became the world's first brassiere.

236

THE SILVERWHITE WAY

Let us take along our sceptical mind and this conclusion: every description contains more of the author than the subject. Everything can be read, everything must be decoded.

AN EMPIRE LOST IN THE WOODS

On 21 June 921, Caliph al-Muqtadir left Baghdad with his mission, whose secretary was a young man named Ibn Fadlān. Two years later in the spring, the mission was back in Baghdad. Their destination meanwhile was Volga Bulgaria, loosely located at the confluence of the Volga and Kama rivers. The objective was to carry out the request of Volga Bulgarian emperor Almış: to attain victory over the Jewish Khazar state, which was located in the Volga's estuary on the Caspian steppes, he desired Islamic help in modernising (a historian would say, centralising) his state and building a fortress to defend its borderlands. Almış could have accomplished the latter on his own, but 'silver magic' had great power according to beliefs of the time. The fortress was to be built in exchange for 'blessed money', meaning with the support of silver Arab coins. Only then could the fortress be impregnable.

The caliph was incredibly flattered by this request and set aside 4,000 dinars for the faraway Northern ruler – a massive sum. The mission was to acquire the funds on the way from selling a state manor in present-day Karakalpakstan (an autonomous republic in Uzbekistan). Alas, a Christian sabotaged the caliph's orders and the mission continued on its way, hoping the money would be sent later. To this day, those 4,000 dinars haven't reached Kazan.

Their vehicles: After wintering in Khorazm, now the area of Urgench, the mission travelled seventy days to arrive in Volga Bulgaria. They arrived on 12 May 922. They went by caravan. On the Baltic Sea, there were shore rights and harbour immunity. In Central Asia, caravans were off-limits – at least in times of peace. However, Ibn Fadlan's task was of a political nature, consequently his person was not untouchable. In moments of danger, the mission hid 'at the centre of the caravan'. How are we to interpret this?

The caravan was composed of 3,000 horses, 5,000 men, and probably an even greater number of camels: opportunity enough for hiding an entire artillery division. This size of caravan wasn't extraordinary: about a dozen years before the events described, Muhammad ibn Sulayman took 24,000 camels as a prize of war after retaking Egypt.

SILVERWHITE

Additional details: Volga Bulgarian emperor Almış had a personal tailor from a Baghdad fashion house. Over a thousand people resided in the palace. The rooms were decorated with Armenian rugs, the throne adorned with a Byzantine brocade. In that post on the Baltic–Volga waterway, Fadlan encountered an array of merchants, some hailing from as far as Sindh in southern Pakistan. Volga Bulgaria was a site for trade, but also a stop on merchants' journeys to the fur forests and tusk shores of the North.

Who were the Bulgars? Volga Bulgaria was a cosmopolitan state, the eponymous tribe of which spoke a Turkic language. A plethora of Bulgar loanwords can be found in the Komi-Zyrian, Udmurt, Mari, and Mordvinian languages, as well as to a lesser extent in Vepsian, Votian, Finnish, and Estonian: north Estonian õun ('apple'), south Estonian ubina, Finnish olma – cf. Chuvash olma; Estonian mahl ('juice'), Finnish sima – cf. Chuvash sim; Estonian humal ('hops') – cf. Chuvash xumla; Estonian küla ('village') – cf. Volga-Turkic kül, kil, meaning 'inn' (Collinder). The last loanword is perhaps the most telling from a travelogue's perspective.

Let's avoid discussing the history of how certain nations developed – ethnogenesis, in scientific terms. The labyrinth has an enticing gate through which several generations of linguists – not to mention travelogue authors – have entered and disappeared without a trace.

We can settle for less. Fadlan's accounts are ethnographically and ethnonymically accurate, as the Bulgar ruler sent with him a translator who spoke most of the local languages. The Bashkirs are Fadlan's *Bashkorts*. His accuracy declines the further the young man's travels and inquiries penetrate into northern and non-Turkic environments. Some peoples of the Volga refused to submit to Islam. One rebellious tribe was led by a man named Varag (Kovalevsky 1956: 116, 139). The name isn't necessarily Chuvash, even though the word *varas* meant 'captain' in their language. Modern-day Chuvash are linked to the tribal name *Suvaz*, though they do not overlap linguistically or ethnically because playing a role in the fate of the Volga Bulgarian union and state (450–1240s CE) that was equal to the Turkic tribes were also the Volga Finns: the Mari, Mordvinians, Zyrians, Komi-Permyaks, and perhaps even more distant Finno-Ugric tribes, but certainly also the Magyar/Hungarian tribes, other Madjars who were assimilated into the present Bashkirs by the Kipchak Turks, and –serving as intermediaries for cultural contacts – the Baltic Finns. In terms of language and location, the Chuvash are the descendants of that former empire. Fadlan consistently ties the peoples of the taiga to

238

THE SILVERWHITE WAY

the title *ṣaqāliba*. Russian, Hungarian, and Finnish researchers believe that the word was applied to Northern peoples as a whole in Arab literature during that period (cf. Haavio 1965: 66–7; Kovalesky 1956: 85). As a result of Slavic migration, *ṣaqāliba* narrowed to designate the Slavic tribes three hundred years later.

MUHU AURORAS AND ARAB JINNS

Ibn Fadlan arrived in Volga Bulgaria at a time of peak solar activity and witnessed the Northern Lights.

> Look! the spirits of Fiery Fighters
> In their clash beneath the skies
> Flashing with their silver spears,
> Shaking all their golden shields
> Set the ship aglow in red.
> Now the men were losing courage,
> The lads were churning in their pants.
> But the son of the Kalev Heroes
> At this fiery frolic smiled:
> 'Let the Fiery Fighters' fencing,
> Flashing of their silver spears,
> Shaking of their golden shields
> Make for us an arc of fire,
> Where we in the glow of light
> Might our course e'en further see!
> Moon refused to come along,
> Sun has long gone into hiding;
> Uku as a gift of grace has
> Set these Fiery Fighters fencing.'
> (*Kalevipoeg* 2011: XVI, 882–900)

Naturally, it was Ibn Fadlan's first time seeing the Northern Lights, and the natural phenomenon had a profound impact on the young man. He described their flickering dance as follows (Kovalevsky 1956: 134–5):

> Thus, the first night that we spent in this land, before the light of the sun faded [a full hour before sunset], I saw the horizon turn a brilliant shade of red and in the upper air there was great noise and tumult. I raised my head and saw a red mist like fire close to me. The tumult and noise issued from it and in the cloud were the shapes of

SILVERWHITE

men and horses. These spectral men held lances and swords. I could see them clearly and distinguish them. Then suddenly another bank of mist appeared, just like the first, in which I saw men, horses and arms; it advanced to charge the first, as one cavalry detachment falls upon another. [...] They clashed for a moment and then parted, and so it continued for an hour after nightfall. Then they vanished.

We questioned the king on the subject. He claimed that his ancestors said: 'They are the believing and the unbelieving Jinn. They fight every evening and have not failed to do so every night since they were first created.'

Thus, the symbolist system does not come from Fadlan, but the Volga Bulgarian ruler. It is local tradition, Northern folklore heard from ancestors. Experts in Arabic studies have drawn attention to this repeatedly.

Two centuries ago, Matthias Johann Eisen recorded in Karksi a story titled 'Turning Northern Lights', in which the 'clash and rattle' of swords is a distant echo of Fadlan's account. Once upon a time, a man got lost in the woods during winter and came across a lone cottage where he ate food cooked by the Northern Lights. As soon as the man stuck the spoon into his mouth, he rose into flight (Eisen 1894a: 83–5):

He soared high in the sky at the speed of the wind, going far towards the north. On the way, he passed others flying northward as well, carrying burning torches. Sometimes, others shot past him in turn.

After a while, there were so many wind-riders that they started battling with thousands of scutchers, dashing this way and that and turning the sky rather bright. Only now did the man realise he'd been flying as an aurora. The torch-bearers brandishing their scutchers and rushing past one another finally got so intense that clashing and rattling filled the whole sky.

The next morning, the man returned to the Northern Lights' house: 'They insisted over and over again that he not tell a living soul about his trip to Lapland and then showed him the way out of the woods.'

* * *

We should add a few words about terminology.

The concept of the Aurora Borealis doesn't appear often in Arabic, nor can it. Ibn Fadlan was forced to employ a comparable notion, an adequate metaphor. Friedrich Reinhold Kreutzwald called magnetism

THE SILVERWHITE WAY

'the flashing force of gravity' and as late as the year 1880 conveyed the idea of atmosphere as *õhupesa* ('air-nest') in Estonian. Ibn Fadlan called the Northern Lights 'jinns'. A jinn is the most common being in Arab folklore, usually invisible and made of fire and air. Sometimes they are friendly towards humans, othertimes not.

Fadlan's term, as we will see, conveys quite accurately the Baltic Finns' notion of Northern Lights. In Estonian folklore, auroras wave, fence, or shimmer (Virumaa); they warm, blaze, or battle. In Järva-Jaani, auroras strike; in Rapla, they strike upwards; in Tõrva, they fight; in Harju, they battle. The following account comes from Muhu Island (Holzmayer 1872: 49): 'On holy nights, you see the sky close over [...], and on either edge of the split, you see two armoured soldiers. Because they want to battle each other, God split the sky to stop it from happening and keep them apart. Light spills to earth through that gap.'

Explanations of the nature of the Northern Lights were slightly different in Saarde, where they formed from the sparks made by swords striking each other. The fencers were 'the souls of fallen soldiers'. Using fencing, fighting, or battling to explain auroras was widespread throughout Estonia. The legend is thought to have originated from funeral pyres in a time when warriors were cremated in full gear and were believed to battle on as fireballs and meteors. We see how the Kaali event cast pale light into an ever more distant past, all the while feeding belief systems that grew increasingly complex, though their common element remained fire or light that spread across the sky. Belonging to this circle are folkloric Estonian notions of Northern peoples fencing, Northern peoples fighting, showing the Lapu the light, and a 'Lapu war being fought'. One account recorded in Koeru was quite similar to the Arab legend of fighting jinns: two armies face each other while their kings fight on fiery steeds. The soldiers watch the 'clashing and rattling' without interfering. Their leaders are the rulers of two different worlds. If one should defeat the other, the armies will rush into battle and the victorious side will remake the entire world (Loorits 1949: I, 285). Fadlan's clashing jinns already echo much more clearly in this account.

The phenomenon, so narrowly unique to the North, astonished the southerners and enlivened their imaginations to such a great degree that Northern Lights ultimately entered Arab folklore – this in a genuinely Arab form, of course, but while preserving the 'clashing and rattling' of swords. Auroras give an account of their home in Scheherazade's 492nd night: the North, similar to the Karksi man's Lapland: 'The White Land

SILVERWHITE

is our home, and every year Allah orders us to appear and battle the disloyal jinns.' Asked where the White Land lies, the following response is given: 'Beyond Mount Qaf, a seventy-five-year journey away.' This isn't geography but poetry. Still, a grain of rational truth lies within: Islam's Mount Qaf is a parallel form of the ancient Greeks' Riphean Mountains, which, as you may recall, ran from east to west along the taiga as a symbol of distance and impenetrability.

Ibn Fadlan's account of the Northern Lights and the earlier variations of Scheherazade's tales were recorded in the same era, as a result of which researchers rule out the sources having a mutual influence. Consequently, the Middle East must have become acquainted with auroras long before Fadlan, which provides certain temporal and spatial promise to the Persian mints coined between 591 and 628 that were discovered in Åland.

Eastern astronomy and geophysics are interwoven with the distant echoes of Northern folklore. It would be naive to hope for more. And, yet, are we able to check Fadlan's accuracy as an observer? His credibility? That possibility does exist.

THE DAWN–DUSK PHENOMENON FROM AN ARAB PERSPECTIVE

Earlier, we discussed the Arab world's fifth pillar. Now it is time to add an important aspect to that pillar: religious custom required every Muslim to quietly meditate at exact times during the day – moments of prayer to be said facing Mecca, no matter where one travels. Great responsibility fell upon Ibn Fadlan in these matters: on top of everything else, the mission was tasked with bringing Islam to Volga Bulgaria. This should not lead us to believe that Islam was not practised in the lands along the Volga before he arrived. In addition to the Baghdad tailor and numerous servants, the ruler was served by the local muezzin, an aide to the imam. However, Volga Bulgaria was an extremely rudimentary formation in the 920s and the performance of religious duties was just as basic. It might be compared to the lands surrounding Reval during the time of Bishop Hiltinus a century later: clergy lacked sufficient power to enforce the new religion. One way or other, Ibn Fadlan endeavoured to bring order. To prevent any false associations, let it be stated that pre-crusades Islam was immensely more tolerant in religious matters than bellicose Catholicism.

Thus, it was Ibn Fadlan's job to accurately set the times for prayer. This shouldn't make him seem to us like a dour missionary, however. Far from

THE SILVERWHITE WAY

it! His diaries demonstrate easy and effortless conversation, a lively and vivid style, and flashes of humour. Nevertheless, conversation occasionally had to be interrupted for prayer at the correct times, as we see in the following excerpt (Kovalevsky 1956: 135):

> We talked for the amount of time it would take to read less than half of a seventh part of the Qur'ān, while we waited for the call to evening prayer. Suddenly we heard it and went out of the tent. Day was breaking. I asked the muezzin:
> 'To which prayer have you called us?'
> 'The dawn prayer,' he said.
> 'And the evening prayer?'
> 'We say it with the sunset prayer.'
> 'And during the night?'
> 'The night is as you see. They have been even shorter than now, for already they are beginning to lengthen.' [...]
> I observed that in their lands the days are very long and remain so for a certain part of the year and the nights are short. Then the nights lengthen and the days shorten. On the second night of our stay, I sat outside the tent and watched the sky. I saw only a few stars. I think there were about fifteen scattered across the sky. The red glow that one sees before the evening prayer never fades, and the night was not dark – a man can recognise another from a bowshot's distance.

Fadlan's account deserves serious analysis. As always, we must take an unknown factor into account, i.e. the author's character, but also the possibility of subconscious or conscious disinformation. Let's begin with the latter.

The mission did not fulfil the expectations placed upon him. The 'blessed silver' for building the border fortress was never received, and although it was not the fault of the caliph or the ambassador, the Bulgar ruler's attitude towards the mission turned chilly, if not to say acrimonious, afterward. It didn't affect the cheerful, sociable Ibn Fadlan working away at his modest secretarial tasks – over time, he became a conduit between the Bulgar ruler and the mission's diplomats. Almış took him on walks and hunting trips, and they spent long evenings together in conversation – the ruler listening to stories of life in Baghdad and in turn sharing fascinating details about the distant, exotic North. This tickled the young official's spirits, of course, but the mission's overall failure also forced him to constantly emphasise that they had done everything that was in his power

SILVERWHITE

without losing a single day. Considering this, we should understand the exceptional detail of Ibn Fadlan's report – a level that every ethnographer and folklorist should wish to achieve – but also the exaggerated emphasis of the mission's zeal, which ultimately provided conscious disinformation in the account just provided.

The lie is obvious from an astronomical standpoint. Again, the mission arrived in Volga Bulgaria on 12 May following a seventy-day journey by caravan. Ibn Fadlan claims that on the very first night, without a day's rest, he began checking the times at which the muezzin called to prayer. Even if one were a religious fanatic, doubts would be raised – especially in Ibn Fadlan's case. And the muezzin's reply confirms this suspicion in the simplest of ways: '[The nights] have been even shorter than now, for already they are beginning to lengthen.' Consequently, summer had already peaked. Ibn Fadlan only began to review religious arrangements forty days after arriving in Volga Bulgaria. That it is wiser to hide such carelessness in an official document is obvious to anyone who has ever had to draft, with a heavy conscience, a business trip expense report.

At my request, the Estonian astronomer Heino Eelsalu (1930–98) provided commentary on the observations, and his analysis gives us several limit values regarding Ibn Fadlan's exact location. They leave two possibilities:

1. Ibn Fadlan's astronomical and geophysical observations were made at some time just around the midsummer, but must have been recorded north of 55° latitude; or
2. Ibn Fadlan's observations were indeed made prior to midsummer, but were recorded even farther north at around 60° latitude.

This is a sober conclusion, not a hazy insinuation of a possible journey even further North. Ibn Fadlan's travelogue contains echoes of Ob-Ugric folklore, and even references to *ma-hor* – the 'land bull', as the Khanty still call the mammoth. Unfortunately, we cannot deduce from this anything apart from what we already know: there was 'lively traffic' (to be understood in 9th-century terms) along the Baltic–Volga waterway and Finno-Ugric tribes were an important component of Volga Bulgaria. Experiences, accounts, and expressions circulated broadly. Let us take a fragment from the end of Ibn Fadlan's first description of the dawn–dusk phenomenon as proof. The bewildered muezzin explained to him the following: 'For if a pot is put on the fire at sunset, there is no time for the water to boil before the dawn prayer.'

THE SILVERWHITE WAY

In all my travels, I have not yet seen a kettle so great that it would not come to a boil in an hour – not even the largest pilaf pots of Central Asia's most hospitable families. A night lasting only one hour would require the pot to be located between 60° and 65° latitude, just near the Arctic Circle. And that is where the experience undoubtedly took place. However, the statement itself was merely a saying in Volga Bulgaria and points to the fact that they were familiar with the polar day, albeit as an exotic phenomenon: people knew it, and people talked about it. Volga Bulgaria's part in conveying Northern folklore to the Arab world has been proved.

Ibn Fadlan's travelogue would hardly have gained the confidence of his contemporaries and our interest had his observations been limited to astrophysical phenomena alone. No one would have taken the time to copy them, and thus the mission's young secretary would have become yet another of history's nameless authors. This fate was avoided, however, because the most fascinating part of Ibn Fadlan's notes, unique even to Arab literature, is a scene in which he meets sailors on the banks of the Volga. We will review it in two parts. But first, a few words regarding terminology.

Ibn Fadlan consistently differentiates the ancient Baltic vessel (called by the taboo names 'wave horse' or 'sea stallion', i.e. *rhos*) from Arab nautical terms, which is completely understandable: at that time, Arab seafaring culture was the world's most advanced. The Baltic ship-boat was in no way comparable, but neither could it be referred to with a term designating an entirely different and more complex vessel.

We should add that Ibn Fadlan's interpreter was unable to translate the sailors' language, which rules out their being local.

BRASSIERE PROBLEMS

I saw the Rūs, who had come for trade and had camped by the river Itil [Volga]. I have never seen bodies more perfect than theirs. They were like palm trees. They are fair and ruddy. They wear neither coats nor caftans, but a garment which covers one side of the body and leaves one hand free. Each of them carries an axe, a sword and a knife and is never parted from any of the arms we have mentioned. Their swords are broad-bladed and grooved like the Frankish [a general term] ones. [...]

SILVERWHITE

All their women wear on their bosoms a circular brooch made of iron, silver, copper or gold, depending on their husband's wealth and social position. Each brooch has a ring in which is a knife, also attached to the bosom. Round their necks, they wear torques of gold and silver, for every man, as soon as he accumulates 10,000 dirhams [stereotype!], has a torque made for his wife. When he has 20,000, he has two torques made and so on. Every time he increases his fortune by 10,000, he adds another torque to those his wife already possesses, so that one woman may have many torques round her neck.

The most desirable ornaments they have are green ceramic beads they keep in their boats. They will pay dearly for them, one dirham for a single bead. They thread them into necklaces for their wives. [...]

When they arrive from their land, they anchor their boat on the Itil, which is a great river, and they build large wooden houses on the banks. Ten or twenty people, more or less, live together in one of these houses. Each man has a raised platform on which he sits. With them, there are beautiful slave girls, for sale to the merchants. [...]

As soon as their boats arrive at this port, each of them disembarks, taking with him bread and meat, onions, milk, and mead, and he walks until he comes to a great wooden post stuck in the ground with a face like that of a man, and around it are little figures. Behind these images there are long wooden stakes driven into the ground. Each of them prostrates himself before the great idol, saying to it: 'Oh my Lord [stereotype!], I have come from a far country and I have with me such and such a number of young slave girls, and such and such a number of sable skins ...' and so on, until he has listed all the trade goods he has brought, 'I have brought you this gift.' Then he leaves what he has with him in front of the wooden post, 'I would like you to do the favour of sending me a merchant who has large quantities of dinars and dirhams and who will buy everything that I want and not argue with me over my price.' [...]

If he has difficulty selling and his stay becomes long drawn out, he returns with another present a second and even a third time. If he cannot get what he wants, he brings a present for each of the little idols and asks them to intercede, saying:

'These are the wives of our Lord [ancestor] and his daughters and sons [descendants].'

Then he continues to make his request to each idol in turn, begging their intercession and abasing himself before them. Sometimes the sale is easy and after having sold his goods he says:

246

THE SILVERWHITE WAY

'My lord has satisfied my needs and it is fitting that I should reward him for it.'

Then he takes a certain number of sheep or cows and slaughters them, distributing part of the meat as gifts and carrying off the rest to set before the great idol and the little figures that surround it. Then he hangs the heads of the sheep or cows on the wooden stakes which have been driven into the ground. When night falls, the dogs come and eat all this, and the man who has made the offering says:

'My Lord is pleased with me and has eaten the gift that I brought him.'

(cf. Kovalevsky 1956; Frähn 1823; Hämäläinen 1938; Frähn 1848, etc.)

When writing his works, the encyclopedist and geographer Yaqut al-Hamawi (1179–1229) used sources that have not survived to our day. He had the following to add regarding the sailors gathered on the Volga's bank (Frähn 1823: 4):

Ros, also written as Rs, is a people whose territory borders the Slavs and the Turks. They have a different language and their own faith and ritualistic practices that have nothing in common with others [i.e. with Islam or Christianity]. Al-Mukaddasi [a late-11th-century author following Ibn Fadlan] says: they live on the island of Wabia, surrounded by a lake, which is their fortress against any who attack them. Their numbers are estimated at one-hundred thousand. They have no fields nor livestock. The Slavs carry out raids against them and take away their possessions. When one gives birth to a son, he is tossed a sword and told: yours will be only that, which you acquire by the sword. When their king passes judgement over two arguing parties but they are not satisfied with his decision, he says to them: pass judgement between yourselves with swords. Whoever's sword is sharper is the victor.

The contrasting of religious beliefs and the emphasis on a lust for pillaging point clearly to the source from which al-Maqdisi acquired his information: Christian missionaries' fables of the pagan North, which Arabs believed bordered with the Turkic tribes of Volga Bulgaria. Al-Maqdisi thus gives us the most indistinct coordinates, or even less – the direction in which one should seek the docked sailors' home. Ibn Fadlan's description of the sailors' religious beliefs is unique but offers no geographical footholds. Only folk dress remains: ambiguous, as they are conveyed to us through Arab terminology.

SILVERWHITE

Hailing from the international trading city of Baghdad, Ibn Fadlan was impressed by silver magic. It tickled his patriotic faculties. Silver magic was widespread, prevalent throughout our entire area of interest, which stretches from the Volga to the shores of the Baltic Sea. By silver, we mean the smallest Arab coins: dirhams, which were used both as change and *en gros*; by weight. A dirham, the Greek < *drachma*, weighed about 3.11 grams. Ancient monetary scales have been found in rather large quantities in Estonia as well. Ibn Fadlan began describing local silver magic the moment he reached the Volga Bulgarian border. Here is his first meeting with the Bulgar ruler Almış (Kovalevsky 1956: 131–2, 141): 'When he saw us, he dismounted and fell down with his face to the ground to give thanks to God, the All High, the Almighty. In his sleeve, he had dirhams which he scattered over us.' This ritual was repeated when meeting the queen: 'When I had given her this robe of honour, her women scattered dirhams over her and then we went back to our tents.' In comparison, here is an account of silver magic recorded on the Pärnu shore (Loorits 1939: 200): 'Money is placed beneath the forestem and the mast's nog, usually that of the foremast, so the ship may have good fortune.'

Almış himself gave the most convincing account of the power of silver magic:

> 'I thought', he answered, 'that the empire of Islam is prosperous and that its revenues come from licit sources. For this reason, I made my request. If I had wanted to build a fortress with my own money, silver or gold, it would not have been impossible for me. But I wanted to have the blessing which is attached to money coming from the Commander of the Faithful, and so I asked him for it.'

A sceptical smirk would be totally inappropriate here. Almış's words should be taken as sincere. Silver brought fortune. It did so because it came from a blessed source – a descendent of Allah's prophet Muhammad. Dirhams also bore the holy script. The king's annual salary was 70,000 sable skins, and Baghdad's money was of purely symbolic value beside the immense fortune. Thus we should take it as silver magic.

Ibn Fadlan repeatedly mentions instances of silver magic being practised among the sailors docked on the Volga. Offering sacrifices to his ancestors at the sacred post, a sailor would ask for silver as a vehicle of fortune. We could de-poeticise this as fortune = prosperity = social status, though it would be a half-truth. The notion of silver magic was both broader and more primitive. As one Estonian researcher remarked,

THE SILVERWHITE WAY

'Nowhere else in Estonian folk beliefs or customs has silver acquired such central significance as among the tiny people who live underground. They forge and guard precious metals as blacksmiths, but can be appeased with silver as bearers of illness' (Loorits 1949: III, 80). As early as the dawn of the Common Era, silver took a powerful step into Baltic-Finnic folklore as a novel cultural factor. The taboo euphemism for sacrificed silver in Estonian was *valkea*, meaning 'fire', similar to the word for 'white'. Sacrificing itself was called 'throwing white', 'casting white', 'placing white', 'whiting', or even 'enlightening'. Coins placed with the dead were called 'light'. Using the word 'white' in place of the taboo 'silver' persisted for centuries in medieval Tallinn, and even crossed over into the local Reval German dialect.

However, not even this can help us pinpoint more precisely the homeland of Ibn Fadlan's sailors. We can only conclude that the Baltic shores cannot be ruled out. But are there any more specific footholds to be found?

We now come to the 'circular brooches' that the women wore on their chests and, as a result of five hundred years of diligent study and interpretation, gave us the comical legend of metal 'brassieres'.

Hang a traditional Estonian brooch on a wall and ask a Portuguese merchant to describe it. The result would coincide with Ibn Fadlan's uncertain wording. We can ask for nothing more. It is important to keep in mind that later annotators, proceeding from their own anatomical experiences, turned single into plural. In Ibn Fadlan's text, there is only one such mysterious item on a woman's chest.

Many Estonian folk songs refer to young women as *rahakaelad* ('money-necks') and *taalderrinnad* ('plate-breasts'). In the context of silver magic, we should add that brooches, silver coins, and the pendants called *rinnalehed* were essential components of women's outfits. A woman wore silver in the form of a coin necklace or brooch on her breast, rings on her fingers, and 'silver chains' on her hips to keep them healthy and strong. Men wore a brooch, rings, and earrings to protect against eye-aches, earaches, and headaches. The jewellery's magical task was even more extensive. One's transition to the afterlife was a long journey, and the deceased were equipped with everything they needed: a symbolic or genuine boat, footwear, food, tools, and coins. (Let's keep this in mind when reading Ibn Fadlan's next excerpt.) Brooches were given the job of a boat, to convey one to Toonela, the afterlife, and its pin was used as an oar: 'a brooch for rowing, a *[rinna]leht* for flying', as a song titled

SILVERWHITE

'The Drowned Brother' goes. A brooch was meant to ferry the seeker to the right place and could also be used to 'row' through the sky to the afterlife. Jewellery's symbolism was such a great aid to the deceased on their journey that plenty was provided alongside everyday necessities.

Ibn Fadlan was amazed that the sailors also prized cheap 'green ceramic beads they keep in their boats'. The beads remained in high fashion among the Baltic Finns for a very long time, and were naturally referred to in folk songs as well.

Baltic Finns' ancient customs offer plenty of support for Ibn Fadlan's silver magic, brooches, and coin necklaces. But how are we to understand his 'brooch [that has] a ring in which is a knife, also attached to the bosom'? Early Iron Age Baltic-Finnic women's clothing and jewellery have been reconstructed on the basis of archaeological grave digs. The manner in which a knife is attached corresponds precisely to Ibn Fadlan's description.

The similarity of ancient Baltic-Finnic culture to the clothing and customs of Ibn Fadlan's sailors is not yet proof, strictly speaking. They are still just examples that can be used to refute a claim, but not prove one. We should keep in mind that proof is not put forth between the covers of this travelogue. So again in this case, we can merely state: the possibility that Ibn Fadlan described Baltic-Finnic sailors cannot be ruled out. Is that so insignificant?

THE MYSTERY OF THE SHIP KING

Author: Ahmed ibn Fadlan.

Main characters: a ship king and his wife.

Secondary characters: attendants, wedding guests, funeral guests, interpreter.

Setting: the Baltic–Volga waterway.

Year: 921 CE.

Act I: wedding.

Act II: funeral.

Note: there is no intermission between Acts I and II.

Baltic-Finnic tribes used the word 'king' to signify a chief or leader before the Common Era.

One day I learned of the death of one of their great men. [...] If the dead man was poor, they build him a small boat and place him in it and set it on fire. [...]

THE SILVERWHITE WAY

When a great man dies, the members of his family [his entourage of closest relatives] say to his slave girls and young slave boys [his entourage who joined under sworn allegiance]:

'Which of you will die with him?'

One of them replies:

'I will.'

Once they have spoken, it is irreversible and there is no turning back. If they wanted to change their mind, they would not be allowed to. Usually, it is the slave girls who offer to die.

When the man whom I mentioned above died, they said to his slave girls:

'Who will die with him?'

One of them answered:

'I will.'

Then they appointed two young slave girls to watch over her and follow her everywhere she went, sometimes even washing her feet with their own hands.

Everyone busies himself about the dead man, cutting out clothes for him and preparing everything that he will need. Meanwhile, the slave girl spends each day drinking and singing, happily and joyfully.

When the day came that the man was to be burned and the girl with him, I went to the river where his boat was anchored. I saw that they had drawn his boat up on to the shore and that four posts of birch or other wood had been driven into the ground and round these posts a framework of wood had been erected. Next, they drew up the boat until it rested on this wooden construction.

Then, they came forward, coming and going, pronouncing words that I did not understand [meaning, even with the help of the interpreter], while the man was still in his grave, not yet taken out.

Then they brought a bed and placed it on the boat and covered it with a mattress and cushions of Byzantine silk brocade. Then came an old woman whom they call the 'Angel of Death' and she spread the bed with coverings we have just mentioned. She is in charge of sewing and arranging all these things, and it is she who kills the slave girls. I saw that she was a witch, thick-bodied and sinister.

When they came to the tomb of the dead man, they removed the earth from on top of the wood, and then the wood itself, and they took out the dead man, wrapped in the garment in which he died. I saw that he had turned black because of the coldness of the country [in summer!]. They had put mead in the tomb with him, and fruit and a drum. They took all this out. The dead man did not smell bad and

251

SILVERWHITE

nothing about him had changed except his colour. They dressed him in trousers[?], socks, boots, a tunic and a brocade caftan with gold buttons. On his head, they placed a brocade cap covered with sable. Then they bore him into the pavilion on the boat and sat him on the mattress, supported by cushions. Then they brought mead, fruit and basil which they placed near him. Next they carried in bread, meat and onions which they laid before him.

After that, they brought in a dog, which they cut in two and threw into the boat. Then they placed his weapons beside him. Next they took two horses and made them run until they were in lather, before hacking them to pieces with swords and throwing their flesh onto the boat. Then they brought two cows, which they also cut into pieces and threw them onto the boat. Finally they brought a cock and a hen, killed them and threw them onto the boat as well.

Meanwhile, the slave girl who wanted to be killed came and went, entering in turn each of the pavilions that had been built, and the master of each pavilion had intercourse with her, saying:

'Tell your master that I only did this for your love of him.'

On Friday, when the time had come for the evening prayer, they led the slave girl towards something which they had constructed and which looked like the frame of a door. She placed her feet on the palms of the hands of the men, until she could look over this frame. She said some words and they let her down. They raised her a second time and she did as she had the first time and then they set her down again. And a third time and she did as she done the other two. Then they brought her a chicken. She cut off its head and tossed it away. Then they took the chicken and threw it onto the boat.

I asked the interpreter what she had been doing. He replied:

'The first time they lifted her up, she said:

'"There, I see my father and my mother."

'The second time, she said:

'"There, I see all my dead relatives sitting there together."

'And the third time she said:

'"There I see my master sitting in Paradise [a green garden, Toonela, the afterlife] and Paradise is green and beautiful. There are men with him and young people [entourage], and he is calling me. Take me to him."' [Here, the interpreter who understands the sailors' customs is a commentator, not a translator. During the ceremony, Fadlan spoke with the interpreter in Arabic, the interpreter with the sailors in an unknown (intermediary) language, while the spells and laments sung by the sailors and the girls were in a third language.]

THE SILVERWHITE WAY

They went off with her towards the boat. She took off the two bracelets that she was wearing and gave them both to the old woman who is known as the Angel of Death – she who was to kill her. Then she stripped off her two anklets and gave them to the two young girls who served her. They were the daughters of the woman called the Angel of Death. Then the men lifted her onto the boat, but did not let her enter the pavilion.

Next, men came with shields and staves. They handed the girl a cup of mead. She sang a song over it and drank. The interpreter translated what she was saying and explained that she was bidding all her female companions farewell. Then they gave her another cup. She took it and continued singing for a long time, while the old woman encouraged her to drink and then urged her to enter the pavilion and join her master.

I saw that the girl did not know what she was doing. She wanted to enter the pavilion, but she put her head between it and the boat. Then the old woman seized her head, made her enter the pavilion and went in with her. The men began to bang on their shields with staves, to drown her cries, so that the other slave girls would not be frightened and try to avoid dying with their masters [the explanation is inaccurate: at the moment of entering the pavilion, the girl symbolically became a wedded woman]. Next, six men entered the pavilion and lay with the girl, one after another, after which they laid her beside her master. Two seized her feet and two others her hands. The old woman called the Angel of Death came and put a cord round her neck in such a way that the two ends went in opposite directions. She gave the ends to two of the men, so they could pull on them. Then she herself approached the girl holding in her hand a dagger with a broad blade and plunged it again and again between the girl's ribs, while the two men strangled her with the cord until she was dead.

Next, the closest male relative of the dead man came forward and took a piece of wood, which he lit at a fire. He then walked backwards towards the boat, his face turned towards the people who were there, one hand holding the piece of flaming wood, the other covering his anus, for he was naked. Thus he set fire to the wood that had been set ready under the boat. Then people came with wood and logs to burn, each holding a piece of a wood alight at one end, which they threw onto the wood. The fire enveloped the wood, then the boat, then the tent, the man, the girl and all that there was on the boat. A violent and frightening wind began to blow, the flames grew in strength and the heat of the fire intensified. [...]

SILVERWHITE

> And indeed, not an hour had passed before ship, wood, girl and master were no more than ashes and dust.[2]

I can't deny that I find the account gripping. I feel as if I am present, a mute participant in a cruel and cryptic mystery. I see the handsome landscapes of the Volga; the bluish-wall of forests stretching to the water's edge at the low bank; the giant silverwhite river and the distant opposite bank, towering like the wall of a cliff. I hear chanted spells and laments; I sense the solemn peace of the funeral-goers and feel the roaring flames painfully baking my face, forcing me to retreat step by step into the dim undergrowth while the wheel of the sun slowly disappears below the horizon.

Let us hold this image before our eyes for just a moment before we begin breaking it down.

The sun disappears below the horizon and the sky burns red, sighing. Now, the scents of the forest erupt, the wind delivers the smell of marsh rosemary, the damp chill rising from the river creeps over the marshes. The final embers still crackle, a vortex of sparks flies into the darkening sky.

FROM WEDDING TO FUNERAL

Ibn Fadlan's account weaves together a wedding and a funeral, two central events in one's life. A 'great man', a captain, passed away. He needed a vessel for his journey to Toonela, the afterlife. He was undoubtedly a wealthy man, because he was given all he needed for the voyage.

First, there was a wedding and symbolic rites – placing the coif upon the bride. This was followed by a funeral and cremation. Generous sacrifices accompanied both the former and the latter.

Estonian folklore offers plenty of support for Ibn Fadlan's account, which we will now present in order of the events.

Engagements, as the etymology of the English (French *engager*, 'to pledge') and Estonian (*kihlus*, a contract) words suggest, were important social events that brought a new member into a large family or community, and that were sealed with pledges – a system of mutual rights and obligations. It was primarily because of this importance that Estonian wedding rituals retained several ancient customs and rituals, chants and runic songs, up till the early 20th century. These practices dated back to time immemorial and had no peers in other systems of ritual.

How far into the past did they reach, exactly?

THE SILVERWHITE WAY

When discussing Tacitus, we mentioned that the boar 'figures' of the Aesti found their final haven in Estonian wedding traditions and made it to the threshold of the 20th century. 'When entering [the bride's] home, the groomsmen found a boiled pig head on the table – one was supposed to cut off an ear. But a nail had been driven through the ear so the man's sword would have trouble' (Tedre 1973: 41, 45). One might think this was a chance or isolated fact. Such was not the case. The man who helped to 'pay off' people playfully pestering or attempting to steal the bride on the way from her house to the groom's was called a 'pig-man' or 'snout-man'. Wedding traditions also preserved the memory of the crucial roles played by swords and ships in the distant past. The bride's dowry chest was decorated with a flag and accompanied by a guard of sword-carrying men (Schroeder 1888: 107). One 19th-century author indeed expressed his astonishment that swords aplenty could 'still today' be found in every Estonian village. According to one rather detailed account from Põltsamaa, a long pole topped by a flag was attached to the bride's dowry chest. At the groom's gate, the receiving elder asked the leader of the procession whether the ship that docked in harbour was for sale. 'It is,' the man on horseback replied, 'but only to him who pays in cash, for the ship carries precious cargo that is rarely found anywhere in the world' (Tedre 1973: 33). What's more, the key to the chest was called the 'ship's anchor'.

Throughout western Estonia, a bride's dowry chest was called a 'ship'. We alluded to the reason earlier: rowers on ancient Baltic ships sat upon their cargo chests. The Latin figure of speech is *pars pro toto* – a portion of something taken to represent the whole. The Estonian word *alus* in its sense of 'vessel' also originally meant the boat's hearth: a limestone slab upon which a fire was built. The semantic trunk–ship connection is beyond doubt. One Muhu superstition alludes to the immense 'power' of a ship and a chest in olden days: as late as 1872, the islanders believed that simply seeing a ship could make a girl pregnant. The frequency of ship nails in ancient cremation burials also speaks to its central role in life (Anu Lõugas to the author, 20.3.1973). We should mention that the nail itself seemingly possessed an ideological function independent from that of the ship. The nail joined planks, but also generations. As we may recall, the Sámi drove one nail into the world post and another into the head of the supreme deity Tooru. In Estonian wedding traditions, a nail was driven into the head of a pig. After the last few chapters, I hardly need to stress the significance of this striking analogy. If your value judgements aren't sufficiently flexible, then imagine a gold nail. Iron is only iron in

SILVERWHITE

the eyes of the chemist – it has had different values at different times. But returning to the question at hand, the oldest elements of Estonian wedding traditions reach back to the days of Tacitus and allow us to draw a smooth line of comparison to the traditions so colourfully recorded by Ibn Fadlan in the year 921.

Monogamy and polygamy alike were practised by Estonians in that time. Henry of Livonia provided evidence in 1223: 'They took back their wives, who had been sent away during the Christian period. They disinterred the bodies of their dead, who had been buried in cemeteries, and cremated them according to their original pagan custom' (XXVI). Describing the defeat of the Lithuanians and the Estonians in 1205, Henry adds: 'A certain priest named John who at that time was held captive in Lithuania reported that fifty women had hanged themselves because of the deaths of their husbands, without doubt because they believed that they would rejoin them immediately in the other life' (IX).

CHASING THE WOLF'S TAIL

By the time Ibn Fadlan arrived on the bank of the Volga, wedding preparations had been going on for days. The bride's relatives were busy arranging the dowry and collecting wedding gifts. Accompanied by two bridesmaids, the bride went from house to house 'to feel joy for the future'.

In Estonian wedding tradition, this part of the preparations was called *hundihänna ajamine*, 'chasing the wolf's tail'. Though first recorded in the 17th century, the ritual is much older, possibly extending back into the most ancient days of Estonian family customs. The origin of the expression – *hänna alla saama* ('to get under the tail') or *hundi käes olema* ('to be in the wolf's clutches') – probably need not be explained. The groom's family had a certain right to the bride in an economic, social, and even biological sense. The primary function of a wedding was to guarantee descendants and pass down the seed of life. Both the bride's and groom's immediate families, extended families, and later the entire community were directly connected by children, future labourers, and bearers of their way of life. Procreation magic was joined by dispelling magic: an array of means and expressions aimed at protecting a person from evil words, envious gazes, and bad thoughts in the most decisive moments of their lives.

One custom intended to instil the bride with courage – *juulutamine* – was recorded on Saaremaa in the early 20th century (Kruusberg 1920b: 65):

THE SILVERWHITE WAY

> Folk on Saaremaa say that there was this custom in the old days, one no doubt from ages and ages ago. During a wedding, the young man would be put lying face-down and bare-bottomed on a table, and the 'male' (the local term they use there) wedding guests would each take turns striking him on the uncovered part with their *membro virili* [or *iduhänd*, 'sprout-tail'], saying: 'This is what you deserve!' People also say that a man from Ülemaa happened to be at a Saaremaa wedding. But he took the custom a little differently... and so the Saaremaa men started beating him. [...] The married men carried out the *juulutamine*.

Representing the groom's guests in Estonian wedding tradition was a 'sword-father' with a white scarf over his shoulders and three brooches on his chest, a 'sword-mother', and groomsmen. The bride's guests included bridesmaids, the bride's brother, and a maid of honour. There were three high points in her transition from maidenhood to married life: departing her family home, arriving at the groom's family home, and placing the coif on her head – ritualistic womanhood. Naturally, these were surrounded by procreation magic and dispelling magic. Ibn Fadlan, without knowing the purposes of the customs, conveyed their visual aspects in outstanding detail.

SHOEING THE BRIDE

The sailors were foreigners. The bride was missing her home gate. Therefore, they erected a symbolic structure 'which looked like the frame of a door' and before which the young woman was lifted three times. In folklore, this is called a rite of passage: the girl is lifted out of maidenhood. This triple lifting, done by the sword-father and groomsmen, was called *piiramine* ('surrounding'), *rautamine* ('shoeing', as with a horse; a metaphor for placing the coif), or *karastamine* ('tempering'). As they lifted the bride, the men chanted (Tedre 1973: 51):

> Surround, surround, groomsmen,
> surround the witches in the heath woods,
> in the oaks behind the house,
> surround the birds in the brush,
> the bad in the fenceposts!

The gateway erected on the bank of the Volga carried other symbolic functions: the bride was not allowed to enter the groom's home along the same path used for carrying out the dead. The jambs, and especially the

SILVERWHITE

head jamb, were ascribed a direct connection to procreation magic – they were seen as a euphemistic symbol of the 'sprout-tail'.

Crossing through the gate, the girl stood upon the territory of the groom's family. A traditional wedding procession wasn't possible in the Volga sailors' camp, as the ritual was held in a cramped space, and the bride's and groom's homes were merely symbolic.

Arriving at the groom's home was the second culmination: a moment swathed in especially thick dispelling-magic protection in Estonian wedding tradition. The bride's foot was forbidden to touch the ground, a custom found throughout Estonia. Laid before her was a rug, blanket, or old coat; sometimes, she was carried into the house. Ibn Fadlan's account, both poetic and tragic, is astoundingly accurate in terms of this custom: the bride places her feet in the palms of the men to be carried to the ship.

THE MAID OF HONOUR

The bride handed her first gifts, two bracelets, to the old woman tasked with her death – a figure whom Ibn Fadlan, as a Muslim, dubs the 'Angel of Death'. Naturally, one could not expect the Arab to know the precise folkloric or ethnographic terms. A bride customarily handed the first gifts to her mother-in-law. This, however, takes place during a voyage, and it would be strange to expect all family members would take part. We also know that the old woman was tasked with preparing the nuptial bed. Usually, this would be one of the maid of honour's responsibilities. The bride handed the next gifts to both bridesmaids, who were 'the daughters of the woman called the Angel of Death'. We should consequently regard the old woman more as the bride's maid of honour. However, from beginning to end, the wedding ceremony on the bank of the Volga was interwoven with funeral customs. Ibn Fadlan's 'Angel of Death' – 'Old Lady Death' or *surmaeit* – could hold a clue tying her to a mythical Estonian figure who faded during the Middle Ages but who is still remembered in our language as *murueit*, 'the Old Grass Woman', who was said to appear on clearings at night with her beautiful daughters. This is pure speculation, of course. However, the blending of characters, the contamination of two simultaneous and opposite rituals, was inevitable and is no longer speculation, but is drawn from Ibn Fadlan's account. During the lifting rite before the door frame, the bride first bade farewell to her mother and father, her home; then, to her family; and, lastly, she greeted her groom and his companions waiting on the green grass of Toonela.

THE SILVERWHITE WAY

THE COIFING

The third culmination of the wedding ceremony was the coifing of the bride, to which Ibn Fadlan only alludes in little detail: the young woman 'put her head between [the pavilion] and the boat. Then the old woman seized her head, made her enter the pavilion and went in with her' at the same time as the men, her 'family', began banging their staves against their shields. The description has been nearly inscrutable for annotators. It's been assumed that the girl was simply drunk. However, the Arab's description leaves no doubt that she intentionally lowered her head. According to ancient beliefs, a person's power resided in their hair and nails. Slaves' hair was cropped short in hope of their obedience and submissiveness. Binding the bride's hair served approximately the same goal in Estonian wedding tradition. And as Ibn Fadlan's next sentence describes the old woman grabbing the bride's head (but from where, if not the hair?), which the onlookers' thunderous banging of shields underscores as the ritual's culmination, we can perceive this as the moment of coifing. From this point forward, the bride is a woman.

Throwing the bride into the nuptial bed in the company of witnesses was also a part of Estonia's oldest wedding traditions (Schroeder 1888: 107). According to Ibn Fadlan, the bed was adorned with several types of flowers and scented plants. We find parallels in Estonian folk wedding songs as well:

> Dearest women, dearest darlings [...]
> Let's go looking at the bed [...]
> Has the groom made hay,
> Struck the ox's head well,
> Cut the berry's stalk,
> Cuddled with cornflower,
> Groped the globeflower,
> Milked the meadowsweet [...]

DUAL SACRIFICES

Baltic Finns had a wide number of possible sacrifices to choose from, ranging from a few morsels or a couple drops of beer to three grains of rye, a belt or ribbon tied to a tree, and even an ox or a horse (Hupel 1781: 227–9). Meagre sacrifices being made alongside entire horses, the crown of wealth, suggest the conservatism of traditions, where plain foraged

SILVERWHITE

berries later turned into the products of agriculture, craftsmanship, and trade.

In Setomaa, a horse was sacrificed to the water god when ice receded in spring. Thanks to Henry of Livonia, we know that water was accorded cleansing properties: the invaders' holy water was rinsed off with pure water, or, in other words, Estonians washed one water with another, not realising that both sprang from the same imaginative source. Water magic was joined by several sauna myths and one of Estonian folklore's finest and most poetic motifs: a man defiled a lake, so the insulted lake left its home.

Wedding sacrifices stem from a very ancient cultural stratum, as fire-guarding was one of the traditions: as soon as the bride was brought into the groom's home, one member of the family was required to sit in front of the hearth and guard the fire from the evil eye. This custom elucidates Ibn Fadlan's fire-bringer who walked backwards towards the ship: the man, not the corpse, was subject to the fire's protection – a custom that Baltic-Finnic Votians still practised in the 20th century (Öpik 1970: 160). Attitudes towards the dead were contradictory in a patriarchal society's customs and mythology: the dead were feared and one tried to render them harmless. At the same time, one needed to maintain a good relationship with the ancestors so as to seek their protection in moments of need.

Earlier, we discussed the beliefs surrounding the cremation of the dead: the soul needed to be freed of the body and set off on its long journey to Toonela. They were to be dispatched in their own ship with necessities for the voyage, according to their social status – in this case, even accompanied by a young woman.

Cremations were moved from the sacred grove to the clearing, the village green, and especially to burn-beaten land, from which the souls could rise as smoke and sparks to continue their earthly life 'on the other side'. 'Whoever perishes goes home,' was the saying in Tarvastu. As August Wilhelm Hupel asserted in 1777: '[Estonians] are never especially mournful over a relative's death.'

Several sacrifices are listed in Ibn Fadlan's notes. It's hard to say which belonged to the wedding traditions and which to the funeral rites. Roosters, hens, and dogs played an important part in both wedding songs and laments. The oldest domestic animals were also sacrificed when their master died: a dog would be drowned in the well or beheaded so it could accompany the man to Toonela, where it would carry on performing its duties.

THE SILVERWHITE WAY

Sacrificing roosters and hens was a widespread practice at funerals and when visiting a new-born infant for the first time. It has been described abundantly and older people can still remember it, myself included. My grandmother told me that when her father-in-law passed away, her mother-in-law took a rooster and at the moment the coffin was carried out of the door, cut its head clean off. My grandmother was from Harju County. No, I don't mind you quoting me. (Aliise Moora to the author, 30.6.1973)

If a Seto cottager had nothing else, then they'd sacrifice a rooster, which wealthier folk mockingly called 'a poor man's ox'.

Still, there was a common rule: roosters were sacrificed to men, hens to women. And here is our thread. According to Ibn Fadlan, altogether eight animals were sacrificed. First, a dog: it was to be the first to reach its master's 'home gate' so that the man could be greeted upon arrival. That funeral sacrifice was purely for the sailor. Two horses, two horned creatures, and the rooster and hen allow us to deduce that they were sacrificed to the man and woman equally – in other words, it must have been a stallion, a mare, an ox, and a cow. Half of Arabic specialists have translated the pair as oxen, half as cows (Kovalevsky 1956: 251). Folklore, as we can see, helps to bring us a little clarity. The hen sacrificed by the young woman herself was apparently a symbolic gift to the groom's household, to her new home.

THE HUMAN SACRIFICE AND AFTERWORLD GRASS

Were human sacrifices common among Baltic Finns? Written sources leave no doubt that they were. Adam of Bremen claimed that the Estonians sacrificed humans. According to Henry of Livonia, the Estonians intended in the year 1190 to sacrifice Brother Theoderich, who later became Estonia's third bishop: 'The people were collected and the will of the gods regarding the sacrifice was sought after by lot. A lance was placed in position and the horse came up and, at the signal of God, put out the foot thought to be the foot of life' (I). The right foot was considered the 'foot of life'.

Henry of Livonia also writes candidly that in 1223 Estonians in Tartu debated whether to sacrifice an ox or the priest Hartwig, 'for he was equally fat'. At about the same time, the Saccalians (Sakala) sacrificed the Danish magistrate Hebbus, eating his heart together 'so they would be made strong against the Christians' (XXVI): a ritual that is practised

SILVERWHITE

to this day in the Catholic Church, the intention being to transfer the spiritual and physical power of the sacrificed man to those taking part in the ceremony. Even in this sense, Estonian folklore is not at odds with Ibn Fadlan's account.

Therefore, we can also regard with certain credibility the young woman's final words spoken during the third time she was lifted: 'There I see my master sitting in Paradise [a garden] and Paradise is green and beautiful. There are men with him and young people, and he is calling me.' The land of the dead beyond the waters is a grassy meadow on the seashore. It is likewise associated with the Estonian legend of Old Lady Grass (*murueit*) and her daughters, who are regarded not only as the rulers of Toonela, but guides to the afterlife. Old Lady Grass is in no way like the mythical elves and fairies that were created by Friedrich Reinhold Kreutzwald on the basis of German archetypes. Consequently, we could ask ourselves: shouldn't we see Ibn Fadlan's untranslatable 'Angel of Death', Old Lady Death, as a personified Old Lady Grass? The question remains unanswered.

In north Estonian belief, the deceased could ride a Toone stallion on Toone's grass: a hero's horse with a white spot on its forehead. In Estonian folklore, this spot or blaze was called *Kalevi lauk*, 'Kalev's blaze'.

We have returned from yet another voyage and the loot isn't too shabby at first glance: Ibn Fadlan's account of the captain's funeral does not conflict with Baltic-Finnic folklore, traditions, or mythology. It holds no proof. But it does hold possibility.

THE GULF OF FINLAND WEDS THE VOLGA

If Ibn Fadlan truly encountered Baltic Finns on the bank of the Volga in the year 921, then we should also be able to find reports of the Baltic Sea, perhaps even Estonia and the Estonians, from other Arab travellers. Unfortunately, it's not at all as one might expect. Ibn Fadlan's mission was an extraordinary one, and, just like anything extraordinary, it has been met with sceptical gazes and ironic smirks, starting with those from his fellow idealist al-Hamawi (who in his geographical dictionary – 3,894 pages in six volumes of Latin script – did save the young secretary from oblivion, but also placed the crown of disbelief upon his head) and ending with Josef Markwart, a 20th-entury German Arabist who declared him a liar … And the name may have stuck (Markwart worked the steamroller with gravity and tenacity), had an unexpected event not taken place in

262

THE SILVERWHITE WAY

Mashhad at that same time: in that scorching provincial Iranian city where thousands of devout Muslims gather at the grave of Harun al-Rashid every year, the Bashkir researcher Zeki Velidi Togan published Ibn Fadlan's manuscript in its entirety. True, it wasn't written by Ibn Fadlan's own hand, but the Mashhad manuscript turned out to be extremely close to the original in both its level of accuracy and the period in which it was recorded (Hämäläinen 1938: 207–27).

Ibn Fadlan's credibility should not be diminished by the fact that his journey was so exceptional. And, luckily, it wasn't after all.

* * *

Over the next few pages, we will change course and move backwards through time. Why? We're accustomed to knowledge becoming more exact in tandem with the flow of time. This is not always the case, of course. Let us recall Erik the Red's son Leif on the coast of America (at about 1000 CE) and the continent's 'rediscovery' in the days of Christopher Columbus. Regressions occasionally take place in geographical knowledge. The field reflected practical needs, and knowledge disappeared along with them.

For reasons that we will discuss later, the Silverwhite waterway from the Baltic to the Volga began to dwindle in around the 11th century and finally lost its importance in the 12th. This is reflected in Arab geography. The knowledge possessed by Abdulfida (1271–1331) was already quite hazy. But the closer we get to Ibn Fadlan, the greater definition the shores of the Baltic take. We can be sure of this (Frähn 1823: 178; Krachkovsky 1957: 386–94): 'The Varangian Sea stretches from the northern world ocean to the south. It has true length and width. The Varangians are the people on the shore of that sea' (Abdulfida).

'Varang' is the Arab word for the Baltic Sea. The Chuvash word *võras/varas* used in the sense of 'chief, captain' is apparently tied to the toponym, and from the west it is joined by several Estonian expressions, all conveying might and wealth: *raudvara* (literally 'iron store'; 'reserve'), *kaasavara* (literally 'accompanied store'; 'dowry'), *tagavara* ('back-up, reserve'), *varakas* ('wealthy'). We can add *varal käima*, 'to (go) acquire wealth'. When a ship was heavily laden and sat deep in the water, the gunwales were made higher with added boards called *varad*. And, finally, the name *varakandja* is used for no other than a *pisuhänd* ('fireball'), *sädemesaba* ('spark-tail') > *lohemadu* ('dragon')[?] (Wiedemann). This would allow us to interpret *varaline* as a member of a crew, a companion tied to the crew not by blood

SILVERWHITE

but by an oath of 'soul and blood', known on the Baltic as *vaering*. This is speculation. All that has been proved is that the Arab 'Varangian Sea' is taken from the notion of *vaering*. The name, as we will see, is written in various ways:

> Here is a great arm of the sea that is called the Varangian Sea. The Varangians themselves are a people who speak an indecipherable language, of which almost no words can be understood. They live next to the Saqaliba people [Northerners]. This arm of the sea is the Twilight Sea in the North. Five islands lie near its shore. (al-Dimasqi (1256–1327); Frähn 1823: 190–2; Krachkovsky 1957: 382–6)
>
> And when this sea has passed the territories of the Varangians, a tall and hostile people living on its shore, it turns to the east and extends past impenetrable mountains and [un-?]inhabited lands to the borders of the Sini [Chinese]. Its length beginning from the west is 180 degrees, its width from at the north 5 degrees. (Persian Nasir al-Din al-Tusi (1201–74), through another author; Frähn 1823: 186; Krachkovsky 1957: 386–94)

Yet the closest we get to Ibn Fadlan is, of course, the famous Khwarazmian, great figure of Uzbeki culture and author of 113 works, Abū Raihān al-Bīrūni (973–1048). He writes (Frähn 1823: 182–3; Krachkovsky 1957: 329–42, 248):

> The sea [...], which in the western part of the settled world bounds the shores of Tandźa and al-Andalus, is called the All-Encompassing Sea. The ancient Greeks called it Okeanos. No one dares to venture out to open sea; sailing is done along the coastline. From those lands, the sea extends north to the territory of the Saqaliba [Northerners]. And branching from the northern part of Saqaliba is a large channel or bay, and it extends quite close to the state of the Muslim Bulgars. This [bay/channel] is known as the Varangian Sea [*Var varank*]. The latter in turn is a people who live on that same coast. Next, it [Okeanos] turns past it to the east, and between its shore and the most distant territory of the Turks [Bashkirs] can only be found unknown, desolate, unsettled lands and mountains.

The Varangian Sea stretches almost to Volga Bulgaria. These lines are still very significant. A talented person's mistake is worth more than an untalented person's truth. Let us see the accuracy of the whole behind the inaccuracy of the fact! How could the Gulf of Finland's ties to the lands

264

THE SILVERWHITE WAY

of the Volga be expressed more vividly than by al-Biruni's conviction that both are immediate neighbours. Or do we need to allude to al-Biruni's credibility?

He pronounced his own name 'Beiruni' and never failed to stress his Khwarazmian origins with characteristic bluntness. 'Son of Ahmed' suggests a child born out of wedlock in the Arab tradition. Al-Biruni's heritage is dim, but not his childhood: as a wunderkind, he would collect plants, hurry to the city's sole educated Greek, and have them identified, thus learning the Greek language. Wealthy patrons came not long after. He received an outstanding education and tasted the bitter bread of tyrants. It would be easier to list the fields in which al-Biruni was not active: his prowess extended from poetry to mineralogy, his favourite subjects included mathematics and astronomy, and to this he added pharmacology in the final years of his life. Later generations remembered him only as a geographer, though he was an encyclopedist in the most rigorous sense of the word, comparable perhaps to Leonardo da Vinci. Focusing on Greek geography, he strove to tie it to Arabs' practical needs and added a wealth of new information to knowledge of the North. He distinguished the Baltic Sea from the Arctic Sea, identified Varangians and Swedes among captains, and had heard of a man who sailed so far north that the sun no longer set.

Although al-Biruni only prized the Arabic language, Persian and Syrian were becoming increasingly important in his era. Times were uncertain and the caliph's power didn't extend much further than the area of Baghdad, but travel was unhindered nevertheless, even to 'pagan-Christian' Byzantium. Al-Biruni's shine had not faded before Mahmud al-Kashgari, one of the first Turkish authors, recorded rich Arabic observations of Turkish tribal migrations, folklore, ethnography, and neighbouring peoples (1072–4). The central point of his circular T-O map was no longer Mecca, but Yettishar – unconscionable impertinence from a barbarian! It was not missing the Volga, Volga Bulgaria, or the Gulf of Finland, and among the numerous tribal names was a mysterious people called the Sumal. Sámi?

Al-Biruni and al-Kashgari ruled the 11th century. Not before or after did the Middle Eastern gaze peer so far north. It was a favourable time. The foothills of the Alps were controlled by the Saracens, the blockade of the Mediterranean Sea was ongoing, Spain was blossoming as the western bud of Arab culture, Charlemagne (748–814) had died without delving fully into the subtleties of the art of writing, the unity of the Christian

SILVERWHITE

empire was crumbling into legend, and ascetic Aachen turned its dour gaze from south to north. By al-Biruni's day, the Silverwhite waterway had become the continent's main highway. Shortcomings turned into virtues. The ancient tribal route running along Europe's furthest peripheries suddenly became the sole corridor maintaining a connection between the Occident and the Middle East, conveying goods that neither side wished to relinquish.

What kinds of goods? And who was conveying them? Ibn Fadlan was interested as well.

NOSE AGAINST A WALL

'The king told me that beyond his country, three months' march away, there is a people called the Wisu among whom the night lasts less than an hour.' Ibn Fadlan brings up the Wisu on four occasions.

Once, a frightening stranger came down the river to Volga Bulgaria – a wild man of whose origins the king himself became curious. He decided to find out more, and shared these findings with Ibn Fadlan (Kovalevsky 1956: 138–9):

> I wrote to the people of Wisu, who live three months' distance from us, to ask for information about him. They wrote to me, informing me that this man was one of the people of Gog and Magog.
>
> 'They live three full months from us [the Wisu]. They are naked, and the sea forms a barrier between us, for they live on the other shore. They couple together like beasts. God, All-High and All-Powerful, causes a fish to come out of the sea for them each day. One of them comes with a knife and cuts off a piece sufficient for himself and his family. If he takes more than he needs, his belly aches and so do the bellies of his family and sometimes he even dies, with all his family. When they have taken what they need, the fish turns round and dives back into the sea. They do this every day. Between us and them, there is the sea on one side and they are enclosed by mountains on the others. The barrier also separates them from the gate by which they leave. When God, All-High and All-Powerful, wants to unleash them on civilized lands, he causes the Barrier to open [...]

The Bulgar king is corresponding with the ruler of the Wisu? Unbelievable. Even so, Ibn Fadlan's claim deserves serious attention.

After the Hungarians, the oldest Finno-Ugric alphabet and written language is that of the Komi-Zyrians. It was used by Russian Orthodox

THE SILVERWHITE WAY

missionary Stephen of Perm to translate church texts in the 14th century and was practised by Moscow iconographers even much later than that, until the Church's pious priests banned it and burned the holy books. However, Ibn Fadlan visited Volga Bulgaria in 922. What was used for correspondence in those times? We can assume that birch bark was the surface: the academic Peter Simon Pallas discovered birch-bark scrolls in Altai's Ablaikiti grave mounds in the year 1771 (see Finsch 1879), and rich post-war discoveries in Novgorod allow us to regard birch bark as the oldest paper of the taiga. Parchment was also used, or at least skin, a clue to which comes from the recently rediscovered Izhorian word *tshirjanahk*, 'writing skin' (Öpik 1970: 193). Izhorians used the word as late as the 18th century, at which time it was also recorded in writing. So, what strange alphabet was being practised in the distant backwoods of northeast Europe? Phrased in such a Eurocentric way, the question was destined to be left unanswered. So it was reworded: what alphabet might have been used by the traditional trade partners of Volga Bulgaria and the Middle East? The answer now received confirmed Ibn Fadlan's credibility: it was a north Iranian alphabet, once also a starting form for Georgian script, which would also explain the correspondence of some Georgian and Komi-Zyrian letters. Here there are no contradictions with Ibn Fadlan's account.

But what about the barrier at the end of the world, beyond which lives a 'beast'-like people? Doesn't this story recorded in the year 922 resemble another written in the early 19th century?

The epic hero Kalevipoeg decided to embark on an expedition and

> To the great world's end to travel
> Sail out to the northern borders,
> Where before no one had ventured,
> Ne'er a road had yet been made,
> Where the vault of all the heavens
> To the earth was fastened firmly. (XVI, 5–10)

A 'wise raven' was able to tell Kalevipoeg that 'gateways hidden away' could be found at the world's end. And in some instances, they would open:

> 'Where you see the bright blue water,
> Broadest billows of the waves,
> See if on the bank rise rushes,
> Iris by the water's edges;

SILVERWHITE

> Stamp there firmly with your right foot
> Mighty heel against the shore, –
> Then the earth its mouth so secret.
> All the gateways hid away
> Will come gaping up before you,
> Where you'll gain the end of the world.' (VI, 48–59)

The *Kalevipoeg* cannot be used as a historical source. But as August Annist, an annotator of the epic (Annist et al. 1961: 486), remarked, its compiler, Reinhold Kreutzwald, was said 'to have also heard "Mary of Magdalene", "Järvamaa", and "the Viru shore" when told about the quest for the end of the world and the barrier before it'. Thus, poetic symbols within the information that arrived from Wisu also featured in north Estonian myths.

Correspondence was held with the Wisu. The Wisu lived far from the people of Gog and Magog. The Gog and Magog survive on fish. The fish are extremely large, possibly schools of spawning salmon, possibly walruses or whales. And that's it.

Let's not be too quick to shrug. All will become clear in time. Perhaps the reason why the Bulgar king is so ambiguous lies in Ibn Fadlan's following statement (Kovalevsky 1956: 138): 'There are many merchants among them who go to the lands of the Turks and bring back sheep, and to a land called Wisu, from which they bring the skins of sable and black foxes.' Merchants guarded their trade secrets.

The Mashhad manuscript, to which we referred earlier, was not written by Ibn Fadlan's own hand. Yaqut al-Hamawi's writing comes the closest to Ibn Fadlan's original. In his geographical dictionary, he dedicated entire articles to Wisu, Bulgar, and the Itil. They contain information that had been lost from Mashhad's manuscript. However, al-Hamawi replaced the Wisu with the *Disuri* people in his text (Frähn 1823: 210–11): 'On this river [the Itil/Volga], merchants sail as far as the Disuri and bring back many furs such as beaver, sable, and squirrels.' Could sons of the desert recognise a beaver? Zakariya al-Qazwini (1202–83) left no doubt that the Arabs' *qunduz* was our beaver: 'The beaver is an animal of land and water that resides in the large rivers of Isu [also Aisu] and builds a den for its purposes on the riverbank.'

Everything having to do with furs is extremely precise. So why the geographical ambiguity? Legends, wandering topics? The mythical Gog and Magog, which not even the extremely critical al-Hamawi left out? 'The people of Gog and Magog lie a three months' journey away and are separated from them by a sea.' This odd national pair, or rather trio, is a

THE SILVERWHITE WAY

wandering topic of folklore. Gog, Magog, and Rōs Khros appear in the Old Testament as general terms for northern tribes. Through Byzantine writing, *Khros > rus* took hold as a general name for eastern Slavic lands, completely separate from the taboo name for an ancient Baltic ship (Popov 1973: 54–5). In the oldest Eastern literary tradition, the three ethnonyms all represented distant, unknown, and dangerous peoples. As civilisations' knowledge of the world expanded, Gog and Magog were pushed further and further northward. In the Koran (c. 644–56), they were as far northeast as one could travel. Alexander the Great (4th century BCE) was said to have built his mighty wall to repel them. The legend is of Syrian origin. Locating both peoples and searching for the wall have played an important part in the development of geography. One of the most famous Arab expeditions in search of the Gog–Magog wall was undertaken in the time of Harun al-Rashid's grandson, half a century prior to Ibn Fadlan. The traveller was known as Sallam the Interpreter. Based on later retellings, we know that he headed northeast. From this, we could conclude that by the time of Ibn Fadlan's journey, there was only room for the folkloric Gog–Magog in the periphery of northeast Europe – the northern Urals. When the 'end of the world' was finally reached, the people found there inevitably had to be Gog and Magog. They were given all of the pair's external features. However, an accurate ethnonym came in place of the mythical duo: *yura*. In the year 1030, al-Biruni knew of only three nations who lived in the North (Frähn 1823: 210–11): 'Few peoples live on that side [of the seventh] climactic zone, only those such as the Isu, Varangians, Yura, and others like them.' Al-Qazwini clarifies this picture (Hennig 1961: II, 255):

> The Bulgars bring Islamic swords to Wisu. They have no hilts or any type of adornment, being bare blades as they come from the forge. [...] If you hang them from a ribbon and flick them, they ring. Such swords are suitable for export to the land of the Yura, who spend great amounts to purchase them.

Flick them! Isn't that worth its weight in gold? The statement has the business-like tone of a merchant, the fastidiousness of a buyer, characters and relationships, but above all precision. Where did al-Qazwini get this? He doesn't conceal his source: 'Abu Hamid al-Gharnati tells us ...' he acknowledges at the beginning of the passage. If we could only find al-Gharnati's manuscript ...

It was found in an archive in Madrid.

SILVERWHITE

THE TWILIGHT ROAD TO TWILIGHT LAND

One moment. What are we searching for, anyway? Ourselves. What have we found?

1. First, the Bulgar lands.
2. A three-months' march away from Bulgar in an unknown direction is a land called Wisu, Isu, or Disur, from which furs are brought.
3. A three-months' march away from Wisu in an unknown direction is a people called the Gog-Magog, aka the Yura, who survive on fish or large sea creatures and purchase unfinished swords at exorbitant prices.

Let us present Abu Hamid's name in its entirety, as it has been shortened to various forms in literature and confused with other authors. Brace yourself: Abū Hamid Abū 'Abdallāh Muhammed ibn 'Abd ar-Rahīm al-Mazinī al-Kaisī al-Garnātī al-Andalusī al-Uklīshī al-Kairuvānī (1080–1169). Henceforth, he will simply be Abu Hamid, traveller.

He is curious, wealthy, pragmatic like a merchant, and has read rather little. As one can tell from his name, he was born in Andalusia, Granada. Starting from the age of 26, he studied theology and Islamic law in Alexandria, Cairo, and Baghdad. As a grown man, he travelled across Persia and the Caspian to the Volga and purchased a home there. He lived in Volga Bulgaria, where one of his sons died. Fifteen years later, he travelled to Volga Bulgaria again. From there, he went to Hungary, where his oldest son was married to two Muslim Hungarians, and bought another house there. A few years later, the restless man moved back to the Volga, and from there crossed the Khwarazmian Empire and Persia to Mecca. At the age of 82, in the city of Mosul, he embarked upon a career that is nowadays called 'writer'. He knew no prototypes, rules of genre, or stereotypes. He simply spoke in layman's terms about himself in the first person instead of genteelly referring to 'the lines of writers to whom the Prophet entrusted the quill', and with downright uncouth frankness acknowledges whom he heard this or that from every single time. This type of story did not fit the canon of a single existing genre. Rather, a new genre was born: the synthetic or cosmographic travelogue, which gained rapid popularity, created a whole school and imitators, and rescued several of Abu Hamid's passages from oblivion thanks to their numerous copies. But it was only in 1953 that César E. Dubler discovered the manuscript itself, long thought lost. It had ended up in Madrid.

THE SILVERWHITE WAY

Truly, what an odd man! Few before him, and even fewer who followed, could claim to have seen the Pillars of Hercules at Gibraltar, the famous Lighthouse of Alexandria, a burial chamber at the Great Pyramid of Giza, and, to top it off, have lived in the Khwarazmian Empire, on the banks of the Volga, and in the taiga. One house that he acquired was in Saqsin. Although the exact location of the city remains unknown, we do know what his house was like: pine logs, frigid cold outside, and sheepskin nailed to the door. These everyday details were within his faculties to describe. When the Volga iced over, he rushed to measure the river: 1,840 paces wide! No other southerner would have come up with such an idea, act, or statement. It would be like measuring length in litres! He delighted in everything, like a little boy. His was a charming personality.

Abu Hamid saw Bulgar as a giant city. He was amazed by the pine-log houses, the oak-log city 'wall', and especially the long summer days and short winter days. Summer was exceptionally hot, but the evenings were cool and the nights cold, at least for an Andalusian. In winter, the king organised raids to retrieve horses and women. The Bulgar residents could tolerate the cold because they ate great quantities of honey, which was said to be quite inexpensive in that region.

How far did the limits of Bulgar extend? Its area of influence depended on a range of factors. The Norwegian Harald Eriksson ravaged the land forty years after Ibn Fadlan's visit, and Bulgar struggled on as 'a tiny little town' (Krachkovsky 1957: 202). Apparently, it had recovered by the time of Abu Hamid. The traveller himself gave this answer (Mongait 1959: 169–81): '[Bulgar's] limits of power extend to the areas that pay it tribute. They are one month's distance from Bulgar. One of them is called Wisu, another Arv. There, they hunt beaver, ermine, and squirrel. The day there lasts 22 hours in summer. From there are brought very good, wonderful beaver skins.' This corresponds with earlier reports. Taking civil and nautical twilight into account, a 22-hour day would place Wisu at 59° latitude, or even a little south. According to the Koran, night begins from the moment that black thread can no longer be told apart from white, and Abu Hamid, as we can recall, had studied the Koran.

But what about 'Arv'? Dubler, who discovered the manuscript, surprised readers by stating that the name 'Arv' should be interpreted as the aristocratic Scandinavian name Arvastus. Slavicised forms of Scandinavian names often appeared in the Novgorod chronicles, but at a different time, in a different place, and in a different form.

SILVERWHITE

Still, Abu Hamid's reports of a mysterious trade partner do not stop there. A little later, he says that he was able to meet people from Wisu and Arv in Bulgar that winter. The account contains incredibly delicate details sketching their visages and appearances: 'They are pink-faced and have light blue eyes and flaxen hair. They wear linen clothing to protect against the cold, and some also cover themselves in skins made from very fine beaver pelts, the fur worn outward.' The foreigners from Wisu and Arv drink 'barley water', which, according to the Muslim who abstained from alcohol, tasted sour but caused a cascade of warmth to spread through his body. A knowledgeable reader will have no trouble identifying this description as beer or recognising the devout traveller as a covert beer-lover.

Abu Hamid's information about the regions neighbouring Bulgar are accurate. He describes ancient Mordvinian beliefs, the city of Kyiv (which he calls Gurud Kuyaw), and Hungary. Particular credibility is evoked by his brief remark that the Unkuria people live in Bashgird. These are, of course, Hungarians or, to be exact, the Magyar tribes that did not participate in the migration over the Volga to the lands beyond the Dnieper and the Danube, gradually assimilating with the Kipchaks and laying the foundation for today's Bashkirs. Connecting Bashgird and Unkuria underpins the traveller's credibility even when he speaks of neighbouring peoples and their endonyms.

Attempts have been made to decode Arv as Udmurt. We could place a greater degree of trust in Abu Hamid. He writes of Bashgird and Unkuria in tandem, and likewise Wisu and Arv appear side by side. Making a note of their white nights leaves no doubt that in his mind, both lie in the North. Would it be too bold to translate 1135–50 Arv as Estonia's Järvamaa? It would be plausible if we could prove that the name Järvamaa had spread internationally.

Here is the evidence given by the *Chronicle of Novgorod*: Järvamaa residents were known as Chud *Nereva*, *Ereva*, or *Erva*. True, the chronicler submits this proof a century later and *Erva* still isn't *Arv*. But in the 12th century, the pronunciation of the Baltic-Finnic word *järv* ('lake') was much closer to *jarv* or *arv*. We must wait for more evidence to equate Järvamaa to Abu Hamid's 'Arv'. But regardless of the results, we can already presume within the bounds of probability that the Arab conveyed a Baltic-Finnic toponym. We could place it in Järvamaa, but also in Valgejärv or Ladoga, critical hubs on the Silverwhite waterway. None other than encyclopedic al-Masudi himself disqualifies the latter possibility: Arabs already knew Ladoga by its Baltic-Finnic name two hundred years earlier

272

THE SILVERWHITE WAY

(Frähn 1823: 210–11): 'The Rus comprise many different peoples. One of them is called the Ladoghia [Ladogas], and it is the largest. They trade in Spain, Rome, Constantinople, and Khazaria [in the lower Volga]' (al-Masudi). And the Persian al-Hamadani, who in the 12th century held Ibn Fadlan's original manuscript in his hands (or a text very much like it), adds: 'The Rus are a populous people who live on an island. It is very damp there. [...] Every woman has a box[!] made of gold or wood attached to her chest, and the box is circular' (Nadzhib al-Hamadani, incorrectly also Ahmed Tussi, 12th century).

We must limit ourselves to this knowledge for now. Wisu/Rus/Ladoghia/Arv coincides with 59° latitude and a few other details that we need not reiterate. In one respect or another, all the names are tied to sailing. This doesn't allow us to draw new conclusions but confirms the old ones: ancient Baltic seafaring, i.e. what we have somewhat ambiguously referred to as 'maritime culture', was not a national phenomenon, but a societal one in which many coastal peoples took part, regardless of geographical advantages and the location of the waterways. But to be more specific?

Let us continue with Abu Hamid.

Up to this point, we have envisioned the Silverwhite route over several eras – the waterway a glinting ribbon from the Baltic to the Volga. Andalusian Abu Hamid adds a new dimension: the winter route. The following passage is not the first-ever description of skis, but it is the most accurate (Mongait 1959: 173):

> The road that leads to the Yura people is perpetually covered in snow. For movement, people use short boards that are made in various ways and attached beneath their feet. The length of a board is one *ba* [about 1.5 m] and their width one full foot. Both ends of the board curve slightly upward from the ground. The foot is placed at the centre of the board. There are openings through which strong leather straps are threaded and bound to the foot. The boards bound to the feet are connected by a long strap like a horse's reins that is held in the left hand. Held in the right is a long staff, the height of a man. At its base is a ball that is woven from wool, the size of a human head but very light. The rider pushes the staff into the snow and draws it back like a rower in a boat. In this manner, he travels quickly over the snow. If it weren't for this clever method of movement, then no one would be able to walk that road, as the snow is like sand; you could never wade through it. (al-Qazwini's travels in Bulgar, 1135–6 and 1150)

SILVERWHITE

Why should we be concerned about skis? What is the connection between skis and the history of Baltic exploration? Only an indirect one, unfortunately. The Baltic Sea was, as we may recall, the link between western Europe and the Middle East – a difficult and unavoidable leg that involved the greatest risks. From Eastern and Occidental perspectives alike, it lay at the furthest limits of actual geographical knowledge. This is simultaneously the source of relative ambiguity in historical accounts, which might drive us to desperation if we do not shape it into a virtue: concentric circles emanating from Eastern and Occidental cultural centres met on the Baltic Sea, forming cross-waves in turn, amplifying and sharpening each other, joining or separating in their oscillatory phases, reflecting or distorting in local folklore, languages, toponyms, and ethnonyms. The East and West lived in different traditions, spoke in different languages, and used different terminology. Simply through indirect data, mundane details, clothing, food, customs, weaponry, and trade items, it is possible to determine the location of a 'country', or at least its position in relation to a third 'country'.

Abu Hamid's description of skis is ethnographically accurate. This demonstrates his personal attentiveness or at least his conversations with merchants who knew the way to Yura. And what did merchants seek from Yura? According to Abu Hamid:

> Yura brings many different goods. Each merchant positions his goods separately and adds his personal stamp. Then, the men leave, and when they return, they find other goods that are demanded by their country. Each merchant finds something next to the goods he left behind. If it suits him, then he takes the traded good for himself; if not, then he takes the goods he previously left behind and leaves behind a new item, and this is done without a single violation of the rules. You couldn't say who is the seller and who is the buyer here.

This manner of doing business is called 'silent trade'. The Estonian saying *kaup on koos*, meaning 'the deal is done' (literally 'the goods are together'), dates to this period. From Abu Hamid, we know that the most sought-after goods were hiltless Arab swords made in Zanjan, Tabriz, or Isfahan – cheaper than those made by the Franks or Rus. The latter were forged from iron and cemented with steel, meaning they could be curled and would spring straight again. The complex technology of tempered Damascus steel, which sprang from Celtic forging traditions, peaked in Rhineland in the 8th–9th centuries and reached Estonia soon after: every

274

THE SILVERWHITE WAY

third 10th–13th-century spear tip found in Estonia is composed of an iron core with a Damascus-steel socket and blade (Selirand 1973: 114). Arab steel swords did not bend but broke. This had less to do with quality than dissimilar properties. Arab swords were preferred in certain cases; in others, Frankish. Unfinished Arab swords were exported to Yura, where they found a competition-free market and an entirely new application: 'It is said that if the Yura do not cast a sword, the type of which I spoke before, into the water, then they do not catch a single fish' (al-Qazwini).

Do you recall Ibn Fadlan? The Gog-Magog are given a great fish from the sea every day, and each 'comes with a knife'. And al-Qazwini: if you 'flick' the sword, it rings. Al-Qazwini shouldn't be criticised for twisting the causal connection in his travelogue. Unfinished swords were in high demand in the North for several centuries because they were used to make harpoons for hunting whales, and especially walrus.

We must applaud Abu Hamid for his unprejudiced spirit. How simple and tempting it must have been to ignore the episode about casting swords into water entirely! The travelogue would have only benefitted in terms of credibility, the author's reputation would have been boosted, and in that sense 11th-century Arabs weren't all that dissimilar from 20th-century authors. We must appreciate the man's attention to detail, because it contains the author's self-sacrifice, his self-loyalty. What's more, we should also recognise Abu Hamid's talent, for he preferred the extraordinary and unbelievable fact over the ordinary and believable. The swords were utilised as harpoons, but naturally also as weapons in battle. In 1472, during a long and intense war against the pagan Komi-Zyrians, the Orthodox missionary Stephen of Perm seized a mass of Asian swords – elegant confirmation of Abu Hamid's accuracy. Komi-Zyrians, the immediate northern neighbours of the Bulgars and eastern neighbours of the Vepsians, were indeed intermediaries in that lucrative trade with Yura, who were called the *Yugra* on the eastern bank of the Pechora River, as I recently (1970) confirmed. *Yugra* is an official word for the Mansi.

Before us is a chain that stretches from the present to the past without a single missing link: Mansi–Jögra–Yura–Gog-Magog. Vaygach Island is separated from the mainland by the narrow Yugorsky Strait: *Ugri väin* in Estonian and *Yugorsky Shar* in Russian, which is a Komi-Zyrian loanword from Mansi. There is no basis for connecting *shar* to the Scandinavian *skär*, as Ernst Kränkel recently attempted in his charming memoir – the similarity is coincidental.

SILVERWHITE

The preservation of Yura–Yugra–Ugri almost intact throughout the centuries leads our thoughts back to the Komi-Zyrian alphabet – a written tradition that seems quite fantastical in the endless primeval forests. Yet the Russian chronicle's oldest, temporally undetermined section already speaks of the 'White Ugri': their first wave crossed the Dnieper to the lands of the Danube in the early 7th century, and at about the year 898 the chronicle names 'Black Ugri' who reached the present-day homeland of the Madjars after crossing the Dnieper. To date, it is unknown what paths the Ugri name took to reach the pages of the chronicle and Old East Slavonic at such an early date.

Active traffic along the Silverwhite trade route and well-developed trade ties are likely the foundation, without which we cannot explain the existence (and preserved texts) of the Komi-Zyrian writing system.

Still, the Komi-Zyrians were not the only intermediaries in Northern trade. Arab authors also speak of a direct route from Wisu to Yura. It was not a transit route through the Volga Bulgarian city: the distance from Volga Bulgaria to Wisu was just as long as that from Wisu to Yura (which we will call 'Ugria' from this point forward). What we have is consequently an isosceles or even an equilateral triangle: the Volga Bulgarian city–Ugria–Wisu. That places Wisu at the mouth of the Gulf of Finland; the route from Wisu to Volga Bulgaria matches our Silverwhite waterway. Henceforth, we will call the flow of trade from Volga Bulgaria to the northeast the 'Ugria Road'. What's left is the triangle's northern line from Wisu to Ugria. We could call this the Twilight Road in accordance with the general Arabic word for Northern lands and assuming that such a route ever existed.

There is no reason to doubt the latter. The oldest description of the Twilight Road was given in the Austrian diplomat Sigismund von Herberstein's travelogue *Rerum Moscoviticarum Commentarii* (1549). The waterway ran from the Viena (northern Dvina) down the Vychegda River to the upper Vym, and from there to the Pechora, northeast Europe's last great river before the stone wall of the Urals. According to the analysis of Sergei Bakhrushin (1882–1950), a professor of history in Petrograd, both cases constitute an old Finno-Ugric waterway that the merchants of Novgorod started using for their trade expeditions to Ugria starting from the 12th century (Bakhrushin 1922: 18).

Is this the world's distant, terrible end?

The people living behind the wall edged further and further away for Arabs and Baltic peoples alike. There was still space for it at first, as well as time ...

276

THE SILVERWHITE WAY

Long ago, many people went to where the land and sky touched. They often went on towards the morning. There, they found a great, towering wall and could go no further. The wall was so tall that you could've easily gotten into the sky from it.

Once, several men went there, wanting to find out what there might be behind the wall. They were all skilled climbers. One climbed up to look. He got up on top of the wall. He still spoke to the others while climbing, but once he'd gotten to the top, he didn't speak another word. He just spread his legs and disappeared behind it.

Another went up to look, but the very same thing happened to him. The others no longer dared to climb.

It was believed that dog-headed people lived beyond the wall. They were beasts like no other. The front half of their body was like that of a man, the back like that of a dog. They had the sense of a human for three days, then the sense of a dog for three. Whatever they'd done in the first three days, they destroyed over the next three. When they had the sense of a dog, they went over the wall to kill people and eat them. (Tapa, in Eisen 1894: 125–6)

As the sailors and travellers pressed further, they made the Twilight Land less enchanted (Frähn 1823: 230–2):

In Bulgar, I was told of the land of darkness [Twilight Land], and certainly had a great desire to go to it from that place. The distance, however, was that of forty days. I was diverted, therefore, from the undertaking, both on account of its great danger, and the little good to be derived from it. I was told that there was no travelling thither except upon little sledges, which are drawn by large dogs; and, that during the whole of the journey, the roads are covered with ice, upon which neither the feet of man, nor the hoofs of beast, can take any hold. These dogs, however, have nails by which their feet take firm hold on the ice. No one enters these parts except powerful merchants, each of whom has perhaps a hundred of such sledges as these, which they load with provisions, drinks, and wood: for there we have neither trees, stones, nor houses. The guide in this country is the dog, who has gone the journey several times, the price of which will amount to about a thousand dinars. The sledge is harnessed to his neck, and with him three other dogs are joined, but of which he is the leader. The others then follow him with the sledge, and when he stops they stop. The master never strikes or reprimands this dog; and when he proceeds to a meal, the dogs are fed first: for if this were not done,

SILVERWHITE

they would become enraged, and perhaps run away and leave their master to perish. When the travellers have completed their forty days or stages through this desert, they arrive at the land of darkness; and each man, leaving what he has brought with him, goes back to his appointed station. On the morrow they return to look for their goods, and find, instead of them, the fur of the sable, squirrel, and ermine. If then the merchant likes what he finds, he takes it away; if not, he leaves it, and more is added to it: upon some occasions, however, these people will take back their own goods, and leave those of the merchant's. In this way is their buying and selling carried on; for the merchants know not whether it is with mankind or demons that they have to do; no one being seen during the transaction. (Ibn Battuta, 1829)

Sápmi and dog-headed men, a Great Wall and the end of the world, the land of souls and a perilous voyage back to one's home shore, all have, in one way or other, left a mark on Estonian folklore, shaping our ancient worldview. It would be pointless to search them for specific geographical footholds: time has washed them away, the centuries worn them smooth, and only the poetic metaphors have reached us in spite of being weathered, filled with ancient scents and colours. The following description of 'the end of the world' was recorded in Saarde in the late 19th century. Swept along by the gripping fantasy, the storyteller himself unconsciously became a character in the events of long ago:

In the year 1790[!], a ship set sail into the great world sea. Its captain had decided to sail until they reached the place where the edge of the sea meets the sky. They had seven years' rations on board. So, they sailed across the waters. Several months they sailed without their eyes resting on more than sky and water. One day, a great gale rose and tossed the ship about like a splinter. When the weather calmed, the sailors realised the ship was spinning in a circle and had stopped going forward. The captain saw the ship was caught in a whirlpool and wouldn't get out unless a shark helped. Red fabric was wrapped around an anchor, which they chained fast to the ship and tossed into the sea. After a little while, a shark bit and the ship started moving swiftly forward. They cut the chain and sailed onward.

After several days of sailing, we[!] saw a mountain in the distance. Boats were let down and six men rowed closer to take a look and they soon came back and said the sea was deep enough for the big ship to sail closer. The ship sailed to the place they pointed out and they let

THE SILVERWHITE WAY

down anchor. Now, they took a look around. On the other side of the mountain, they saw white people far below the opposite side of a wall. The men immediately decided to lower somebody down. A sailor was tied to a rope and lowered. The white people untied the rope from around him and it was pulled back up empty. The same thing happened with a second sailor. A chain was padlocked around a third and he was lowered down. Those people sure stroked and kissed and pulled, but couldn't get it off. The sailor was pulled back up, but he was mute [...].

The ship flew swiftly onward towards where the sky was said to touch the land. For several months, we saw nothing but sky and water. The Sun started getting shorter, and the day kept on getting shorter and shorter. Finally, the light went away entirely. Our hearts were filled with fear and we asked the sky for help. Our supplies were running out and no one knew where home was. The darkness kept getting thicker and thicker. The captain wouldn't come out of his cabin and the sailors were all in desperation. Several weeks went by that way. Then, the sailor in the crow's nest called out: 'Light! I see a city!' [...] The anchor was cast into the sea, a boat lowered to the water, five others and I the sixth [!] got in and rowed towards the illuminated place. [...] When we entered, we saw lots of stakes stuck straight into the ground. People had been hung from those stakes: one by the hands, another by the feet, a third by the tongue, a fourth by the chest, a fifth by the neck, and a sixth by the navel ... (Saarde, 1898)

The notion of a vortex in the sea where water spun down into the underworld at a terrible speed was common throughout the North. It belonged to common Baltic folklore and also appeared in Kreutzwald's *Kalevipoeg*:

> The whirlpool was about to swallow
> Craft together with her cargo.
> The Lappish sage he took a barrel,
> Varrak took a little cask,
> Covered with some felt of red
> All the outside of the cask,
> Bound it with the strongest fetters
> Hanging from the side of the ship,
> Should some fish then eye the bait,
> Quickly it would come to take it.
> On the waves appeared a whale

SILVERWHITE

Bent on catching that red bait;
In its mouth it gulped the cask,
Took itself off in a hurry,
Hauled the ship out from the whirlpool,
Saved the boat from the mouth of hell,
From the door of the underworld

(Kalevipoeg XVI, 638–56)

The Near East and the Baltic Sea? Two separate worlds?

The discoveries of physical Baltic and Arab objects also weave with poetry: every third coin found in Estonia has been of Arab origin. Every third coin has a path behind it ten times longer than that of the first two. We should speak of coin-kilometres in the fashion of tonne-kilometres to correctly assess the weight of the Silverwhite waterway and the proportions of coin discoveries. Nearly two-thirds of Arab coins have been concentrated along a coastline 25 kilometres long, demonstrating the waterway's importance in daily lives in ancient Estonia. Christian Martin Frähn, an outstanding 19th-century Arabist, determined that the coins were minted between 765–976. True, ten times more Arab coins have been found in Sweden, but they constitute an amount equal to Estonia's in relation to population. In addition, the silver treasures found in Sweden over the last two centuries have been concentrated not on a narrow strip of shoreline, but on the island of Gotland.

Further evidence of the harmony of Baltic maritime history, folklore, and trade routes is the fact that not one Byzantine coin has been found in Estonian or Swedish archaeological sites dating to that period: the former Finno-Ugric route remained dominant, by then the international Silverwhite waterway joining the Baltic Sea to the Near East via Ladoga, Valgejärv, and the Volga-Bulgar. Alas, its days were numbered.

VII

'THEIR GREATEST MIGHT IS IN THE SHIPS'

(VERSE 366 OF LIVONIA'S OLDEST RHYMING CHRONICLE)

FLYERS OF THE EARLY IRON AGE

'*Kalid* are logs used as levers when rolling ships along or prying up stones,' folk in Emmaste (Saareste) once said. In Harju County, ship levers were also called 'oxen', which is tied to the history of Härjapea ('Oxhead') River in one way or another. Although it is now channelled through underground sewers, Tallinn once stood on its banks. Over time, the name of the tool name broadened to refer to sailors in general. It seems that the term *kalimees* ('*kali*-man') or *kalev* was originally used for crew members who had pledged an oath of mutual aid and loyalty. If this is true, then one shade of meaning of the word *kalevipoeg* (literally 'son of *kalev*') could be 'guild boy'.

Kalevipoeg's stories are concentrated in Virumaa and have spread out from there. On the other hand, tales of sacred groves (Estonian *hiis*, Finnish *hiisi*) are rare or missing entirely among the Livonians, Latvians, and Lithuanians, in quite sharp contrast to Estonian folklore. Both can be explained by the Silverwhite waterway's location.

> 'Flyer' was the name they gave her,
> For in flight she cut the waves,

wrote Reinhold Kreutzwald in the *Kalevipoeg*. The ship's name – in Estonian *lennuk*, today 'airplane' – sounds overly modern with its modern association. In fact, it is the opposite. *Lennuk* was a colloquial word that vividly stressed speed and agility, flying from wave to wave. In Estonian, you say a busy person is 'in flight' (*lennus*), overlooking the dual meaning. If we recall the comparison of a ship to a village swing soaring high, then

Kreutzwald's creative wordplay even acquires the character of a genuine folk tapestry.

A ship flying on bird's wings ... Poetic metaphor for speed and lightness? That and much, much more.

The word for 'flying' was associated with a range of powerful natural phenomena and magical fancies. Variations are used, for example, in expressions ranging from a person achieving success to a sudden, inexplicable disease that strikes an animal. *Tulelendav* ('fire flying') is a folk word for 'meteor', and *vana lendav* ('old flyer') was another term for the mythical Estonian *lohemadu*, the Latin 'dragon'.

The idea of a flying ship is plainly connected to the belief system and taboo words that we discussed earlier. *Lohe* was a downpour, *lohemadu* a waterspout, and *tuulelohe* a whirlwind, but only *lendmadu* ('flying snake') signified a dragon itself. The line between symbol and reality was rather hazy and often infinitesimal in those times. When setting out to hunt, you could draw an ancient bull in the sand and then spear the image: the animal had been killed, you just needed to find it. A verbal symbol was similarly equivalent to reality. Merely speaking something aloud could conjure the undesired entity. To avoid manifesting a dangerous dragon (*lendmadu*), one would say 'old long', 'long-tail', 'mountain eel', and in even older writing (1638) 'whiteman' or 'flyman' (Stahl 1638: III, 198; IV, 44). Even so, it was most common to replace the dragon with the words *uss* ('snake, worm'), *huss*, or *uisk*.

In ancient beliefs, snakes were not repulsive or dangerous but creatures worthy of respect, keeping sickness away and bringing fortune to a household. People were fascinated by snakes' deadly poison and death-like hibernation, which allowed them to be associated with the land of the dead – the afterlife where one's ancestors carried on their daily activities. Snake worship, which sprang from ancestral honouring, was practised in full force even as late as the 19th century: the north Estonian *kodu-uss* and *maja-uss* ('home snake' and 'house snake'), Mulgimaa *maja-uisk*, and southwest Estonian *maja-tsiug* enjoyed general care and protection. House snakes were fed milk – giving water could bring misfortune.

Ibn Fadlan described (in 922) signs of snake worship along the Volga with astonishment. Centuries later, travellers visiting Old Livonia shared his sentiments (Laugaste 1963a: 72–3): 'Finns also share with Estonians the practice of feeding home snakes' (Thomas Hiärne, 1638–78).

Even so, Estonians' snake worship was not a 'pure' ancestral cult. There were additional influences. In Estonian tradition, the snake also acquires

'THEIR GREATEST MIGHT IS IN THE SHIPS'

and protects property. Adders were kept in grain chests. From wedding traditions, we recall that a bride's dowry chest was called a 'ship' in western Estonia. Ships were sailors' home on long voyages. Naturally, they wished to take their ancestors' power, protection, and might along with them. Any kind of relocation in those times was associated with much greater dangers than we can possibly imagine today. A person left the protection of their 'friendly' home spirits; they were suddenly helpless and alone. Therefore, a home snake was also taken to sea. Saaremaa captains brought a bag of sacrificed snake heads on board. Two centuries ago, a live snake would even be kept in a chest. On the other hand (Põldmaa 1973):

> Saaremaa captains once used adders as compasses. They'd take one to sea in a bottle. An adder always holds its head to the north. You could also use it to predict the weather. [...] Sailors said that as long as they had an adder on board (it was kept in a little crate), there was no fear of shipwreck. The snake brought fair winds and protected them from misfortunes.

The actual and symbolic participation of home snakes in nautical voyages was so important that it even left a mark on ship construction: bows were carved into representations of snake heads. And thus the name: *draakonlaev*, drakeship, dragonship. The foreign loan is misleading. As we've discussed, *lendmadu* was the Estonian term for 'dragon'. It was a taboo word at sea, so *uss* ('snake') was used instead. *Uss/uisk* gradually became the Estonian word for an ancient ship navigating the Silverwhite waterway. *Uss* meant 'ship', *uss* meant 'dragon'. And so, we return to where we began:

> 'Flyer' was the name they gave her,
> For in flight she cut the waves

Uss/uisk is still used to refer to a certain type of boat on the Väinameri Sea today — a mysterious monument to ancient associations that shaped the Estonian language, symbols, beliefs, and worldview.

We could have refrained from this diversion had it no connection to the Twilight Road, Wisu, Yura, and sources of written history.

THE CITY OF INTERNATIONAL FRIENDSHIP

Uisk spread from Estonian to the Russian language as a general name for a certain type of vessel. What's more, it appears repeatedly in the Early Middle Ages on the Twilight Road, on voyages to enigmatic northeast

283

SILVERWHITE

Europe, to the 'end' of the world, to the legendary Great Wall before the land of the dog-headed people.

Eastern Slavs moved north along the Dnieper waterway, settling on the shores of Lake Ilmen and becoming neighbours to the Baltic peoples. They came from the south, and the new environs were foreign to them. They borrowed a plethora of words to refer to the Northern forests, birds, agricultural methods, waters, and almost every species of fish. One of the very first words they borrowed was *laev*, 'ship', which only appeared in writing with the *Chronicle of Novgorod* in 1143: *loyva*. Based on phonetic history, we can determine that Russian borrowed the word *loyva* much earlier than that. The word was used simultaneously and later in Russian for a large two-masted vessel, usually with a foredeck. *Laevnik*, 'captain', was borrowed into Russian as *laybochnik*: 'a sailor, captain, or ship owner' (Kalima 1919: 60).

This was a ship with sails, of course, as linguistic history also documents. *Puri*, 'sail', appeared in Nestor's *Primary Chronicle* in the year 907. Conveying the Baltic-Finnic *u* with the symbol pronounced *yer* shows that the word entered Russian along with the very earliest Baltic-Finnic loans.

The hub where these nautical terms were borrowed appears to be Lake Ilmen: the crossroads of the ancient Silverwhite waterway and the new Dnieper waterway, along which a favourable trading post soon developed – the centre of future Novgorod. Large-scale archaeological studies undertaken in the 1960s and 1970s produced several discoveries comparable in importance to the ancient finds at Helgö. It has now become clear that Finno-Ugric peoples, Balts, and Slavs all played a part in Novgorod's birth. The fortress town comprised three different linguistic communities that were also equally represented in the famous Novgorod *veche* – the governing authority until 1478. This also explains the reasons why Novgorod appears under various names on older world maps.

The *Primary Chronicle* underlines the role Baltic peoples played in the development of the Russian state. It's as if the chronicler remembers he is speaking of north Europe's oldest inhabitants and places them next to the Slavs: 'In the share of Japheth lies Rus', Chud', and all the gentiles' (*Primary Chronicle* 1953). Against this backdrop, we should assess the information of varying weight and tendencies that concerns Baltic-Finnic captains and voyages.

The oldest and partly legendary stories are from the year 882. A prince named Oleg, known in Norse chronicles as Helgi, joined Novgorod's lands to Kyiv and thus subjected the entire Dnieper riverway to one rule. It was

284

'THEIR GREATEST MIGHT IS IN THE SHIPS'

a remarkable event that the meticulous chronicler appears to remember well. Many 'Chuds' participated in Oleg's military campaign, listed immediately after the Scandinavians. When his fleet reached Kyiv, Oleg posed as a merchant en route to Constantinople. It was a military tactic. Merchants apparently enjoyed immunity. In this way, he managed to lure its rulers Askold (Höskuldr) and Dir (Dyri, also Tiuri, Turi) to the riverbank, where they were killed (Nissilä 1972). We needn't trust the legend's fine details, as the chronicle was written three centuries later, but certainly its spirit: the author regarded a mercantile voyage of Scandinavians and Baltic Finns to Constantinople as an entirely plausible excuse. Another raid on Constantinople was made during the reign of Prince Igor/Ingvar, pillaging the shore 'as far as Heraclea and Paphlagonia' and ending in 944 with a peace treaty 'henceforth and forever'. The treaty is important from a seafaring standpoint, as it presents statistical material on Baltic-Finnic navigation. The forty-five envoys who signed the document represented the prince of Kyiv and merchants of the Dnieper waterway. These are the names recorded: Ivar, Vefast, *Isgaut*, Slothi, Oleif, *Kanitzar*, Sigbjorn, Freystein, Leif, Grim, Freystein, Kari, Karlsefni, Hegri, Voist, *Eistr*, Freystein, Yatving, Sigfrid, Kill, Steggi, Sverki, Hallvarth, Frothi, and Munthor; and the merchants Authun, Authulf, Ingivald, Oleif, Frutan, Gamal, Kussi, Heming, Thorfrid, Thorstein, Bruni, Hroald, Gunnfast, Freystein, Ingjald, Thorbjorn, Manni, Hroald, Svein, Styr, Halfdan, Tirr, Askbrand, Visleif, Sveinki, and Borich. In 1941, Russian historians wrote that from the list, Isgaut and Kanitzar were certainly Estonians, and Eistr was a Finn. We will allow linguists to decode the names further. As mentioned, the chronicle was written three centuries after the events in question, and even the Scandinavian names were greatly distorted in the original text.

The credibility of the chronicle has been reinforced by archaeological coin discoveries: Byzantine coins started appearing in Estonia from the year 950. All earlier Eastern coins are of Arab origin and were minted in the Near East, North Africa, Samarkand, Tashkent, and Bukhara (Selirand and Tõnisson 1963: 181; Lewis 1958: 214–19). The line is quite sharp and temporally coincides with the events described in the chronicle.

Can linguistics and chronicles be used to detail Estonians' nautical culture?

From the *Chronicle of Novgorod* in the year 1190, we find another type of 'coastal Estonian' vessel in addition to the *loyva*: the *shneka*, as the chronicler calls it. This is the well-documented Norse *snäkkia*, 'the galley's little sister'. We'll speak of its appearance later.

285

German nautical historian Paul Heinsius (1956: 144) claims that ancient Baltic ships lacked any cooking facilities. Linguistics proves this false. Estonian lent to the Russian language *alus/alas* > *alazh*: a place for cooking at the bow of a ship. The 'ship' meaning of the Estonian *alus* indeed came from shared meals, which used to accomplish a great ritualistic task. Eating together brought the crew together, tying them by common 'flesh and blood' as the Catholic ritual still holds true. *Alus* may have originally meant a family, then a crew, and metaphorically also the ship itself, just as we might interpret the word 'family' in a narrower nuclear sense and 'farmstead' in a broader sense. Lending the name of a cooking nook to Russian appears to suggest the true gravity of a seemingly insignificant detail.

The Volkhov River connecting Novgorod to Lake Ladoga is incredibly turbulent and difficult to navigate. One 12th-century illustration depicts towing a boat upstream: a horse has been hitched to the bow and its sail is billowing backwards. This was not ignorance on the artist's part: sails were used for steering, keeping the vessel in the middle of the river or near the bank where the current is weaker at a bend. Horses were inseparable from ships during the ancient era of Baltic seafaring. Their importance can be seen from the wide distribution of grazing islands along ancient waterways. The animals were crucial in pulling a boat against the current, but could simultaneously signify the danger of attack: mercantile voyages turning into raids. Limitations applied. Special grazing islands were located near larger trading harbours. If a captain had peaceful intentions, then he left the horses behind. Some known grazing islands might include Hobulaid (literally 'Horse Islet') at the mouth of the Bay of Haapsalu; Russarö (also 'Horse Island'), three miles south of the Hanko promontory; likely Naissaar (Russian *Konevets*, *kon'* > horse) at the mouth of the Bay of Tallinn; and finally Konevets Island in Lake Ladoga. A horse-worshipping ritual practised by the Votians and Izhorians, one in which only men took part, may also date back to the earliest days of seafaring.

Ships were steered with a long tiller that even in modern Estonian dialect is called a *saps*. This extremely important term was subsequently lent to Russian: *sopets*. Tsar Peter I used the word when drafting his navigational regulations: 'all gunners and sailors must each guard his place while standing by the wheel or *sopets*' (Kalima 1919: 46). The Baltic-Finnic term for tiller spread across the vast Komi-Zyrian territory and ultimately worked its way into the Irtysh-Tatar lexicon. Sound unbelievable? Just recall the ritual wedding oars of the Khanty! The symbology of taiga boats naturally focused

'THEIR GREATEST MIGHT IS IN THE SHIPS'

on the tiller, which guided the fate of the vessel and its people. Here, we can also understand why the word for 'tiller' is native to Baltic languages, but not that for 'helmsman'. There were no parallels to the Norse skipper in the distant past or the newer era. Skippers' rights were broader than those of today's captains, but they were not the owners of the vessel. The ship, as I have said earlier, was common property of the community, village, or family. The crew were simultaneously stevedores. The helmsman's share in the cargo might have been greater, and with his knowledge of distant lands and languages, he probably also had representational duties, thus leading to the adoption of the international term.

> Skipper, boy,
> hold the reins, turn the ropes,
> our seas here are shallow,
> our reefs here are sly,
> our capes four-cornered!
> (Haljala, Selja)

We only listed terms that can be used to draw certain conclusions about ancient nautical culture. These conclusions are buttressed by a rich selection of more minor terms that disseminated from ancient Baltic seafarers into the Russian language: altogether over sixty words used in navigation and ship construction. The Baltic-Finnic *lämsä*, originally 'lasso' or 'reins', was lent to Russian as *lyamka* and adapted to mean 'traces' (in the equine sense). The Komi-Zyrian and Udmurt *lamka* likewise mean 'ship traces'. When the bard in the song we just cited appeals to the skipper to 'hold the reins, turn the ropes', we interpret it primarily as a poetic symbol for the tiller. The Estonian expression 'hold the reins in your own hands' equates to 'direct your fate'. A seaman would have a somewhat more specific interpretation: the sheet. Yet, using draft horses to move against the current and, ultimately, the appearance of 'reins' in nautical terminology offer tempting opportunities for contemplating questions such as how a specific object can pale into an abstract idea in folklore, then densify into poetic symbol in turn and thus in turn resist the eroding flow of time in the bard's memory, all without the singer himself necessarily being aware of the original connection. If archaeology is the skeleton of distant history, then linguistics and the decoded symbols of folklore can be regarded as prehistory's blood and breath.

Thus, we have circled back around to dragons and flying ships. This time, they soar onto the Twilight Road.

287

SILVERWHITE

The third linguistically documented type of Estonian vessel is the *uss/uisk*. It can be found in the *Chronicle of Novgorod* in 1320 as *ushkuy* and in 1473 as *oskuy*; it also appeared in the *First Pskov Chronicle* in 1463, and was even used by Tsar Peter I himself (Kalima 1919). This is the most significant loanword from a perspective of geographical exploration, as the 'snake ship' led to a unique class of sailors in Russian: *uss*-men or *ushkuynikud* (e.g. in the *Chronicle of Rostov*, 1371). For several centuries, they ventured along the Twilight Road and in the land of the dog-headed people. Among others, the Russian chronicles have provided us with the name of one such *uss*-man: Samson Kolyvanov, who in 1357 was killed along with his companions in distant Ugria. Here, we are interested in his compound surname: Kolyvanov, 'of the Kalevs'. Instances of *kali*, *kalev*, and *kalvi* are in no way rare in the Russian historical tradition. 'Russian law' itself mentions the Kylfings on three separate occasions. They enjoyed a legally advantageous position, regardless of their paganism: whereas ordinary people were required to produce two witnesses in court, a Kylfing needed only one. Eleventh-century Byzantine sources call the Kylfing '*koulpingoids*' and the oldest Norse sources '*kulfings*', differentiating them from the Varangians. The Kylfings of *Egil's Saga* demand tribute from the Sámi around the year 885. Their territory of Kylfingaland is situated near Sweden and Varangia. Linguists believe the word 'Kylfing' derived from *kaliväge*, 'forces of Kali' (Saareste 1951). The Baltic *a* turned into an *o* in the oldest Russian loanwords (e.g. *laeva* > *loiva*, *Karjala* [Karelia] > *Korela*), and Estonia's *v* corresponds to the Russian *b*. The Russian toponym *Kolyvanskaya zemlya* meant north Estonia, the coastal lands of the Silverwhite waterway. The Moscow historian N. Gordeyev arrived at the same conclusion from analysing northern Russian toponyms, in which the ancient *kalev* has been preserved to this day (Kolbyagi, Kolbezhitsy, etc.) (Gordeyev 1973: 67). A 12th-century birch-bark record discovered in Novgorod that mentions the Kylfings appears to confirm these conclusions as well.

So, Russian historical tradition recognises Baltic-Finnic seafarers by two names: *kalevs* and *uss*-men. Let us add colour to the landscape. In a dissertation defended in Uppsala in the year 1734, Peter Niklas Mathesius says the following about the ancient *kalevs*' activities in Lapland (1734: 18–20):

> Long ago, there lived a giant named Kaleva. He had twelve sons. Three sons began building a fortress here on the lands of Päralaht. One, who was called Soini, was said to have built his castle in the middle of the

288

'THEIR GREATEST MIGHT IS IN THE SHIPS'

lake where Limingo Church now stands; the second built his of granite and earth in Pühametsa. And although the forest has buried it, you can still make out the embankment and two gates. The third, who was named Hiisi, built himself a towering fortress 10 miles due east of the Kajaani fortress, and it was an extremely frightful and terrible sight. He built it of massive stones and earth amid the bogs and marshes, steps leading up to it, a *süld* of land [2.13 m] between each, and with two big gates. He called it Hiiulinna, Hiiu Fortress. That's why even today, when the people of Kajaani are angry, they say, 'Go to Hiiu!' just like some in Sweden say 'Go to Blåkulla [Witch's Hill].' Soini once went to visit Hiisi, rowing 13 miles through tremendously great rapids in a single day. He pulled himself up through all the rapids, sitting, until the roughest of them all that is called Pelli, where he had to push with his knees. The place where night found him is now called Soini Cape, and even today you can see the traces of the boat, as big as those on a tar boat. Now, not even five of the strongest men could manage it in such a boat over four, or even five days, but he came upstream in just one.

The 1114 entry in the *Primary Chronicle* records a voyage of the *uss*-men heading towards Ugria. Do you recall the 'Wisu' ruler's letter to the Volga Bulgarian ruler Almış about the people beyond the mountain? 'They are enclosed by mountains on the [other sides]. The barrier also separates them from the gate.' The description of the *uss*-men is remarkably similar. The features of the land beyond the mountains hadn't changed notably over the course of two hundred years (*Primary Chronicle* 1953: I):

'I sent my servant', said he, 'to the Pechera, a people who pay tribute to Novgorod. When he arrived among them, he went on among the Yugra [Ugrians, Mansi and Khanty]. The latter are an alien people [speaking in a strange language; consequently Komi-Zyrian was intelligible?] dwelling in the north with the Samoyedes. The Yugra said to my servant, "We have encountered a strange marvel, with which we had not until recently been acquainted. This occurrence took place three years ago. There are certain mountains which slope down to an arm of the sea, and their height reaches to the heavens. Within these mountains are heard great cries and the sound of voices; those within are cutting their way out. In that mountain, a small opening has been pierced through which they converse, but their language is unintelligible. They point, however, at iron objects, and make gestures as if to ask for them. If given a knife or an axe, they supply furs in return."'

SILVERWHITE

The nautical traditions of the Baltic peoples and eastern Slavs interwove in the 'New City', rapidly developing Novgorod, but it was still the era of the Silverwhite waterway. A certain type of Volga Bulgarian jewellery travelled past Narva and spread there along with its foreign name, *tallukar*, giving evidence of the direct Turkish/Tatar cultural influence on the northeast Gulf of Finland.

Perhaps al-Biruni was right when he spoke of contact between the Gulf of Finland and Volga Bulgaria? It's not impossible that even the name 'Narva' is a memory of an international trade route.

Norse captains dubbed the Strait of Gibraltar 'Norvasund'. It isn't difficult to recognise the root word *narwa*, which in Norse languages meant a small opening, incision, or scar, but primarily an access, strait, or narrow, just as the last-mentioned English word reflects today. Lauri Kettunen has associated 'Narva' with the Vepsian word *narva*, *narvainen*, meaning a threshold or shallow place in a river. In 1972, the Finnish linguist Viljo Nissilä pointed out the high frequency of Finnish toponyms bearing the *narva* root, ruling out Vepsian origins. He concluded that their origin remained the Norse *narwa*, which spread in turn to Vepsian and northern Russian dialects. Thus, Narva could have several renowned namesakes: in addition to Gibraltar, Norway's Narvik – 'narrow bay'.

Apart from Arab coins in plentiful archaeological sites, we on the Baltic have also preserved a memory of the Sanscrit *rupya* – *robi* in Estonian dialect and *ropo* in Finnish dialect (Paul Ariste to the author, 9.4.1969).

The ancient Finno-Ugric waterway functioned as the Silverwhite trade route for two and a half centuries. By 1060, it had peaked and begun its decline, destined to ultimately disappear. The decline of the Silverwhite waterway reflects the rise of Novgorod and the Slavs' arrival on the shores of Lake Ladoga.

Perhaps Estonian folklore has preserved a memory of those dramatic events as well:

> The silk ships sailed past,
> the coin ships went along ...
> (Haljala, Selja)

GEESE EAT POLISHED SILVER

The two and a half centuries of the Silverwhite route was also seen as a long period of time back then. Coastal peoples on the Baltic grew accustomed to monetary circulation.

'THEIR GREATEST MIGHT IS IN THE SHIPS'

Row, boat, go, boat,
row, boat, to the land
where the roosters eat gold,
roosters gold, hens tin,
geese polished silver,
fine birds pennings,
crows old coins.
(Jõelähtme, 1888)

A reduction in the import of Arab coin forced Baltic peoples to begin minting their own. Denmark began in 930, followed by Sweden, Norway, Poland, Schleswig, and Sigtuna in central Sweden. Even so, only the Danish currency evolved into a national system. Is it mere coincidence that the inflow of Arab coins into Estonia lasted the longest (Rasmusson 1963: 135–51)?

Just ten years, twenty at most, after the cessation of Arab silver imports, Gotland rose to become the Baltic Sea's coin hub.

Visby's successful coin initiative speaks to the island's leading mercantile and organisational position in the Baltic Sea. Unlike coins minted among its neighbours, Gotland's were disseminated widely: to Sweden, Norway, Denmark, northern Germany, and, of course, Estonia, though it didn't last long. Revala became Visby's rival and, ultimately, its successor. Not three centuries passed before Queen Margaret I was forced to pledge the city of Stockholm to Tallinn as surety.

Coins have a direct tie to seafaring.

The relationship between Gotland and Estonians was stamped by a trade agreement, Estonians' first international legal contract, of which some details have been preserved (Moora 1968: 525). These contracts do not so much reflect closer ties as they do a simple need to conclude pacts – in other words, danger.

The Baltic's pivot was not painless. When the Silverwhite waterway floundered, a flurry of attempts was made to establish new waterways in other estuaries.

THE SAGAS' POETRY

The Norse sagas leave no doubt that their authors firmly believed Estonia was directly on the Silverwhite waterway, which Scandinavians knew as the 'east road', *austrvegr*. Mentions of the Estonians reach back into the 6th century. According to the *Gutasaga*, Goths attempted to settle

SILVERWHITE

on Hiiumaa, where the populous settlement on the Kõpu Peninsula had died out by the early Common Era. Somewhat more elaborate was 13th-century Icelandic historian Snorri Sturluson, who wrote the *Ynglinga Saga*, which was based on earlier sources. The Estonians, men of the Silverwhite way, conducted multiple raids against the Swedes. In about the year 600, Uppsala king Yngvar (*Yngvar konung*) decided to exact revenge and set out on a counterraid. It ended in failure:

> Certain it is the Estland foe
> The fair-haired Swedish king laid low.
> On Estland's strand, o'er Swedish graves,
> The East Sea sings her song of waves;
> King Yngvar's dirge is ocean's roar
> Resounding on the rock-ribbed shore.[1]

Sturluson naturally refers to the Baltic as the East Sea, *Austmarr*, and other names are also in agreement. The Estonian forces are those of the *eistneskr* and he refers to Estonians as *eistr*, *aestr* – the same ethnonym used by Tacitus. Preceding the song is a block of prose which contains other significant evidence.

Prior to Yngvar's raid of vengeance, the Swedish dominions had been sacked by *Dönum ok Austrvegsmönnum* – Danes and Men of the East Road. Who were the latter? Snorri Sturluson answers in the next sentence.

First, Yngvar made peace with the Danes. Then, he turned his warriors against the East Road, bringing him to Estonia, *Eistland*. In Snorri's eyes, the ideas of the East Road, Estonia, Men of the East Road, and Estonians overlap. The famous Icelandic historian lived in Norway for several years. We should treat his chronology and tendency to venerate kings with scepticism, but not his geography or the kings' shameful defeats. Snorri's saga, which is often also referred to as *Heimskringla*, 'the circle of the world', after its opening words, contains, in addition to genealogical and mythological constructions, a wealth of ethnographic and geographical information. This information is believed to have come from travellers and is, in any case, the most accurate part of the saga.

King Yngvar fell in battle in Estonia. According to Snorri, the battle happened in a place called *al Stein* – 'near the stone/cliff'. The king was then buried 'close to the seashore under a mound'. Both the mound and the cliff were located in *Adalsýsla*, which has been identified as the northwest coast of mainland Estonia. Yngvar's strategic plan was aimed at the Men of the East Road's most critical East Road stronghold. It

'THEIR GREATEST MIGHT IS IN THE SHIPS'

would seem highly unfounded to tie the event to Ridala or Lihula (Kruus 1924: 14). Rather, we can read from the description a reference to the cliff at Toompea, the mightiest natural stronghold along the Silverwhite waterway. Nevertheless, closer determination of the semi-legendary king's burial site is of lesser importance. What matter much more are Snorri's general geographical beliefs.

The *Heimskringla* is not the only work that identifies Estonia with the East Road. In a history of Norwegian kings called *Fagrskinna* ('fair leather', after its parchment), King Eirik invaded 'Eistland, and plundered in many other countries around the Baltic. He also raided extensively round Sweden and Gotland. He went raiding north in Finnmork and all the way to Bjarmaland' (Finlay 2004). The toponyms and conclusions of this passage agree with the story of Yngvar. The *Fagrskinna*'s Estonia is *Eistland* and its East Road *Austrvegr*. Estonia is a narrower concept and the East Road broader; Estonia is part of the East Road. The passage also connects Estonia to Finnmork (Finnmark, Sápmi) and Bjarmaland (the northern Dvina basin). Here, the logic is also sound: both lay along the Silverwhite waterway.

Yet another source speaks of a connection between the Baltic Sea and Bjarmaland: the Icelandic monk Nikulás Bergsson, who died in 1159. Having stated that the ocean connects to the Baltic Sea near Denmark, that north of Denmark lies Norway, and to the east of Norway lies Sweden, the learned man continues as follows: 'East of Denmark is Lesser Sweden, then comes Öland, then Gotland, then Hälsingland, then Vermland and the two Kvenlands, located north of Biarmaland. From Biarmaland stretch uninhabited lands northward to the limits of Greenland'.

This text is anything but simple. Hälsingland is a Swedish province about 150 kilometres north of Uppsala. Vermland, however, lies to the west of Uppsala. The monk's pointer must have traced the Swedish shore to the Far North and then arced back down to Uppsala's latitude. There, it leapt east across the sea: *Kvenland* is Finland. Adam of Bremen called it the land of the Amazonians and placed it just slightly north of Estonia. From there, he drew a brisk line to the east and northeast: over Biarmaland 'to the limits of Greenland'. Ancient seafarers believed that Greenland was connected to the northern European continent via Novaya Zemlya. Nikulás's pointer had quite the tempo! But what interests us lies elsewhere: why did he divide Kvenland, Finland, into two?

The monk's list appears to confirm the ancient waterway. There could be no other way to order the geographical knowledge and secure it in

SILVERWHITE

memory. From the latitude of Sweden's Mälaren, he crosses the sea to 'both' Finlands. This corresponds to a nautical course running along the shore of southwest Finland, across the Gulf of Finland, and onward to the east along Estonia's coast. We should remember that Finland was an unimportant region in the 12th century: its population was nearly four times smaller than that of Estonia, and the Baltic Finns' linguistic divergence was not yet significant. A foreigner might not even notice the difference. In this case, the 'two Kvenlands' would mean Finland and Estonia. But both are north of Biarmaland? The claim isn't incorrect if you follow the Silverwhite waterway further to the southeast. And only then can we interpret the next apparent contradiction in the monk's text, simultaneously claiming that the territories to the north of Biarmaland are uninhabited. This should be understood as the tundra belt along the Arctic Sea east of the lower Dvina, upon which a seafarer's fantasy projected a non-existent link to Greenland.

Unexpected support for the 'two Finlands' hypothesis (it can't be seen as anything more) comes from a document drafted at the papal palace at Rome in the year 1120. It lists the bishoprics of Sweden and notes that the 'counties' of *Findia* and *Hestia* – Finland and Estonia – are under the authority of Sigtuna. The Church proceeded from the practical needs of missionary work. An intimate knowledge of local language was crucial in the eyes of clergymen. Was that why Nikulás abandoned a former naming tradition and spoke of two Finlands instead of Eistland? Could this point to his or his informant's personal knowledge of local languages?

In distant Rome, and even in nearby Bremen, *Findia* and *Hestia* were imagined as islands. Even later authors believed Estonia was an island or a cape. Clearly, this reflects Estonia's favourable position at the mouth of the Silverwhite waterway.

Estonia is part of the East Road; Estonians are Men of the East Road; Estonia and Finland have probably borne a common name or been considered a pair; raids to Estonia continue on to Sápmi and Biarmaland. These conclusions match Swedish gravestone statistics: on forty different occasions, 120 runes mention the toponym *austr*, 'in the east', as a site of battle. This above all encompasses Estonia and Finland, but also the entire Silverwhite waterway to the distant southeast and south. On twelve occasions, runes name Estonia, Courland, or other Baltic sites; Russia eleven times. The closer to the silver crisis, the more frequently. The number of toponyms culminates in 11th-century runes (Melnikova 1973: 60–1).

294

'THEIR GREATEST MIGHT IS IN THE SHIPS'

This time of unrest is best described by the imprisonment of Norwegian crown prince Olaf Tryggvason. Several variations of the story are preserved in the sagas. All except for one say that Olaf's captors were the Estonians, but apart from the Thiodrek version, none name the site of the prince's imprisonment. Thiodrek's manuscript, which was written two hundred years after the event, suggests Saaremaa as the place (*Antiquités russes* 1850: II, 85, 416–17).

The legend goes as follows. The prince sailed with his mother and men to take refuge from his father's killers with the prince of Novgorod. Estonians captured the ship. Olaf, his mother, and the crew were imprisoned and sold into slavery. Six years later, Olaf was bought free by a tax collector and taken to Novgorod.

The most credible parts of the saga, written to eulogise the Norwegian king, are those that speak poorly of Tryggvason. At the end of the passage, the young heir to the throne slays his enslaver, whom he encounters by chance at a Novgorod market years later. The moral 'evil will be repaid' is overly conspicuous. However, the elements of his enslavement and purchased freedom are credible and allow us to make certain generalisations.

Snorri Sturluson describes the episode as follows: 'As they sailed out into the Baltic, they were captured by vikings of Eistland, who made booty both of the people and goods, killing some, and dividing others as slaves' (Sturlason 1907). The nautical route from Scandinavia to Novgorod was too well-known to require any more precise geographical details. A general eastern course needed no explanation. Yet the fact that Snorri brought attention to it anyway means he must have had a particular thought in mind; an explanation for what followed; a causal tie between maintaining an eastern course and the Estonians' arrival.

Ancient coastal rights explain these circumstances as follows. Navigating eastward, the Norwegian ship found itself in coastal waters that would nowadays be called territorial. Their extent was estimated as the line of sight from the shore to the horizon, approximately 12 miles. Tryggvason's companions did not fulfil the requirements of common law and were forced to pay a dear price. The young man was separated from his mother, and his old foster father Thorolf was slain before his eyes. Understandably, the events left a deep imprint on the young man. Later, as the first ruler of all Scandinavia, he signed an agreement with the Anglo-Saxon king Aethelred to nullify shore rights. The pact was rather limited, just as their authority was at the time. The rulers agreed not to sell into slavery the crews of ships that ended up in their personal possession or landed within their realm. Instead, the sailors would be allowed to leave as free men and take from the ship whatever they could carry upon their backs (Niitemaa 1955: 44).

It's not clear from the saga where exactly the Norwegian ship was captured and its crew sold into slavery. In a document dated 1254, two of the five mentioned harbours on Hiiumaa are Sõjasadam (Sottesattema) and Orjaku (Oryocko). The latter is believed to have been a hideaway where slaves were kept until they could be sold. Tryggvason was exchanged between three masters. Ultimately he ended up in a family where he was kept 'long and well' and was 'greatly loved' by his master. The first two transactions happened shortly after his capture and must have been made on slave markets like that at Orjaku.

Tryggvason's liberation is a miraculous, fairy tale-like episode in the saga. It is described colourfully and reverently, fitting for a future king. Olaf Tryggvason's 'remarkably handsome' features were said to have caught the attention of one Sigurd. 'Sigurd came as a man of consequence, with many followers and great magnificence.' He was there to collect taxes and rents from King Valdemar of Holmgard (Novgorod). Their meeting at the market was incidental. What's more, it turned out that Olaf was Sigurd's nephew. It all fits the bill of a folkloric travel tale.

However, Olaf was recovered at a price nine times higher than that of any ordinary slave. This is in accordance with the overall royal accolade, but also contains more mundane details: firstly, it makes clear that harbour or market peace applied in the place where Olaf resided – a law that not even 'a man of consequence, with many followers and great magnificence' dared to violate, even if it involved freeing the heir to the throne. Secondly, the passage describes the market's environs: 'Sigurd asked him (Olaf) to go to his master', or, in another version, 'the farmer went home after (meeting at the market).' Thirdly, it turns out that the master was very aware of his slave's high status and didn't burden him with heavy labour. All aspects totally correspond to the rules of shore rights and market peace. Shore rights reflected the idea of national sovereignty. Sigurd did not regard the capture of the Norwegians' ship as an illegal act. That very same notion of sovereignty also allowed harbour peace to be enforced in certain places, at certain times, and under certain conditions, supplanting shore rights. Foreigners like Sigurd were allowed to visit such places, as they were untouchable in the harbour and at its market.

Thus, at the heart of Tryggvason's story lies a ship that violated shore rights. The guilty parties were held hostage and freed in exchange for

'THEIR GREATEST MIGHT IS IN THE SHIPS'

a ransom appropriate to their status. We see a system of shore rights and harbour peace by which both parties abided. We see a marketplace governed by international law and the permanent settlement around it. Perhaps a dozen sites around Estonia corresponded to these conditions in the late 10th century, including Tallinn – or was it primarily?

According to legend, Copenhagen was founded by Absalon, bishop of Roskilde and later archbishop of Lund. In approximately the year 1180, the prelate invited a 30-year-old man named Saxo who had studied liberal arts in Paris to come and work for him. Saxo later became known as 'Grammaticus' in handsome Latin, lauded by another master of style – the Renaissance scholar Erasmus of Rotterdam.

Saxo Grammaticus's *Gesta Danorum* closes the chapter of ancient Estonian seafaring. The work comprises sixteen books. The first nine present a history of the Baltic Sea, based on oral tradition. The last seven are based on existing documents and accounts of the time. His esteemed informant was Theoderich: the third bishop of Estonia and a fluent speaker of Livonian, who was killed in the Battle of Lindanise a year before Saxo's own death. The bishop contributed several episodes from the Baltic Finns' past to the historical chronicle, including the adventures of Saxo Grammaticus's favourite hero, who brought together all the most noble qualities of the ideal seaman: a man known by the Latin name Starcatherus, Starkad, or Starkather: 'it is recorded that he came from the region which borders eastern Sweden, that which contains the wide-flung dwellings of the Estlanders and other numerous savage hordes' (Saxo Grammaticus 1979). Starcatherus was a genuine Norse Odysseus who voyaged on the Arctic Sea and the Volga-Bulgar waterway alike.

If a true historical figure did lie behind the mythical depiction, then Saxo Grammaticus, in any case, painted him in more of a folkloric light. Starcatherus was a symbol of the era, not needing to be its fact. He conveys to us important information, all the same: in the eyes of his contemporaries, 'Viking' was not a national but a social concept; one which encompassed every tribe along the Baltic Sea that had achieved sufficient skills in shipbuilding technology and navigation.

THE FIRST PACTS

Two competing Baltic camps had developed by the time of Olaf Tryggvason.

Strengthening to the south was the North Sea Empire, binding Denmark and England beneath a single crown while establishing close

SILVERWHITE

family connections and alliances with Poland. Denmark's mortal enemy was central Sweden, which had established equally close family connections and alliances with the rulers of Novgorod.

The Estonians' first pacts were also composed according to the two coalitions' spheres of influence. Nineteenth-century St Petersburg historian Ernst Bonnell suspected an alliance formed between Revala and Cnut the Great somewhere between 1025 and 1030. Northwest Estonia, the key to the Silverwhite waterway, was the natural target of the Swedish–Novgorod coalition. Maintaining Revala's independence would only be plausible if there existed an alliance with Danish Cnut, who was keeping a concerned eye on rising Swedish influence on the Baltic Sea. At that time, opposition to the Danes came in the form of the Swedish king Olof Skötkonung and his future son-in-law Yaroslav, the famous Jarisleif the Lame of Norse sagas. In Sweden, Olof Skötkonung was criticised for neglecting the Silverwhite waterway. Having bound their families through wedlock, Olaf and Yaroslav subsequently began a campaign to control estuaries on the Gulf of Finland. Yaroslav conquered Tartu in the year 1030, occupying the city for thirty years. At the time, Tartu was already a noteworthy market for trade and craftsmanship with roots stretching back into the 5th century. Several raids were conducted on the Häme region, Harju County, and the 'Iron Gate', which has been interpreted as the narrow strait between Aegna Island and the Viimsi Peninsula. Cnut wrought destruction upon his adversary's fleet in turn, and the Norwegian ruler was forced to seek refuge with Yaroslav after his next campaign (1020) – just as Tryggvason had with Vladimir a quarter of a century earlier. The dynastic interweaving of Sweden, Norway, and Novgorod is too complex to unravel here. However, an entry made in the Danish chronicle in the late 12th century reads: *Canutus rex duxit exercitum in Estoniam* – Cnut led his army to Estonia. Does this reflect a former union?

The details of the Estonians' compact with the Goths are somewhat clearer. According to Henry of Livonia, crusaders accused the Goths of being in a pact with the 'pagan' Estonians in the year 1203. Men from Saaremaa had set sail in sixteen ships to raid the Öresund/Øresund Sound and pillage the county of Listerby near Lund on the coast of western Sweden, which belonged to Denmark at the time. When the Estonians docked in Visby along with their booty and Christian slaves, they met German crusaders on their way to Riga in the harbour. According to Henry of Livonia:

'THEIR GREATEST MIGHT IS IN THE SHIPS'

> After a few days, the Esthonians approached with all their rapine. The pilgrims, seeing them sailing, blamed the citizens and merchants for permitting the enemies of the Christian name to cross their harbour in peace. The merchants and citizens pretended to agree but really wished to enjoy peace with them. And so the pilgrims went to their bishop and requested permission to fight the Esthonians. The bishop, therefore, knowing their wish, endeavoured to turn them aside from the project. (VII)

The pact between Gotland and Saaremaa was so strong that not even the pope's legate William of Modena could break it a quarter-century later: 'He came to Gothland and sowed the Word of God and displayed the sign of the holy cross for the remission of sins to all who bore the Christian name, that they might take revenge upon the perverse Oeselians. The Germans obeyed and took the cross. The Gothlanders refused' (XXX).

As if the Goths' rejection of the legate's command weren't enough, Pope Honorius III had to place Visby's German merchants under his special protection (17 January 1227) from them, still in connection with the planned raid on Saaremaa. The contractual relationship with Saaremaa appears to have endured even after the raid and Saaremaa's subjugation, as one year later Gregory IX appealed to the clergymen of Linköping and Gotland, demanding that the Goths not sell the pagans weapons, horses, or ships. Today, this is called an embargo. But the appeal apparently went unanswered, as the pope was forced to renew the embargo in a letter to the archbishop of Uppsala and the bishop of Linköping one year later.

The structure of the pact between Saaremaa and Gotland casts some light on earlier practice. It was based on two principles: 'eternal' or 'pledged' peace and canonical 'divine peace'. The 'eternal' peace was confirmed by a pre-Christian ritualistic oath. It was a practice that predated Christianity itself and was consequently more solemn, revered, and placed first in order (Niitemaa 1955: 41–3). This structure reflects the age of ancient maritime law, and shows especially how tenaciously the traditions endured. We know that in the 13th century, Saaremaa's county chiefs concluded contracts with their own seals (LUB I, 369 and reg. 321; also Johansen 1933: II, 748).

Common-law agreements could only remain valid through mutual adherence. The Estonians must have applied the same rights they enjoyed in Gotland to any Goths in Estonian harbours. This meant protection in roadsteads, harbours and markets; the immunity of foreign merchants. It

SILVERWHITE

was rather business-like administration, as we can also see from the story of Olaf Tryggvason's imprisonment and release.

THE SHIP IN LANGUAGE AND MEMORY

What did the Estonians' vessels look like?

We know three names from the Estonian language, chronicles, and sagas: *laev* (Novgorodians' *loyva*); *snäkkia* (*shneka*), which survives in the Häädemeste region's traditional pontoon-like vessel called *nikki*; and *uss/uisk*, which was recorded in Russian history as *ushkuy* and in maritime history as *pyratica*.

Archaeological ship discoveries are extremely rare in countries along the Baltic Sea. In 1872, a vessel found on the shore of Koiva was believed to be an Estonian *uisk* or *pyratica*. Its 'hull is still relatively well-intact, its deck made of double planks, between which is jammed well-preserved thick wool cloth to keep water from seeping into the vessel. The site where the ship was unearthed and its manner of construction allow us to conclude it is very old, and it is thought to be an Estonian ship' (*Revaler Zeitung*, 29.11.1872). The vessel's closed deck compels us to believe it was a merchant ship or perhaps even a much later barge.

Historically, a *laev* was probably used for mercantile purposes. As we said before, ancient merchant vessels had small crews of up to eight rowers and close planking. Its tonnage was increased at the cost of space, which in turn increased the draught (up to 1.2 metres) and reduced speed.

We are now able to describe the *snäkkia* or *nikki* in greater detail thanks to the Roskilde finds, where it is called 'Wreck no. 5'. The *snäkkia* was 17 metres long and only 2.5 metres wide. The uppermost plank was peppered with rowlocks: thirty men rowed the ship. Its slim, almost arrow-like shape (width : length = 1 : 7) and large number of rowers leave no doubt that the *snäkkia* had relatively low tonnage but was a highly specialised, swift and easily manoeuvrable warship. The *Chronicle of Novgorod* appears to confirm this: coastal Estonians portaged seven of their ships past the Narva rapids and sailed from there to Lake Peipus. The Roskilde *snäkkia* was built of oak up to the waterline, then mainly pine to the gunwales. Its stem was hewn from a single timber, the tip arced slightly back, its sides carved with imitation clinker planking. Leaving aside its proportions and the number of rowlocks, the *snäkkia* was no different from an ordinary merchant vessel. Wreck no. 5 was sunk to block Roskilde Bay sometime around the year 1050. By that time, it was no longer used as a warship:

'THEIR GREATEST MIGHT IS IN THE SHIPS'

the ship hauled cargo for its final years (or decades), of which later-added ribs are evidence (Olsen 1963: 20–34).

Novgorodians saw coastal Estonians' *snäkkia* as a ship of war. Henry of Livonia, who held a certain bias, simply called Estonian warships 'pirate ships', *pyratica*. Two different terms were apparently used for one type of vessel. A late-13th-century *snäkkia* could carry a large crew and was counted among the largest ships. We can probably regard both the *snäkkia* and the *pyratica* as the Estonians' *uss*. The term was adopted into Russian as a loan (*uss, uisk > ushkuy*), but was translated into Scandinavian languages: *uss = snäkkia* (cf. *uss*, 'snake', and modern Swedish *snäcka*, 'snail'). But the Estonian *uss* was not quickly retired: as late as the 14th century, *snäkkias* were used in the Bay of Tallinn to commandeer and patrol (Heinsius 1956: 201). The ship embodies the continuity of ancient Baltic seafaring and the medieval Revala naval tradition.

The German naval historian Paul Heinsius described the Estonian *snäkkia / pyratica / uss* as follows (1956: 53):

> On the other hand, there is much we can already say about the vessels used by Saaremaa Estonians and Curonians, who alongside the Scandinavians at the turn of the 12th and 13th centuries played a decisive role on the Baltic Sea primarily as pirates, and who until the appearance of the Hansa probably held factual superiority in the eastern areas of the Baltic Sea. No one could best them or, apparently, withstand their raids [...]. The early-13th-century Estonian pyratica should be understood as a 13-manned open vessel that was powered by both oars and sails. The ships were large enough to hold, in addition to the crew, loot, church bells, prisoners, or even 'an innumerable amount' of sheep on four vessels. They must have had gunwales approximately as high as those on the Hansa auxiliary vessels, though they were lower than those of large cogs. In one naval battle in 1210, the Curonians were able to trim their pyratica such that their bows went high above the Germans' auxiliary ships. From this, we can conclude that the sides of the Estonians' and Curonians' pyratica curved upward rather sharply at the nose.
>
> The rigging comprised a single sail that could be hoisted by eight men. However, an eight-man crew was likely not enough to bring the ship safely into harbour: in 1203, the Saaremaa Estonians destroyed their own ship whose crew was too small. From this, we can tell that several men were needed for rowing and sailing alike, and conclude that the ship had old-fashioned rigging.

SILVERWHITE

If you account for the fact that the Estonians of Saaremaa and elsewhere could overpower almost any enemy until the Germans' arrival, this gives us a certain starting point for assessing the size of other Baltic ship types in the early 13th century.

Several of Heinsius's assessments are now outdated.

The overall picture of ancient Baltic vessels is brought to life by history's most vivid source: modern language.

The ship was a living being to Estonians, imagined to be formal and like oneself. Its keel – rather small, like that of all Nordic vessels – was called the *emapuu* (literally 'mother-tree') or simply *ema* ('mother'). Attached to the keel was the ship's *kest* ('shell'), *nahk* ('skin'), or *kehalauad* ('body-boards'). The broadest part of the body was the *magu* ('belly'). A row of side planks in the plural was called *laed* ('roofs'). One could ask: 'How many roofs high is the ship?' The answer described its tonnage. Ships called *seinkorralaevad* were built whose planks were placed horizontally against one another and formed a smooth wall (*sein*; carvel-built). Using the clinker technique, there was also a *kordlaev* ('layer ship') where planks overlapped like shingles. Both shipbuilding techniques could be utilised in a single vessel: carvel-built below the waterline, clinker-built above. When the ship's frame was complete, it was reinforced with *ribid* ('ribs'). Perhaps an even older name for them is the Baltic-Finnic *põõn*, which was also borrowed by Russian at a very early age: *peny*. Attempts to anthropomorphise vessels did not end there, but included *jalad* ('legs'), *sääred* ('shanks'), and *tallad* ('soles'). In six coastal Estonian counties, the word for 'shank' was used synonymously with 'anchor'. A 'sole' was a protective plank that was loosely attached to the keel. Low-draught Nordic ships cruised in shallow bays and rivers, and would land directly on the shore in harbours. If a vessel ran into a shoal at high speed, the 'sole' would break off and free it. This clever solution demonstrates the ship's universality.

A vessel's strength, seaworthiness, and speed depended greatly on the shape and construction of the bow. The bow was the pride and honour of a Nordic ship, but also its weapon, its threatening sword when ramming an enemy ship to sink it. It is where ancient seafarers believed the ship's character was concentrated, and the names of its various elements leave no doubt: *rind* ('chest'), *pea* ('head'), *nina* ('nose'), *kulm* ('cheek'), and *keel* ('tongue'). The stem, into which the ends of sideboards were notched, was hewn from oak, birch, or maple. There are much fewer

'THEIR GREATEST MIGHT IS IN THE SHIPS'

names at the stern or 'tail' of the ship, even though it was the skipper's or helmsman's position: gripping the tiller, he would sit on the starboard side, later creating the Estonian word *tüürpoord* – 'tiller-side'. As we know from Henry of Livonia, the right was also Estonians' lucky side. The ship's mast was supported freely on the foot, which was fixed to the keel. The mast was held up by shrouds, *harused*, which were initially tied to the bow and stern. Attached to the mast was a large four-cornered red sail called a *raa*, also in dialect a *ravila* or *peel*.

One could bravely go to sea on such a vessel. In Estonian, ships can 'sail', 'run', 'run on a sail', and 'fly'. Sailing against the wind requires one to *kaubelda* ('barter') or *loovida* ('tack'). One needn't fear losing one's way:

> The Starry Wagon, Bear of Sweden,
> The Northern Nail, the son of stars,
> Guided him with sparkling eyes
> Pointing from the sky his path
> To the traveller on the waves
> *(Kalevipoeg* IV, 63–7)

Captains felt at home even on the night sea, seeing in the Milky Way a *taevalaev* ('sky-ship') or *tähesadam* ('star harbour'). The sky wasn't anonymous in that era. It was certainly majestic, but close; endless, but could be grasped. Stars guided the ship and *tähele saama*, 'getting to the star', meant arrival. One was to aim for a certain star – *tähele panna*, 'place on the star' or, in modern Estonian, 'notice'. Johann Wilhelm Ludwig von Luce described an incident (1827) in which a gale struck sailors on a pitch-black night. Yet a star shone through a hazy cloud momentarily – long enough for them to find their way home. When the right course couldn't be found, it was said one *ei saanud tähte ühtegi* – 'didn't get a single star'. Determining time is critical for navigation. In daytime, there was the sun for its determination, then switching to the Great Wheels at night, also known as the Great Wagon – the Big Dipper. The brightest star in the heavens, Sirius, was called the 'Slave Star', though this was likely in an ancient meaning of the word that later faded and found another interpretation. Ancient nautical routes reflect from the sky today: in Finland, the Pleiades are known as the 'Viru Sieve'; the Little Dipper is Estonia's 'Swedish Horse'.

The familiar sky rested upon an equally familiar sea, supported by the four directions like planks on a boat frame. The directions – referred

SILVERWHITE

to as 'sky arcs', 'world halves', or 'world arcs' – divided the world into sixteen parts and guided seafarers who were far from shore. This was rare, however. Even during the next century, sea routes were still angular, following the shore from cape to cape. This type of sailing was called *nina ninalt sõitma* – travelling from nose to nose. Perhaps that is one of the oldest navigational terms that Estonian has preserved.

There is an Estonian nature myth that dates to this era of seafaring: about a time when the wind disappeared. A spider found the wind asleep above the world sea, but a speedy fly brought word to the land and claimed the honour of having found it – that's why the spider takes his revenge on flies to this day. There were many instructions for finding the lost wind. 'Wind knots' were taken to sea to undo in the event of a calm. Untying one would bring a gentle fair wind, two would bring a strong wind, and three a gale (Kuusalu). In the *Ynglinga Saga*, Snorri Sturluson complains of the Baltic peoples being able to whip up favourable winds whenever it suited them. Legend has it that heroes used 'wind sieves' to turn against the wind. Using a whistle or simply whistling through one's lips, copying the living wind, would awaken the *tuulejumal* ('wind god'), *tuuletaat* ('Old Man Wind'), *tuuleema* ('Wind Mother'), or *tuulehobused* ('Wind Horses').

THE SHIP IN PLOUGHING AND DEFENCE

The first written accounts of Estonians' naval tactics date to the 12th and 13th centuries. In violation of Visby harbour peace, the crusaders prepared to attack Estonian ships in 1203. According to Henry of Livonia: 'When the Esthonians heard this, on the other hand, withdrawing eight of their pirate ships [*pyratica*] a little from the others, they thought that they could surround the pilgrims as they came in between and so capture the ships prepared against them' (VII). The Estonian fleet was composed of sixteen vessels. Eight were withdrawn 'a little from the others'. The other eight were 'loaded with bells, sacerdotal vestments, and captive Christians' – at least that is what the hostile Henry of Livonia claims, having to somehow justify his violation of harbour laws. Half of the vessels appear to have been *ussid*, each with thirty men aboard, and the other half heavily laden merchant vessels with smaller crews. The crusaders outnumbered the Estonians and the attack took place during a lull in the wind, compromising the ancient ships' advantages. Was it early morning? The wind only picked up halfway through the battle, and then 'the sail was raised on high by the eight men who survived and, when wind filled the sails, this same man was taken

'THEIR GREATEST MIGHT IS IN THE SHIPS'

captive, and when the ships were brought together he was killed. That ship was burned by fire because of the scarcity of men' (VII).

The last remark isn't convincing. Rather, we should see setting the ship alight as a cremation ritual at sea and the killed crusader as a funeral sacrifice. Based on the Roskilde wrecks, we now know that a normal-sized warship could be manned by eight rowers and used as a merchant vessel. The speed diminished, but the Visby naval battle was limited to the raid just outside the harbour. The Germans refrained from sailing out to sea, not to mention from pursuing the Estonians, so the ship's reduced speed consequently wasn't the cause of its burning.

More details regarding the immunity of territorial waters and defence are found in the next passage from Henry of Livonia's chronicle (1204): 'Thereupon the before-mentioned knights laboured long with their companions in the struggle with the rough sea and at length came to a region of Esthonia. The Esthonians, wishing to take their lives and their possessions, attacked them with ten pirate ships and twelve other ships' (VIII). Again, the chronicler is speaking of two different types of ships. Their ratio is similar to the previous episode: out of the twenty-two vessels, nearly half are warships: Henry of Livonia's *pyratica*, Estonians' *ussilaevad* or *uisud*. The twenty-two ships were manned by at least 420 sailors. This could not have been the fleet of a mere two or three coastal villages, but points to the coastal dwellers' permanent patrol and rapid summoning of a larger force. The crusaders sailed in cogs, the particularities of which we will discuss later. For now, let us just note that the cogs' ratio to the ancient Baltic ships was about the same as that of a destroyer to a torpedo boat today: the first has strategic dominance, the second tactical. The cog was a newcomer to the seas, heavy and clumsy, but its captain was even more vulnerable. The Baltic Sea was a foreign environment to the Germans. Those mentioned in the given passage had departed the Daugava 'before the Nativity of the Blessed Virgin Mary' and arrived in Visby on the eve of St Andrew's Day. The first is on 8 September and the second day would be 29 November – a full 88 days spent on a journey of 240 miles! The chronicler acknowledges that the crusaders spent 'especially many days in hunger, thirst, and cold'. The scant details give us little to go on in determining their route and the site of the battle. They picked up 'fifty shipwrecked Christians standing on the shore', which corresponds approximately to the crew of one cog, and a storm drove them 'among some very dangerous rocks out of which they came with great fear and difficulty'. Most likely, a storm coming out of the southwest blew them

SILVERWHITE

into the coastal waters of northeast or north Estonia, and on the way back they came upon the perilous and legendary Hiiu Shoal (Neckmansgrund) near Tahkuna, Hiiumaa island.

One naval battle fought in 1214 is painted in much richer detail because Henry of Livonia was himself present. A fleet of nine cogs pulled up anchor in June and set sail from the Daugava towards Gotland. 'On the [second] night there was a contrary wind and thunder and they suffered a great storm through the whole day. At last they were forced into the new port on Oesel [Saaremaa; *Portus Novus* in Osilia]' (XIX). It is impossible to determine a more exact location of Saaremaa's *Portus Novus*. A south-by-southwest wind struck the fleet after about 42 hours had passed, consequently in the Irbe Strait, and battling the storm all day could have flung the ships to any harbour located between the southern tip of the Sõrve Peninsula and Kihelkonna on Saaremaa's western coast.

When the Estonians of Saaremaa learned that the ships had come from Riga, they threatened war. This is a substantive admission that refutes the pseudo-romantic legend of Saaremaa Estonians being professional pirates. The fleet sought shelter from the storm in harbour, and the harbour was obliged to offer such haven. However, it became clear that the fleet was not mercantile. What's more, the ships had left from Riga, with which the Estonians were at war, and, as Henry of Livonia personally acknowledges, they were manned by crusaders. The right to safe haven did not apply in these conditions. Indeed, the Estonians revoked the right and apparently acted as law required, 'threatening' them with war – i.e. demanding that the fleet leave. The storm raged on, and the captains did not dare risk venturing out to sea in an unfamiliar place, so the Estonians closed off the harbour: '[they] built structures of wood on the seashore. These they filled with rocks, seeking to block off the harbour, whose entrance was narrow.' It was a common practice that the Saaremaa Estonians had used to close off the mouth of the Daugava near Riga in the spring of that same year:

> The Oeselians came with a great naval force to Dünamünde [Daugavgrīva], bringing pirate ships and brigantines with them. These they filled with rocks and sank in the depths of the sea at the entrance of the river. They built wooden structures which they similarly filled with rocks and they cast these into the mouth of the Dvina [Daugava] to close the route and the harbour to anyone who was coming. (XIX)

This could have been the end of the matter and the Estonians would have allowed the cogs to leave the harbour in exchange for ransom, but then

306

'THEIR GREATEST MIGHT IS IN THE SHIPS'

they imprisoned eight of the Germans, who confessed that among the men on the ships were no other than Bishop Philipp of Ratzeburg and Bishop Theoderich of Estonia. The die had been cast in favour of battle.

'At the first light of dawn the sea opposite us looked black with their pirate ships.' The German fleet was trapped by the storm and blocked the harbour exit. So they chose the simplest strategy to repel the Estonians: 'Our ships were all gathered together, so we could defend ourselves more easily from the enemy.' The Estonians attacked with fire, catapults, lances, and arrows. Henry of Livonia acknowledges their effectiveness: 'Some of them launched three big fires, kindled from dry wood and animal fat, and set upon structures made of huge trees. [...] They steered it straight toward the midst of our ships. [...] The flames of this fire [...] were taller than all of the ships.' As the cogs held raised structures at the bow and stern (*castellae*), the height of the fire must have been 6–7 metres.

Even more fascinating is the chronicle's next sentence, which describes the exact attack formation and tactic: 'Other Esthonians, meanwhile, were rowing around us and they wounded many of our men with their lances and arrows. Still others returned, took the same route around us, and threw their stones and staves at us [and shot from *ballistae*].' The nine rafted ships constituted a 30–35-metre-long floating castle. The bow *castellae* were shorter, the stern *castellae* taller. The procession of Estonian ships rowed past the cogs' bows, shooting lances and arrows, then turned around and used catapults to attack their tall stern *castellae* from the opposite direction. Henry of Livonia refers to these weapons as *ballistae*, contrasted with the heavy trebuchet that could throw boulders or barrels filled with rocks – the distant predecessor of shrapnel – with a weight of up to 150 kilograms. We do not know the construction of these particular *ballistae*. Estonians and the soldiers of Prince Vetseke (Vyachko) used a weapon of the same name to defend Tartu. It couldn't have been very large, as that would have required the ancient ships to have some kind of platform or closed deck, of which we lack any information. The chronicler's illogical repetition of throwing stones and shooting *ballistae* appears to imply that the *ballistae* were used in addition to hurl flammable projectiles made from a mixture of tar, pitch, and animal fat.

Saxo Grammaticus conveyed the account of an Estonian (and Curonian?) raid on the island of Öland, from which we find out that the ancient ship could also be used on dry land in case of emergency (Laugaste 1963a: 15–16):

SILVERWHITE

When the barbarians saw themselves surrounded by such a great fleet, they then, taking courage from desperation, started preparing no longer to flee, but fight. They built, partly from their own ships and from the enemy ships they had seized, partly from logs and stumps, a fortress-like stronghold. It had two very narrow entrances that were therefore difficult to access, through which they climbed or wriggled as if through rear doors. They also fortified the side exteriors, assembled from ships, winding sails around them several times for defensive purposes in the manner in which fibres of fabric are joined [...]. They additionally sharpened stakes for jabbing, gathered stones from the shore suitable for catapulting, and, so it wouldn't appear as if their morale had fallen, feigned the gaiety of song and dance like bacchants while the Danes spent a miserable night in silence.

Finally, a few words about ancient ships' speed.

In the year 875, Ohthere of Hålogaland sailed around Scandinavia to the White Sea. His ship was still equipped with old-fashioned rigging that could not be used to sail by the wind. The naval historian Walter Vogel has estimated Ohthere's average speed at 3 knots; 5.5 kilometres per hour in land terms.

The ships in Adam of Bremen's day were much better equipped. The chronicler divided the Baltic sea routes into day-journeys, *etmal*. It is questionable to attempt to calculate average speed based on this ambiguous unit, as the captains sailed 'from nose to nose'. Nevertheless, the speed of such vessels (in 1073–6) is estimated to be an improved 4.5–6.8 knots, i.e. 8–12 kilometres per hour.

The principles of shipbuilding were outstandingly homogeneous around the entire Baltic Sea, stemming from the manner of constructing sewn taiga boats. Even so, the vessels' tonnage varied noticeably. Length, width, and draught constitute a ship's primary measurements. These do not determine tonnage. Even with equal measurements, the tonnage of a pot-bellied ship can be twice that of a slim ship. Primary measurements are not determined so much by building materials or skills as they are by the nature of the sea: wave height and frequency, wind speed, and, above all, the vessel's universal navigability on sea and river alike were decisive factors. The size of the vessel wasn't a virtue in and of itself. We still know the dimensions of Olaf Tryggvason's royal ship. It was large for its day: the keel was 75 cubits (32–44 metres, depending on the cubit) long, and the width was 28 feet (8.5 metres). It wasn't a vessel meant for the Baltic but for the North Sea, and even there it was used more for representational

'THEIR GREATEST MIGHT IS IN THE SHIPS'

purposes. According to legend, the royal ship was slow and difficult to steer. Every basin shapes its optimal ship type.

The ancient Baltic ship was in use longer than one would generally expect. Its 15–25-tonne cargo capacity could not compete with that of the cogs, which was ten or more times greater on average. Yet as a nimble warship that wasn't limited by the winds, depths, or harbours, its advantages endured for several more centuries. The vessel continued to provide for shipbuilders, seafarers, and Estonian folklore, whose outstanding accuracy was all thanks to the continuity of tradition.

FROM MARKETPLACE TO TOWN SQUARE

Chroniclers do not record standards but deviations from the standards – especially those who were hostile. The early years of the second millennium were not characterised by piracy, but by the rapid development of agriculture, craftsmanship, and shipping, as a result of which the first early towns sprang up all along the Baltic.

Classical historical records spoke of a city's 'founding' and saw in it the will of an individual; the mighty, genius-like sweep of a ruler's arm: here, arise a city! These notions have wandered into the realm of folklore. Reality, as always, was more complicated. Lund's prehistory is one instructive example. Being the seat of the archbishop, Lund later had close ties to Estonia. General belief is that it was founded by Cnut the Great when he returned from his campaign in England and established his first mint. The city's birth is dated to 1019 or 1022, and Denmark's first were indeed minted in Lund in 1018–35. Nevertheless, the town did not spring up because of any one person's will, no matter that historical sources agree in claiming so.

Three kilometres to the east of Lund is a marketplace named Trehögar, whose immunity was protected by 'blessed peace'. However, even during the Middle Ages trading in the 'blessed' site took place according to the 'pagan' calendar, pointing to a pre-Christian tradition. Archaeological digs have confirmed that the Danish king erected his mint next to the ancient trading site, probably in the hope that the market peace of ancient customs would expand to his industry. The early town needed ideological advantages in addition to economic ones in order to flourish: the protective shield of pre-Christian beliefs. Continuity between old worship sites and market peace, on the one hand, and medieval city law and city peace, on the other, is characteristic of all Baltic early towns.

309

SILVERWHITE

Ancient Baltic law developed in Birka, in the Mälaren lake region of central Sweden, and became known as the Bjarkey laws. It organised and regulated relations between foreign merchants and the local population. Over time, Bjarkey laws expanded to cover numerous other trade islands. Over two dozen places bearing some variation of the name 'Birka' are known throughout the Baltic coast (Calissendorff 1971: 29; Niitemaa 1963: 190–203).

But to what extent did these conclusions apply to Estonian harbours and marketplaces? Seafaring and trade are indivisible. During the Silverwhite waterway's golden years, one important Baltic commodity was slaves, for whom there was great demand in the East. Transporting slaves through Christianised western Europe was unthinkable. Regardless of slaves' countries of origin, one primary means for their transport was down the Silverwhite waterway, through the 'pagan' taiga. Although the slave trade declined from the 11th century onward, it still played a noteworthy role on the Baltic Sea (Moora 1968: 518). According to the treaty on the occasion of Valjala's surrender (February 1227), the Saaremaa Estonians were required to liberate their Swedish slaves. Given that the condition was stipulated as a separate point, researchers believe the number of slaves must have been great. One record from the year 1255 addresses slaves' right to inheritance. We spoke earlier about Hiiumaa's Orjaku harbour as one likely slave camp. As late as 1565, one Swedish farming family in Orjaku had the surname Skalk, 'slave'. Estonian common law was later superseded and sanctioned by the Teutonic Order, which in the early 15th century still treated prisoners of war as slaves who could be legally sold (Niitemaa 1952: 355).

The decline of the slave trade was tied to the decline of the Silverwhite waterway. Its consequences were fateful for several harbours and fortress towns, whose economies were underpinned by the exchange of such 'commodities'. Early medieval Birka disappeared from the stage of history in the 970s, leaving only the Bjarkey laws as a legal term. However, Estonia had risen to become the world's northernmost grain producer by that time, and the transition to new trade relationships took place somewhat more smoothly. Estonians' primary exports were grain and furs. They were sold in agriculturally poor Novgorod, where the cereals repeatedly helped to alleviate famine, and especially in south Finland and Gotland. Thanks to the latter, the most common unit of weight throughout the Baltic Sea in the 12th century became the Livonian pound, *Lievespfund*. Estonians' main imports included iron, tin, Swedish and Hungarian

'THEIR GREATEST MIGHT IS IN THE SHIPS'

copper, salt, silver coins and bars, weapons, horses, and luxury harnesses and cloth. In the areas of Revala and Rakvere in north Estonia, silver was turned into jewellery that travelled onward to the north and northeast in turn, all the way to the shore of Varangerfjord and even further.

Consequently, it is highly probable that ancient Baltic law applied in Estonian harbour settlements. This is confirmed by several written records which allow themselves to be projected into a more distant past. In March 1227, Bishop Hermann of Saaremaa along with other feudal lords granted privileges to merchants doing business in Estonia. They allowed for conflicts between merchants to be settled according to Gotland law. However, if a dispute arose between a merchant and an Estonian, then the Estonian was to be compensated according to common law: *secundum iustitiam et constitudinem terrae nostrae, nach dem rechten und gewohnheit unseres landes* (LUB I, 453, columns 569, 570). The privilege bestowed by the master of the Teutonic Order upon Lübeck's merchants a quarter of a century later repeated this verbatim. Merchants were permitted to harvest coastal forests for repairing their ships, replacing masts, and lighting fires. They were allowed unhindered trade with the 'pagans' or Russians even in times of war, and in the event of shipwreck they were allowed to salvage as much as they themselves could manage. In terms of the latter, the merchants' privilege truly included certain rights that they seemingly lacked before. However, conflicts were to be resolved on the basis of common law. This remained effective and solely legitimate, speaking to its deeply rooted tradition. Shaping and instilling such a thriving tradition could only be done in the era prior to Estonia's conquest. Consequently, historical records reflect the system of administration and justice that was enforced in pre-crusades Estonian harbours and marketplaces. One such telling detail is an equine law: only horses that had been brought for sale were allowed in pasture. Did this mean there were other horses on board the ships? That even in the late 13th century, sailors brought horses for towing their vessels upstream? Unlikely. Rather, we should regard this as a traditional element dating back to the days of the Silverwhite waterway, when horses were simultaneously used for trade, towing, and attack. Legal norms have considerable inertia. In one later document, particular emphasis is given to Estonians' right to export horses if they are not intended for warfare (Niitemaa 1952: 355).

Thus, several legal customs that applied in medieval Estonian harbours and marketplaces were established prior to the crusades. Yet regardless of how great legal inertia may have been, such laws could only survive times

SILVERWHITE

of social upheaval if seafaring and trade ties remained uninterrupted. And so, an almost scholastic question comes onto the agenda: when did a harbour and trading settlement become an early town?

Early towns are difficult to define, as they may have been extremely dissimilar on the surface. Schleswig had no city council in its glory years (the 1200s), even though it was already a city. Viipuri (Russian Vyborg) had a city council as early as 1393, though it wasn't yet a city (Niitemaa 1963: 190–203). Still, substantive similarity lies in the apparent difference. Common law was remarkably similar all throughout the Baltic Sea, and here Gotland's leading and organising role emerges.

Estonia's most important ancient harbours were, from south to north, at Häädemeeste, Tahku, the Pärnu/Emajõgi estuary, Virtsu, the Bay of Haapsalu, Rohuküla, Kurkse, Laoküla (near Paldiski), Revala, and, likely from the 10th century, the Pirita estuary and the shores of Maardu and Kolga. From there onward stretched an entire network of harbours commonly called 'Viru', which was connected to the 'Pargasporti' harbour on the opposite Finnish shore. 'Viru Harbour' included Vergi along with Kaupasaar (literally 'Trade Island'), and Toolse and Mahu. Estonia's islands hosted an even larger number of harbours. According to one tally there were up to twenty on Saaremaa alone, as a result of which the island was referred to as one great harbour in the Middle Ages.

Active seafaring sowed Estonian toponyms all along the Finnish shore, particularly in the vicinity of Helsinki: Estnässkat-ön, Estluotan, Estvikshällan, Estvik, and Revalsviken. Based on archaeological data, Helsinki's own development as an ancient harbour can be traced without interruption starting from the time of the Silverwhite waterway. Its crystallisation point was at Santahamina/Sandhamn (Niitemaa 1963: 196–7). Not all harbours developed into towns, not all towns grew into cities, but all Baltic towns grew from their harbours.

HARBOUR LANDSCAPES

> He pushed his boat onto the landing,
> Shoved the tarred to sea,

as a song recorded in Pöide, Saaremaa, goes. On Saaremaa, a landing, *lauter*, is the basin behind or between stone quays where ships are docked. The lines seem to show that the words *lauter* and *valgma/valgam* – a stoneless, reedless beach where boats were pulled ashore – were once

312

'THEIR GREATEST MIGHT IS IN THE SHIPS'

interchangeable. This also leads one to the idea that both words once meant a place where ships were built or kept for the winter and tarred again in spring.

A harbour, *sadam*, on the other hand, has always meant what it means today: a natural, later also artificial, marine area where vessels were protected from storms, high water, melting ice, and attacks; could also be loaded and unloaded without disturbance; and where their rigging, mast, or planks could be repaired in addition to stocking provisions and drinking water. Ships' capacity was quite limited at the time. When a storm flung Riga's cogs to the Saaremaa shore and kept them in place for days, the seamen were forced to row to shore and reap 'the crops with their swords'. Hunger. Over the ten-day journey, half a litre of water and a kilogram of food per man each day would reduce an ancient vessel's tonnage by 4.5 tonnes, a cog's tonnage by even 7.5 tonnes. Harbours were consequently essential and they dotted the shores much more densely than today, and the mutual rights and obligations of seafarers and local residents had to have been regulated by customs.

The words *sadam* and *valgam* come directly from Baltic-Finnic. The first Estonian harbour to appear in written records was *Sottesattama*, *sõjasadam* or 'military harbour', in 1254. As the parchment addresses the dispensation of foresting and fishing rights, it must be legally and topographically accurate. At the same time, the expression *finis portus*, 'harbour border', is used in connection with another harbour. It furthermore concludes: 'All will retain the right to fish and harvest trees just as before, without limitations.' The lack of limitations is an eye-catching exception. Still, it concerns the allotment of Hiiumaa between the Teutonic Order, and the new bishop of Saaremaa and Läänemaa – or, to be exact, between both feudal rulers' vassals, i.e. those living in the parishes of Karja and Pöide. And, indeed, the islanders' indigenous customs were used as a basis for the agreement. At the time, Hiiumaa was a sparsely populated ('*insula deserta*', 1228) island where a lack of limitations did not harm anyone's interests. Tree-felling was unavoidable given the level of shipbuilding at that time. Masts constantly had to be replaced, planking repaired. The demand for timber was great. In more densely populated areas, the right to harvest trees was, unlike on Hiiumaa, limited in the interests of the community. The rules went as follows: trees could be chopped down for fires or to repair a mast or planking, but seafarers did not have the right to build a new ship without permission (LUB I, 291). This is repeated almost word for word in several texts written in the late 13th century,

SILVERWHITE

though they simultaneously refer to old customs: *antiquam libertatem*. Pre-conquest maritime law was adopted in written records. This is also proved by the acknowledgement and adoption of the Pöide and Karja parishes' usage rights in the 1254 record.

So, we know the words *lauter*, *valgma*, and *sadam*. A guest in the latter harbour was allowed to fell trees under certain conditions. Adherence to these conditions required monitoring. The harbour had a border. Hiiumaa's 'military harbour' was differentiated from ordinary harbours, of which four were listed: *Sarwo* (Sarve), *Oryocko* (Orjaku), *Rauky* (Reigi), and *Pylayasari* (Pihlasaar).

We may tentatively transfer these conclusions to Revala's harbour, using the following logic: if harbours on sparsely populated Hiiumaa were divided into military and non-military harbours, if the concept of a harbour did not extend to that of a landing, if a harbour had a border (more legal than topographical), then this most certainly would apply to the densely populated Revala shore. Revala's military harbour was most likely separate from its mercantile harbour, but the harbour area was definitely delineated.

So, where was Revala's harbour located? The question comes as a surprise, but even more surprising is the answer: we do not know. It couldn't have been in the same place as Tallinn's harbour today, which lay beneath the lapping waves in the 12th century. Continental rebound in northeast Estonia is estimated at 25 centimetres each century. This rate is, however, slowing. By even the most modest estimates, sea level was 2.5 metres higher ten centuries ago, likely even more. One could hardly have recognised Tallinn's environs. In places where the coast slants gently, the sea stretched a kilometre inland. Tallinn's roadstead was a placid, shallow bay arcing to present-day Musumägi (Virumägi), and Toompea Hill was framed by the sea on both sides. Waves lapped lazily against pebbles on today's Aia Street, which was a rosehip-dotted beach ridge that extended northward as a bare headland. It took time before the shore could support Pikk Street on its sandy nape. There were only 400–500 metres separating the foot of Toompea from the sea. The hulking limestone fortress looked even more towering, merciless, and menacing.

Tallinn's medieval street network is like a geological diagram showing the Revala cove's gradual withdrawal towards the sea. If the harbour was located at the very back of the cove, then today's Pühavaimu Street would have been the shortest route to the landing site (Miller 1972: 177). The first chapels, monasteries, and warehouses sprouted on the

'THEIR GREATEST MIGHT IS IN THE SHIPS'

harbour's edge – the future city's skeleton. Alas, the cove narrowed, draughts deepened, the harbour was forced to move closer to the sea, and from Pühavaimu's trunk branched off Olevimäe Street, which in 1312 was called *mons arenae*, Sand Hill, and converged with the Small Coastal Gate – the harbour's next location. The continent's continued rise and increasingly deeper draughts finally pushed the harbour even further from the edge of the city, and out of the old trunk sprouted a third branch, the most direct and hitherto longest route to the harbour: *strantstrate* (1362), Beach Street, or today's Pikk ('Long') Street. Dates can be mistaken. They merely represent the very first mention of one name or another in a historical record, reflecting the general direction, not the dynamics, of urban development. The overarching picture is grand and convincing enough without it: three stages, three harbour locations, and three roads connecting them – Pühavaimu, Olevimäe, and Pikk, spreading fan-like from a common trunk, lengthening as the harbour and the shoreline grew more distant. They also drew the future city wall along with them, shaping it into a peninsula running north along the ancient headland to defend the harbour. Structurally speaking, ignoring topography and the harbour's defining role, there would be no reason for having such a long and drawn-out city wall. It took immense efforts to build the wall. As the art historian Villem Raam's research shows, the wall ran along a different path in the late 13th century: Kuldjala Tower was Tallinn's northwest 'corner tower', from which the wall took an almost right-degree turn across Lai Street, continuing approximately down Vaimu Street eastward, encompassing four monastery plots and structures to then join the western defence wall between Bremen and Munkadetagune towers. At that time, the wall did not yet extend to where the Small Beach Gate would later be built. Consequently, the wall reflected the first harbour location. Its estimated construction in the late 13th century is quite accurate. Therefore, Tallinn's oldest harbour, or rather the first stage of its early medieval harbour, was located somewhere in the area between Viru Gate and today's Kalev Spa until around the year 1300.

This was more than a favourable location. Toompea's mighty cliff shielded the cove from southeast and western winds, which are prevalent in Estonia. The tall pebbly ridge stretching to the northeast offered fair places to beach and quieted winds blowing from the northwest and north. The cove's southern shore ran at a smooth incline, was reedy, and gradually transitioned into a coastal meadow.

SILVERWHITE

However, this whole handsome landscape would have evaporated like a mirage if the harbour had lacked drinking water. Where did the ancient ships get their water? One of Tallinn's littlest-known historical monuments is on today's Tartu Road. The little church wrapped in a faded cloak was part of St John's Almshouse, originally St John's Hospital, where lepers were kept. Maintaining a colony for people with infectious diseases alongside a busy highway seems absurd. And no doubt it was.

In the mid-19th century, Tallinn mayor Friedrich Georg von Bunge was poking around the Town Hall cellar when he came across Estonia's oldest historical document hidden beneath a big pile of parchments. It turned out to be a letter written in 1237 by papal diplomat William of Modena, permitting the hospital to accept donations and inheritances. At the time, the institution was called *domus fratrum leprosorum de Rewalia*. Consequently, it was located in its current position even prior to 1237. Hospitals were generally built at a distance from roads and settlements, out of the way. Lepers were treated with downright superstitious fear during the Middle Ages.

And St John's Hospital was no exception. The building's surroundings were totally unrecognisable from what they are today, adding important details to the ancient harbour landscape. Firstly, there was no Tartu Road as there is now. It didn't exist, and couldn't have, as water rippled along its course and Narva Road was deep beneath the sea.

The landscape was complemented by the frothy, turbulent Härjapea River, which flowed into the cove from the northeast. It came at a steep decline of nearly 40 metres along the two-kilometre riverbed. The swift current swept much debris along from the sands of Ülemiste, but gently slowed in the low estuary, the alluvium settling and forming sandy dunes. The mouth of Härjapea River was only 600 metres away from the harbour, perhaps even closer. What does this tell us?

Härjapea River divided the city from lepers. The hospital was located across the river, on the opposite bank of a delta nearly 300 metres wide, on a coast exposed to wind from every direction, outside any danger of contact, and most likely on a headland extending into the sea. Its stone-arched sauna received water through a pipe leading into the river in the year 1380 (Gustavson 1969: 152; Sulev Künnapuu to the author, 4.3.1974).

The clean, swiftly flowing Härjapea was home to salmon. Eels swam upstream into Lake Ülemiste. The lower end of the river flowed to the northwest or the north and was not protected from the sea winds. The

'THEIR GREATEST MIGHT IS IN THE SHIPS'

coast on the cove's eastern shore had a gentle slope, and when the river ran high, sandy alluvium collected at its mouth. Over time, the combination of sea winds, waves, and deposits led to the rise of a low sandbar, which likely accelerated the cove's stagnation. As there have not yet been any archaeological studies of Tallinn's harbour, we can only guess what slight evidence appears to explain the river's strange name.

In contemporary Estonian, *Härjapea* (literally 'Oxhead') means *Härjaneeme* ('Ox's Cape'). Revala's shore offers a convincing analogy: near the Kalamaja district, there is a small headland named *Hundipea nukk* ('Wolfshead Promontory'). It is a very old name and interesting for another reason as well: in 1374, the toponym was *Susipea* instead of *Hundipea*. *Hunt* (> German *Hund*, 'dog') was a later word used to replace the feared, taboo *susi* (archaic 'wolf'). Among taxpayers in the Oleviste parish was listed a fisherman named Zedenpeyke. Back then, *susi* was used in the dialect's genitive case: *suden*. The name's evolution can be traced forward: the fisherman Zuddenpe (1421), four fishermen named Zuddenpe (1424), and Peter Suddenpe (1445) (Johansen 1951: 107). Estonians often used animal names as a basis for toponyms.

'Härjapea' is plausible as a headland, but doubtful as a river. Perhaps I can be permitted to pursue the following logic. The river, which was only two kilometres long in the days of the early town, could have been named after a prominent feature located at its mouth. This could originally have had religious significance as well. M.J. Eisen writes that on Tõnismägi (St Anthony's Hill), the original name of which has unfortunately been lost, swine heads were still sacrificed in the 17th century. The association between marketplaces and holy sites is beyond a doubt, but that would explain the names Lindanise or Revala more than Härjapea. Yet what if *härg* had a different prehistoric meaning that has disappeared from contemporary language?

One such secondary meaning is a 'lever' used for lifting up heavy objects, primarily stones. In the latter event, a lever could also be called a *kivihärg* – 'rock-ox'. Larger and heavier things, such as ships, could not be lifted with a lever. What was done then? Props were placed beneath the vessel and it was rolled up onto land. These were called *pullid*, 'bulls'. A bull is not an ox, but, based on the *susi/hunt* word pair, we can see that the substantive aspect of place names can be more lasting than the word's form. In that case, 'Härjapea' meant a headland where ships were beached; in other words, a landing or *valgma*. The sandy shore at the river's mouth was Revala's winter boatyard. Its name extended to the river, remaining a

317

SILVERWHITE

toponymic monument even when the ancient landing had long since been extinguished from memory.

This interpretation may turn out to be fantasy, but not its conclusion. Härjapea River was routed into sewer pipes in the 1930s. Since then, its underground current is only marked by manholes and street names. Yet when installing the pipes, an ancient ship was discovered on that ancient shore (1932): the first-known find of its kind in Estonia, parts of which were displayed in the Tallinn Maritime Museum as item no. 15 (Estonian Maritime Museum 1937: 3). Perhaps we are beginning to glimpse Revala's landscape?

> Mossy pine trunks.
> The glinting snake of a road.
> Morning mist, juniper arrows, a meadow yellow with buttercups.
> A cove clogged with seaweed.
> A seagull on a rock, alone with its reflection. The sea's distant roar. The splash of an oar. Time stopping.
> A sail.
> A sail.
> A sail.
> The whole world.

NOSE-TO-NOSE TO REVALA

Estonia's oldest nautical guide was written in the early 1200s and is held in the Danish National Archives in Copenhagen. It describes the sea route from Sweden's (at that time Denmark's) Blekinge province across Mälaren to Revala.

In the 13th century and even much later, ships would hug the Swedish shoreline, following the ancient Bronze Age route with which Pytheas himself became acquainted. It was navigated from landmark to landmark – unusually shaped islands or capes, larger trees, groves, shoals, or other more noticeable marks. The piloting book composed for the Danish king only lists the most important ones and is seen as a summary of folkloric sailing – reminders.

Let us stick to Revala's immediate area. Slightly north of today's Stockholm lies Arnholm Island. After sailing from south to north, ships were to take an almost 90-degree turn eastward at Arnholm and cross the mouth of the Bay of Bothnia at the Åland Islands. Not a word of a direct

'THEIR GREATEST MIGHT IS IN THE SHIPS'

route to Revala. The distances from one island stop to the next are given in nautical *penikoormad* (approximately 4.5 nautical miles or 8.3 kilometres).

The points of reference are Arnholm, Lemböte, Korppoo, Aspö, Refholmi, Malmö, and Jurino Island. It recommends sailing north of the last-mentioned. From there, the next point is Hanko Cape, which extends over 30 kilometres to the southwest and was an important landmark for sailors. The peninsula was hard to miss. 'If there is a favourable wind from the west, then you can sail straight from Arnholm to Hanko Cape.'

Next came the Porkkala Peninsula. Although it lies only 48 nautical miles east of Hanko Cape, the guide provides four whole intermediate landmarks. This is quite common for the 13th century, not to mention earlier times. Captains clung to coastlines. A few of the toponyms are Baltic-Finnic, from which attempts have been made to draw conclusions about the author's ethnicity. For example: 'Hanko Cape, which in Finnish is *Kumionpää* [Cuminpe] ... *Oriinsaare* [Horinsarae], which in Danish is Hestö'.

Such frequency of landmarks can be explained by the plentiful shoals along Finland's southern coast. The Åland Islands were also dangerous, albeit unavoidable on the East Road. As soon as the captain had landed safely on Finland's shore, his first concern was to turn his back on the land and continue the voyage along Estonia's coast. Hanko offered fair opportunities for this. So why sail on to Porkkala?

The captain faced a dilemma and, strangely, it holds the key to Revala's prehistory. He could sail 40 miles further eastward through perilous coastal waters and turn south from Porkkala. He could have immediately steered south from Hanko and avoided the many shoals. From our perspective, the second choice would be more prudent, even self-evident: there would be only 14 miles more open sea and 48 miles fewer shoals. It's hard to see the problem.

Yet there was a problem for the ancient captain, and he resolved this problem differently. The guide only permitted sailing directly south from Hanko in exceptional cases and especially with fair winds, which had to blow straight from the north. The traditional sailing route went on to Porkkala and only then was the crossing made to Estonia. Why? Between Porkkala and the open sea is Mäkiluoto Island, and from there to the northern tip of Naissaar is only 19 nautical miles. From Hanko, the crossing would involve 14 more nautical miles on open sea, which posed a greater danger to the pre-compass ancient captain than 48 miles of shoal-pocked waters. It was sensible nautical practice.

319

SILVERWHITE

The distribution of archaeological finds in south Finland corresponds exactly to the navigational guide: their frequency takes a sudden decline east of Porkkala, as if an invisible line were drawn.

Sensible nautical practice did not prevent those very same captains from reaching the White Sea, Iceland, or Greenland (the last two significantly later than it was previously estimated, of course, though navigational techniques were similar to what was described everywhere). Memory took the place of maps, recording the experiences of many generations as long sequences of images: seaside cliffs, capes, shoals, two islands embracing, rookeries, a lone tree ... The landmarks were memorised, worn into one's head, sung on long winter nights, and thus Estonians' rich maritime folklore endured to the turn of the 20th century.

However, a sailor had to see land at least in the morning and at night, which placed great responsibility on Porkkala's southern tip: from there, one could easily spy the northern tip of Naissaar, Virbuotsa Cape. Porkkala was a signpost that directed ships on the Silverwhite waterway to the Bay of Tallinn. This should not be taken metaphorically. As the navigational guide instructs: 'From Porkkala across the Estonian Sea [*mare Estonum*] to Naissaar, six nautical *penikoormad*; from there to the Paljassaar Islands, half a *penikoorem*; from there to the Revala Fortress [Raeuelburgh], half a *penikoorem*; and keep in mind that one sails between the southern and eastern directions [SE] from Porkkala to Revala Fortress.'

Earlier, the conservatism of folkloric navigational skills allowed us to compare the waterway to a highway. Folkloric experience persisted during the age of the compass, competing with nautical charts and guiding the first cartographers' quills. As one medieval Dutch navigational guide cries out: *Wacht u voor Sibbernes!* — 'Not before Tahkuna!' On Ortelius's map (1570), an elongated Hiiumaa stretches menacingly into the west, while Osmussaar – crucial as a landmark – has grown to the width of Lake Peipus. Three centuries after the Danish navigational guide was written, the route still clings to Porkkala and the Åland bridge, despite the nautical technological revolution that had taken place. It was already a highway in Tacitus's day. Was it only a path in Pytheas's?

Funnelling the nautical routes into the Bay of Revala suggests two possibilities. Either the shore of the bay was uninhabited in the era of the Silverwhite waterway and remained uninhabited even after its decline, or one of the Baltic Sea's first fortress towns was already germinating and spreading out roots. There was no third option, because courses could only meet in Revala Bay under one condition: shore rights did not apply there.

'THEIR GREATEST MIGHT IS IN THE SHIPS'

A beach was common community property. Anything that appeared, floated, or was washed ashore – including shipwrecked sailors – belonged to the finder. Outside harbour boundaries, shore rights were universally applied: the land was off-limits to foreigners, the shore hostile and hazardous. Yet sailors needed drinking water, firewood, fresh masts, and grazing for their horses. In the time prior to Ohthere's voyage (of 875), ships sailed at an average speed of three knots. Staying a night in Revala Bay was unavoidable, stopping often inevitable, because from there the course turned at another 90-degree angle: this time to the east. Ohthere, a Norwegian, was yet unable to sail into the wind – a skill that came to the Baltic even later. True, there were oars.

Let us look at this from another angle. Visiting a site under harbour law involved additional costs. Gifts were typical of even the most primitive trade relationships. Not every captain was willing. He could have more compelling reasons to avoid Revala. There may have been an ongoing conflict with his home. Henry of Livonia describes the opposite situation: crusaders came upon a Saaremaa fleet as it was raiding Denmark's Listerby province. The inevitable clash was avoided by the Estonians claiming a peace treaty with Riga. A relatively firm set of rules governed the Baltic Sea. States of war or peace were decisive even far from one's home harbour – they were respected. If a captain on the Silverwhite waterway was unwilling or unable to visit Revala, then he had to have another stopping island as back-up – one close enough to the Porkkala–Naissaar line but still outside common-law territorial waters. Naissaar and Aegna were therefore out of the question. If we could find a stopping island that met all these conditions, then it would be rather considerable proof in favour of early Revala's possibility.

What do our friends the Arabs think of this? By at least the 10th century, the Arabs knew the Baltic and its islands: five, though Gotland was not one of them. Perhaps this can be explained by Helgö's and Birka's leading position in the Scandinavians' eastern trade. There are other unrelated explanations.

Ordering the islands by size would probably be wasted effort, as even Ortelius's map exaggerates Osmussaar to outstandingly sprawling proportions. Only the island's favourable or unfavourable position along the waterway could play a critical role. Going with the Arab conception, the Scandinavian 'island' could have been most important, leaving Denmark in second place. Gotland was seemingly treated as a part of Denmark, Öland even more so. Competing for the three remaining places would

SILVERWHITE

be Bornholm in the south Baltic (an essential target for ship routes), the Åland archipelago, and perhaps also Saaremaa and even Naissaar, given its importance as a landmark.

Already in the 19th century, the Arabist Christian Martin Frähn pointed attention to al-Maqdisi's manuscript in the so-called Rousseau collection in St Petersburg, which mentions the island of Daremusa. Frähn interpreted the name as a distortion of Denmark (1823: 53). Daremusa and Danemarka do sound quite similar. However, the Russian Arabist A. Kovalevsky pointed out an entirely different connection: the classical Arabic *d* is more correctly conveyed as a pronounced *s*, and *t* as a silent *s*. In this case, Daremusa should instead be read as Saramusa, and the description adds discrete facts in favour of Saaremaa:

> West of the land of the Rus is the island of Saremusa, where there are great primeval trees, some of them so mighty that twenty men with outstretched arms still could not wrap around it. Because of the sun's distance and little light, the people light fires in their homes even in the daytime. There are also wild people on the island who are called *peräd*. (al-Maqdisi, 966)

In the passage, we can detect sacred groves and reminiscences of the Estonian word *pered*, 'families', and the island's northern location is notably emphasised. We briefly encountered al-Maqdisi through his colourful descriptions of 'travellers' perils'. As an author, he used informants of several exotic nationalities.

A mere seventy years separate those lines from the Baltic Sea's next mention. The famed Khwarazmian al-Biruni called the peoples on the shores of the Baltic and the sea itself *Var varank*. The name comes from the Varangian seafarers, just like the name of the Kola Peninsula's Varangerfjord. Russian sources also knew the Baltic Sea both as the Viru Sea, *Morye Viryanskoye*, and the Varangian Sea, *Varyazhskoye Morye*. The word is of Norse origin, meaning a crew bound together by a common oath, and does not constitute an ethnic identifier. Thus, Varangian Sea would be a 'social' toponym – a place name derived from profession or status.

Prangli Island's name also appears connected to this, its metamorphoses over the ages being as follows (Johansen 1951: 107): Prangli (1973), Rangilaad (1860), Wrangelsaar (1723), Wrangi (1599), Wrange (1586), Wrangö (1580), Wrangoe (1566), Wrano (1533), Wrange (1529), Rangelisare (1525), Wranghoe (1491), Rango (1397), Varank (? 1030). This is anything but an immaculate progression, but it does force us to

322

'THEIR GREATEST MIGHT IS IN THE SHIPS'

view the island (6.44 km²) in a new light. What stands out are its favourable location on the Silverwhite waterway and ideal condition as a stopping island: neutral waters from a common-law standpoint, but still in eyesight of Virbuotsa Cape at Naissaar's tip. With the same wind and a course change of just a few degrees, a captain could avoid Revala and reach an island that offered drinking water, forest, and shelter from storms.

Prangli Island has not been archaeologically studied. However, on one fine autumn evening, a handful of beads were scattered across a table at the Eduard Vilde Memorial Museum in Tallinn: the same type of beads that perhaps already Ibn Fadlan described on the bank of the Volga. There were more discoveries as well. Former Prangli schoolteacher Johannes Akkatus had unearthed them with his own hands. I asked for more information. According to Helene Akkatus's letter dated 30.11.1973: 'There were bountiful finds, but Johannes's late brother Lex carried them off when he wasn't looking. The older boys on the island would go rooting around, too, in the dunes near Kelmase Harbour. Old islanders said there were once raiders' houses on the dunes, even gold and silver found.' The law protecting archaeological discoveries hadn't been made yet.

* * *

Once, Kalevipoeg came from Finland and stepped on a big boulder, but fell and bit off a piece of the rock. When he stepped on the boulder, the middle sank into the earth and his footprint was left and you can still see it today. (Prangli)

* * *

Kalevipoeg came from Finland, his trouser legs rolled up, but the water was deep and he said: 'Oh, this puddle! Wanted t'wet me trousers!' (Viru-Jaagupi)

The Gulf of Finland, the gateway to the Silverwhite waterway, has carried different names in different languages. In Estonian myths, it is known as the Finnish Sea: 'The Finnish Sea came from the Tallinn Sea. Kalevipoeg helped to make one spot a little deeper with his shovel, that's how there came to be a sea between us and Finland. Wasn't a sea between us and Finland before. Back then, we and the Finnish people spoke the same dialect' (Kohala).

SILVERWHITE

In the early 1200s, the Danes called the gulf *Mare Estonum*, the Estonian Sea. To Finns, it is *Viru laht*, the Gulf of Viru. Captain Solovey Budimirovich sailed towards Prince Vladimir (before 1014) along the Viru Sea, *Morye Viryanskoye*. Could this name have also reached the Arabs, keeping in mind that, according to their geographical beliefs, the Gulf of Finland extended near Bulgar?

Frähn's iron authority has passed down from generation to generation an interpretation that no one has doubted. *Visu* are Vepsians, *ves* in the Russian chronicles. Frähn further quotes the historian August Ludwig von Schlözer, according to whom the people have allegedly 'long since extinguished and turned into the Slavs, not leaving the slightest trace of their name behind' (Frähn 1823: 219). So when the academic Andreas Johan Sjögren came across the Central and Southern Vepsians barely a year later, and subsequently the Northern (Äänis) Vepsians, it became a glowing justification for the interpretation of the Russian chroniclers, and especially of Frähn.

However, Frähn in no way arrived at his conclusions in a straight line, and he described all his doubts and searching with characteristic meticulousness. *Visu* is one of several possible name forms that he derived from Arab manuscripts. The divergences were broad: other forms included *isu*, *vishu*, *disur*, *rasua/rasu*, and even *dalsu* and *valik*. Frähn ascribed these discrepancies to erroneous copies. 'Not without hesitations, but still with a degree of conviction, I dare to hypothesise that it could be the Wes people known from Russia's oldest history [...], whom Nestor, in both his tables of nations, [...] names and whose territory he clearly states as *Bielo Osero* aka *Weisser See* (north of Novgorod).' And from there, Frähn etymologises the Germanic tribes' languages to prove that the name White Lake was the Germanic *Wii-see*!

Identifying the Vepsians as the Visu could very well be correct; however, Frähn's argument is built on an entirely false basis that he took from the 18th-century Russian historian Vasily Tatishchev. There is not one known instance in which Scandinavian seafarers began translating toponyms from Baltic-Finnic into their own language. The area of Vepsian settlement, as we now know, stretched much further east in the days of the Silverwhite waterway. Still, the Vepsians and the Visu cannot be connected by a White Lake.

Applying Frähn's own rules, *visu* could also prove to be *viru*. As we saw earlier, the toponym was widely known and disseminated eastward into Russian folklore. The name Toolse extends back into the middle of the

'THEIR GREATEST MIGHT IS IN THE SHIPS'

1st millennium and highlights the northern Estonian coast's importance on the waterway even in ancient times. The dissemination of Kalevipoeg myths also agrees with this. The Novgorod, Voskresensky, and Sofia chronicles all recognise Estonians by the name Viru (1268 and 1368). To Finns and the Finnish language, we are still Viru (Viro) to this day. The word's entry into Arab geographical literature seems more likely than not.

The Silverwhite waterway shaped an array of urban and provincial harbours on all Baltic shores. As economic ties and ship routes transformed, some turned out to be tragically premature. Hedeby grew to become a city in the 9th century, Novgorod in the 10th, and Visby in the 11th. The others ticked along as villages or disappeared entirely: the thousand-year Helgö, the mysterious Birka. Their dependence on trade is incredibly obvious.

Does this also apply to the harbour at Revala? We know that silver jewellery made in north Estonia found wearers in Ingria, the far reaches of the northern Dvina, northeast Sweden, and even the northern Norwegian coast. It was a golden age for blacksmiths when grain had simultaneously become Estonians' primary source of wealth. Estonia came to be the world's northernmost grain producer. Grain had to be dried in threshing houses because of the climate. The drawback became a virtue, as the dried grain preserved longer and came into high demand (Johansen and Mühlen 1973: 390). The fate of Revala's harbour was not dependent on the whims of transit trade. The continuity of ancient seafaring needn't have been interrupted. There were some hopes for Revala.

Is this vision true? Al-Idrisi certainly believed so.

VIII

SIXTEEN YEARS WITH AL-IDRISI

THE WORLD BROKEN INTO SEVENTY SHARDS

Al-Idrisi divided the world into seven climates that ran in zones parallel to the equator and were numbered I to VII. Estonia lies in the seventh and northernmost.

The climactic zones were divided into sections in turn, numbered from one to ten from west to east. Estonia ended up in the fourth section. Before Estonia, al-Idrisi dealt with the southern shore of the Baltic, Denmark, and the southern part of the Swedish peninsula, which together formed the third section. After Estonia, he moved on to Russia, which formed the fifth. Together with Estonia in the fourth section of the seventh climactic zone were Sweden's eastern coast, Finnmark (both Norway and Sápmi), south Finland, Latvia, Lithuania, the area around Lake Ladoga, and region of Novgorod bordering it. The last-mentioned was necessary for al-Idrisi to tie the Baltic section's coordinates to those of the next, Russia. Having referenced the Novgorod region, he then began describing the Baltic islands.

At least seven later copies of the manuscript of varying qualities have been preserved, each section illustrated with its own map. The first to attempt decoding the Estonian part was Theodor Nöldeke in 1873, followed by the Swedish linguist Richard Ekblom in 1925. The latter's results were corrected and significantly enhanced by the Finnish Arabist Oiva Johannes Tallgren-Tuulio. Andrus Saareste rejected Tuulio's interpretation of Revala. Apart from Uku Masing, no Estonians have made close studies of al-Idrisi (Ekblom 1925, 1931; Tallgren-Tuulio 1930; Tuulio 1934).

Ignaty Krachkovsky believes (1957) that Tuulio's textually critical process established a new chapter in modern al-Idrisi studies. His method

SILVERWHITE

was simple and convincing, proceeding from four of the higher-quality manuscripts. These, he presented as photocopies and gave the Arabic text in Latin script. Every discrepancy between the manuscripts is noted and, if necessary, analysed. As a result, he achieved the closest manuscript to al-Idrisi's original to date. When analysing this text, Tuulio used every toponym found on the maps. Additionally, he utilised the *Little al-Idrisi* manuscript – an Arab's summary of the larger work. Tuulio's method was synthetic. In addition to linguistic analysis, he gave attention to explaining how several of the mistakes arose, restoring the psychological background of the collaboration between al-Idrisi and Roger II of Sicily, and conjecturing what the questions and informants' likely answers were, which provides a basis for determining the informants' possible direction of travel in turn. Only then did he begin decoding the toponyms, dedicating a separate chapter to each. Tuulio was assisted by his brother Aarne Michael Tallgren, who was a professor at the University of Tartu, an outstanding archaeologist, and who wrote the summarising chapter.

Vowels are not marked in the Arabic text. The form of Revala's name that appears on al-Idrisi's maps and in his manuscript requires firstly linguistic and secondly cartographic decoding in order to place it on a network of modern coordinates. This is particularly difficult in Revala's case, whose history has given us a total of nineteen different name forms (Neus 1849, and two more: Ojansuu 1920). In al-Idrisi's *Nuzhat al-mushtāq* (The Excursion of the One Who Yearns to Penetrate the Horizons), he includes a place named Qlwry. The possibilities for pronouncing it are broad and range from 'Flury' to 'Kaleweny'. According to Ekblom's hypothesis, it signifies Revala.

> The fourth section of this seventh climate contains most of the Rwsyyt land and Finnmark land, Häme [Tavast] land, *Estonia* [Estlanda] land, and the land of the Madzhus [those who practise cremation]. [Decoded place names are given in their original form, their approximate pronunciation in brackets. Decoded place names are followed by the al-Idrisi-era place name in brackets. Estonian toponyms are italicised.] These lands are for the most part empty and barren with sparse villages; they are covered by snow for very long and the settled regions do not extend far. As for the Finnmark land, it is rich in villages, farmed arable land, and herds of livestock, yet it has no cultural centres apart from the cities of Turku [Aboa] and Uusikaupunki [Qalmark, Kalanti, predecessor of Uusikaupunki (Niitemaa 1963: 190–203)]. These two are large cities but unfertile lands surround them; the populations of

328

SIXTEEN YEARS WITH AL-IDRISI

both live in poverty because there are not enough essential provisions to be found to cover their needs. The rain there pours endlessly.

It is 200 miles from Uusikaupunki to Sigtuna, going westward.

The king of Finnmark rules land and arable fields on the Norwegian island I spoke about earlier [in section 3].

The distance from the city of Uusikaupunki to the mouth of Mälaren ['the second branch of the Qotelw river'] is 80 miles; from Mälaren to the city of Ulvila [Ragwalda, medieval Ravinankylä (Niitemaa 1963: 190–203)] is 100 miles.

Ulvila is a great and flourishing city that is in the neck of the sea and belongs to Hämeland [Tabast]. Hämeland is rich in farmed fields and villages, but the settled areas do not extend far. This land is hounded by snows even more than the land of Finnmark, and one could say that it is not released by snow or rain for a single moment.

It is 200 miles from the city of *Anhel* to the city of Ulvila.

Anhel is a handsome, noteworthy, flourishing city that is part of the *Estlanda* land. Estlanda's cities include *Qlwry* [Kaleweny], a small city or rather a large fortress. It is populated by farmers whose earnings are small, but who have large herds nevertheless.

It is a six-day journey from the city of Anhel to this place in the southeast; likewise, travelling along the shore, it is 50 miles from Anhel to the mouth of the *Pärnu* [Bernu] river, and 100 miles from there to the *Flmws* [Falamus] fortress, which is far [a certain distance? a short distance?] from the shore. The fortress is desolate in winter because the people flee to caves that are far from [close to?] the sea, where they take shelter, lighting fires and keeping the fires burning night and day for as long as the cold wintertime lasts. When summertime arrives, the mists disappear from the shore and the rains cease, then they continue living in the fortress.

It is 300 miles from there to the city of Mdswna [Madsuna]. Madsuna is a large, homogeneous, flourishing, very populous city whose residents are the Madzhus, who worship fire. It is 70 miles from there to the city of Cwnw [Sortau, Sortavala?] on the coast of the Madzhus' land. But one must count as part of the Madzhus' land, which is far from the sea, the city of Qby [Kainu], a six-day journey from the sea. But from the city of Kainu to the city of Kaleweny is likewise four days. And from Kaleweny westward [Paris MS: southward!] to the city of Gintiyari [Hulmkari, Novgorod?] is seven days.

It is a great, flourishing city on a hilltop, the residents of which have fortified themselves against Rwsyyt attacks. This city is not under the rule of any king.

329

SILVERWHITE

Within Russia's [Rwsyt, Wrusyt] settled regions is the city of Mozir [?], which lies near the sources of the Dniester River. From this city to the city of Przemyśl[?] is a four-day journey southward. Przemyśl is called Twya in Greek. Przemyśl and Mozir are settled places in Russia ['the country of Christians' – Paris MS A], where many such places can be found, travel ye by length or width.

The Twilight Ocean [Baltic Sea] contains numerous unfertile islands. As for the settled islands, two of them are called islands of the Amazonians –Amazonians are the Madzhus. Only men populate the western island; not a single woman is seen there. The other is populated by women, and there no men are found. Every year, the men cross the strait separating them on their ships. It takes place in springtime. Every man finds a woman, lives with her, and stays in her dwelling for about a month. Then, the men return to their island. They remain there until the next year, in the given season, then sail back to the women's island again and repeat the previous year's acts, each man staying with a woman for an entire month. Then they return to their own island. This is done by all. It is their custom, a tradition they have long established.

To reach them by the shortest route, one must depart from the city of Anhel, and the journey is three days. One may also journey from the city of Uusikaupunki, but also from Ulvila. Yet hardly a man who sails to their islands can reach them, so frequent are the fogs on that sea, so long the twilights, and so weak the daylight. (Tallgren-Tuulio 1930: 30–9; Kahk 1960: 16–17)

ONE REVALA TOO MANY?

Seven hundred and fifty words. Enough to inspire quests for answers, starting with Wilhelm von Humboldt, and sufficiently scant to lure seekers hoping to make new discoveries to this day.

Al-Idrisi describes the Baltic Sea but is silent about Gotland, its centre.

Al-Idrisi describes a people called the Madzhus, but the exact meaning of the word is 'a magician, a wizard', and it isn't a great leap from there to 'fairy tale'.

Al-Idrisi describes Estonia but places it west of Tavast, present-day Häme, and Finnmark, Finland, west of that in turn. In this case the entire distance from Estonia's western coast to the eastern coast of Sweden would be incorrectly filled with dry land, with no room for salty sea water.

SIXTEEN YEARS WITH AL-IDRISI

And finally, *Qlwry*, al-Idrisi's Kaleweny, or the supposed city of the Kalevs: our golden nut to crack! The problem is that it's a four-day journey from the city named Qainu to supposed Revala. However, it is a six-day journey from that same Qainu to the sea. The conclusion is obvious: Revala is at least a two-day journey away from the shore. This is how al Idrisi understood it as well, placing Qlwry/Revala far inland on his map. This 'Revala' is not our Revala; it is not Tallinn. However, Russian historical tradition recognises Kolyvany as Revala, and Revala alone; a city that, as we well know, is on the coast.

We can sympathise with the traveller who throws his hands up at this chess problem and declares al-Idrisi's geography flawed, erroneous, and distorted by legends and fancies. Hyper-critique is only human. Many Arab authors have tasted the salt of hyper-critique at various times throughout history, Ibn Fadlan in perhaps the most painful way. But when passions have abated and gazes become keener, it turns out again and again that hyper-critique characterises the researcher more than the researched.

So, how are we to explain the contradiction between the two Revalas – one coastal and one inland? There is one too many. Ergo, al-Idrisi confused Revala with some other city.

Let us proceed with the hypothesis that the city al-Idrisi confused with Revala is the place called Anhel. But, first, let's clarify the Baltic toponyms.

RULES OF THE GAME

Al-Idrisi did not name Gotland. This could be a serious accusation, but it is not. Adam of Bremen was also missing Gotland, even though he lived and worked in Bremen, the gateway to the Baltic Sea, while al-Idrisi was in distant Palermo. We are urged to show even greater caution in our judgements by the fact that Icelandic historical tradition doesn't mention Birka or Helgö. These are outstanding gaps. We must accept them. The relativity of judgements should be self-evident.

Al-Idrisi named the Madzhus as a pagan people settled east of Estlanda. It is a general term, not an ethnonym, and bears a religious quality in this case. Al-Idrisi himself guides us to this interpretation when, in the second section, he describes the people of Nubia as 'Madzhus who believe in nothing'. Here, they are contrasted primarily with orthodox Muslims, but also with the Roman Catholic and Greek Orthodox confessions, which al-Idrisi distinguishes very consistently. Another Arab provided a more evocative description: 'they pray to idols' (Shams al-Din). As late

SILVERWHITE

as the early 19th century, the Kazan Tatars, bearers of the Volga-Bulgars' Islamic tradition, still called the Mari 'Madzhus' (Frähn 1823: 136–7). Thus, the Madzhu fire-worshippers east of Estlanda could be the Votians, Vepsians, Karelians, and perhaps even the Viru. We discussed cremation traditions at greater length in connection with Ibn Fadlan. From Henry of Livonia, we know the great solemnity given to performing these rituals (1223): 'They disinterred the bodies of their dead, who had been buried in cemeteries, and cremated them according to their original pagan custom.' This may have been one of the most important events practised in ancient Baltic-Finnic society, and it is entirely understandable for al-Idrisi to name the easternmost tribes after the remarkable aspect.

But placing Finnmark to the west of Estlanda? This has caused extensive headaches for historians who recognise Finnmark as Finland. The root dates back to the days of Tacitus and, as we remember from Ohthere's voyage in about the year 875, Finnmark was used to refer to all of northern Fennoscandia, Sweden, and especially the furthest reaches of Norway. We could equate Finnmark to Lapland/Sápmi, but only if we interpret the latter as occupying the significantly more extensive historical boundaries that encompassed areas in south Finland, Sweden, and particularly Norway. To achieve total clarity in this issue, we must employ the aid of al-Idrisi's third section, which addresses the region west of the Baltic Sea: Scandinavian lands, in a narrower sense. There, al-Idrisi mentions Finnmark on three occasions. In his introductory statement at the very beginning of the section, he informs the reader that it will 'include the coastal area of Poland, Sweden, the areas of Finnmark, the Danish island [peninsula], and the Norwegian island [peninsula]'. Here, an equals sign shouldn't only be drawn between Finnmark and Sápmi, but also between Finnmark and central Sweden, as Ohthere previously did. In his fourth section, al-Idrisi felt the need to again stress the ambiguous connection between Finnmark and Norway. He did so, as we saw, with the words 'The king of Finnmark rules land and farmed fields on the Norwegian island I spoke about earlier'. When dividing the world into climactic zones, al-Idrisi moved north from the equator; when dividing the climactic zones into sections, he moved from west to east. He remained loyal to these two directions of movement in every individual section and its accompanying map: south to north and west to east. Norway geographically lies west of Estonia, and, in full harmony with this fact, al-Idrisi places Finnmark to the west of Estonia on his map. The Norwegian island (peninsula) lies across from the Danish island (peninsula), thus forming the southern

SIXTEEN YEARS WITH AL-IDRISI

part of Finnmark, Scandinavia's barking dog-head. Between Estonia and Norway/Finnmark is Lake Mälaren, a crucial nexus of Baltic sea routes. However, al-Idrisi is unfamiliar with the name Mälaren. Instead, he uses the 'second branch' of the Göta älv.

So al-Idrisi has the Göta älv flow into the Baltic Sea? He is not innocent of that, either. The Göta älv is a river that flows from the giant lake Vänern to the Kattegat strait. Located at its mouth on Sweden's western shore is today's Gothenburg.

How did al-Idrisi manage to have Göta älv flow in the exact opposite direction and empty into the Baltic Sea near the Åland Islands? The underlying reason was a unified inland waterway and his false deduction that there existed a similarly unified river system. In al-Idrisi's understanding, the sources and headwaters of the Göta älv were located deep within central Sweden, from which the river split in the separate directions of Kattegat and the Baltic Sea. Oiva Tuulio is fully responsible for this decoding of the Göta älv. Supporting evidence can be added from earlier and later history alike. Apollonius of Rhodes treated the Rhine, Rhône, and Po rivers as a single waterway: entering the Po River from the Adriatic Sea, a ship could exit at the mouth of the Rhine! In the early 18th century, the Congo or Niger rivers were believed to be another branch of the Nile. Estonian geography offers supporting arguments as well, and all the more valuable, given that they were written a mere seventy years after al-Idrisi's work. Henry of Livonia consistently calls the Emajõgi, Pärnu, and Little Emajõgi rivers by a single name: *Mater aquarum*, Mother of Water. The reasoning behind this was apparently a notion of the Little Emajõgi branching westward into the Pärnu and eastward into the Emajõgi near Lake Võrtsjärv.

Al-Idrisi's readers have been bothered by the form of Pärnu's name, which is regarded as much younger. Paul Ariste and Finnish linguist Viljo Nissilä have explained that in both Estonia and Finland, there exist toponyms that lack any equivalent in other Finno-Ugric languages: in Finland, Perniö, and in Estonia, Pärnu, to which the linguist Valdek Pall recently added Peipus. The names given to large bodies of water are the most conservative linguistic monuments. Peoples may come and go, but hydronyms remain. We should add that a large number of Arab coins from the years 738–990 have been discovered in the Pärnu estuary (Kruse 1859, see coin tables).

Keep in mind that al-Idrisi basically positioned Mälaren, aka the 'second branch' of Göta älv, correctly: between Estlanda and Finnmark/

SILVERWHITE

Norway, and west of the former. Mälaren, was the intersection of several ship routes, and al-Idrisi endowed it with an important function: in his mind, Mälaren was the link between his previous western section and the section dedicated to the Baltic. He used Mälaren to determine distances to all the closer 'cities' along the Baltic; in other words, their locations. The sections needed to fit together because al-Idrisi was composing a global geography.

What did al-Idrisi call a city? It's not easy to answer this question. The Russian chronicles sometimes refer to a city as a fortress, other times as an actual city, and even the concept of a church is extremely uncertain: according to some sources, Kyiv was said to host 400 churches only a few dozen years after its Christianisation, and even earlier sources claimed that a whopping 700 churches burned to the ground in that same city (Klyuchevsky 1956: I, 172). Stereotypical descriptions are a disease common to all early medieval chronicles and al-Idris alone cannot be blamed. Even so, he attempted to differentiate a village from a fortress, and a fortress from a city. Al-Idrisi's citizens know sciences and religion, and are skilled craftsmen. He counted four cities in Poland and three cities in Finland that fully met these conditions.

THE BALTIC DOESN'T EVEN EXIST!

Al-Idrisi relied upon Ptolemy's worldview. To the classical geographers, the Baltic Sea didn't exist. There was the 'surrounding ocean' and a few islands dotting its coastal waters. Over time, the islands grew and acquired names – Scandinavia, Danemarka – without affecting this overall view. No inland sea sprang from it. In the best case, stretching above the Baltic Sea was a sound that separated the shore from islands and the world ocean. Estonia's coast ran west from west to east in a much straighter line than the shore of the Arctic Sea in old school atlases, and that's precisely how it should be imagined.

Al-Idrisi also envisioned the mainland's northernmost shore as a runway from west to east: located in the 'Twilight Ocean' are the Amazonians' islands, which are a three-day journey from Anhel.

Al-Biruni knew of *Var varank*, al-Idrisi only of the Twilight Ocean. In this sense, he was less accurate and more traditional than al-Biruni. The information that travelled down the river through Volga Bulgaria to reach al-Biruni's ears corresponded more to fact.

SIXTEEN YEARS WITH AL-IDRISI

As a scientist, al-Idrisi relied upon contradictory sources: academic tradition that ignored the Baltic Sea, and an informant *from* the Baltic Sea whose practical experience would inevitably conflict with his theoretical starting data. Theory conflicting with practice is more typical of the Occident than of the East.

What form did this variance take? The only conceivable form, and one that was quite logical from al-Idrisi's perspective. If the 'Twilight Ocean' stood in place of the Baltic Sea, then the Baltic's coasts were consequently oceanic coastlands and not situated across from, but rather next to, each other.

In 1957, Ignaty Krachkovsky directed attention to one of Oiva Tuulio's conclusions, whose importance had not been fully appreciated. From al-Idrisi's foreword, we know that he first composed the maps and only later drafted the texts, which were meant as commentary. Analyses of both confirm this. With one exception, that is: in al-Idrisi's section on the Baltic Sea, the order was reversed. He first wrote the text and then used it to draft his map. For this, he was forced to transcend the contradiction between theory and practice.

He proceeded from the knowledge that the seventh climactic zone was the world's northernmost and also the end of the Eurasian continent. In theory, all sections in one climactic zone were located at the same latitude. Al-Idrisi drew a straight line (mentally or on paper, it makes no difference) and made Finnmark/Norway the westernmost coastal land. He positioned Göta älv slightly to the east, which isn't incorrect. However, the river's second branch was supposed to drain into the sea even further to the east, opposite to Häme or Tavast. He drew 'the second branch of the Göta älv river', actually Mälaren, and to the east of it he wrote: *ard Tabast*, Tavastland, Häme.

When ships sail eastward from southwest Finland, they reach Estlanda. Al-Idrisi placed Estlanda east of Tavast. Sailing east from Estlanda, one reaches the land of the Madzhus, and their easternmost city is Sortau. The distance isn't very long when measured from west to east. Al-Idrisi spread them across three sections. Sigtuna and the mouth of Mälaren concluded the third section.

Then he took the first distances of the fourth section to tie the sections together, as a result of which Sigtuna and the mouth of Mälaren appear in both. He repeated the same process when moving from the fourth section to the fifth. Sortau was made the link, appearing on both maps. The fifth section addressed Russia. Let's leave the details aside. Or shall we?

335

SILVERWHITE

Al-Idrisi worked for sixteen years. He may have written the third section in the first year, the next in the eighth, and Russia's section in the sixteenth. Or vice versa. His informants did not arrive at Palermo's royal court in order of climate and section. And the informants were different every time, except for Russia's section, for which he couldn't find a single messenger.

Let's only take away the two most general conclusions. Firstly, theory and method forced al-Idrisi to incessantly ask: what lies further east of this state, this area, this region? Secondly, practice and experience only allowed the informants to move along ship routes in their answers. Sailing was the common thread to all their experiences. Al-Idrisi pulled it straight across his frame of theory and the Baltic Sea disappeared – suddenly, we are faced with the Arctic Sea, the Bering Strait, Japan, and Korea to the right. We could feel offended, could hyper-critically call al-Idrisi a talentless geographer. But how then should we judge the Danish mathematician Claudius Clavus, who in 1424 depicted the Baltic Sea as part of the Arctic Sea on his map (Dreijer 1960: 67–70; Hennig 1961: II, 369)?

IN PROCRUSTES'S BED

Oiva Tuulio (1878–1941) got to work with a blank sheet of paper. He finished his first monograph in 1930. His archaeologist brother Aarne Michael Tallgren (1885–1945) could only cautiously remark (Tallgren-Tuulio 1930: 135):

> If we truly are to read 'Ragwalda' and place it in any specific location in Ulvila parish, then at least at this given moment, it isn't yet possible to present any archaeological proof in favour of that site's importance. On the contrary, it seems like the population of Häme was striving to stay clear of the coast in that era ...

However, the study of early Baltic towns picked up steam after the Second World War and completely confirmed Tuulio's hypotheses. Uusikaupunki was preceded by a fortress town named Kalanti, al-Idrisi's Qalmark. Medieval Ulvila (city charter: 1365) was preceded by Ravinankylä, al-Idrisi's Ragwalda (Niitemaa 1963: 190–203). An archaeological study in the Aura River valley proved that Turku/Aboa was preceded by Nousiainen and Koroinen: a site that was already actively trading with Estonian areas in the 7th century. All of Finland's *kaupunkit* (cities) appear to date back to the Viking Age, and the Finnish word itself arose in the days of the Silverwhite waterway (Gardberg 1963: 173–89).

336

SIXTEEN YEARS WITH AL-IDRISI

Assessments of al-Idrisi have largely depended on how successfully he is decoded. In terms of Finland and Poland (limiting ourselves to Baltic regions), his authority today is greater than it was in Tuulio's day, which is primarily thanks to Tuulio and Tadeusz Lewick. The Estonian assessment of al-Idris as a geographer is somewhat more sceptical.

In principle, of course, it isn't impossible that Estonia *is* an exception; that al-Idrisi described Estonia more superficially than he did other Baltic lands. We cannot rule out the possibility. In that case, our sceptical judgement would be entirely justified.

In addition to the seventy-section map, al-Idrisi's manuscripts were supplied with a small-format overview map that depicted the entire world. Understandably, only a few toponyms could fit. Out of all Northern place names, only one can be found on the Oxford manuscript's overview map: Estlanda, Estonia (Tallgren-Tuulio 1930: 30). We could conclude from this that at least in al-Idrisi's opinion, Estonia was not a second-rate subject.

Al-Idrisi names 48 toponyms in his 'Baltic' section, 29 per cent of which are Estonian place names. This also does not appear to justify a sceptical attitude. He had quite a lot to write about Estonia. The attention he gives to Estonia is out of proportion to the other sections of world geography and Estonia's geographical area. This doesn't speak to al-Idrisi's accuracy. However, the criticism is aimed at his somewhat artistic theory and method, not the data extracted from informants' experiences. Estonia seemed important to al-Idris. Why?

HANILA'S FIRST PATRIOT

The explanation may lie in the informant's identity; in his person.

Researchers, including those who disagree with Tuulio's interpretations, are in agreement that the Estonian toponyms al-Idrisi conveys are in Baltic-Finnic form. Even so, it has been regarded as only natural that the informant, albeit residing in a Baltic-Finnic environment, was Germanic. Arguments include the fact that the Finnish toponyms Turku and Hämu are given in the Swedish form of Åbo and Tava. However, the informant could have preferred to give internationally recognised name forms, especially to a man composing a book of world geography. Using the little-known river name Pärnu speaks greatly in favour of the informant's Baltic-Finnic origins. One way or another, the first Baltic-Finnic toponyms were recorded between the years 1138 and 1154 at Palermo's royal court, and it was both a remarkable and unusual event.

SILVERWHITE

Another small detail appears to prove that al-Idrisi's informant was a Baltic Finn. In the third, westernmost section, al-Idrisi consistently uses Zwdh, Zweda, to denote Sweden. The messenger who provided details for this section is believed to have spoken a Romanic language (Tallgren-Tuulio 1930: 11). But al-Idrisi becomes somewhat confused in the fourth section, the causes of which can be understood through the *Little al-Idrisi*, which he wrote only after finishing the larger book. In the third section of this condensed version, the geographer writes Rwdh in place of the Germanic Svea/Zweda (Tallgren-Tuulio 1930: 115, n1). One can read this as the Baltic-Finnic name form *Rootsi/Ruotsi* (Sweden), which al-Idrisi could only have received so plainly from his earlier Baltic informant. This was most likely the case. At the very least, it explains the incredibly vexing flaw at the beginning of the fourth section, which reads: 'most of the Rwsyy land and the Finnmark land'. Al-Idrisi emphasises that it is the greater part, as he has already reviewed the smaller, more western part in the previous section under the name Zweda. Most Arabists who had no knowledge of Baltic-Finnic languages have interpreted the toponym as 'Russia' because of its similarity in sound, despite the glaring contradiction: al-Idrisi only addresses Russia in the subsequent, fifth chapter, and there he refers to it as *ardi al-Rwsh*, 'country of the Rus'. Earlier studies even tied 'Roslanda island' to Iceland, failing to realise that Baltic-Finnic languages would allow *Rootsi/Ruotsi* to be found much closer (Frähn 1823: 216).

Introducing his fifth section, al-Idrisi stresses that his description of Russia is not based on anyone's personal observations. Perhaps this is the most substantive nod to our anonymous informant's existence and, consequently, also evidence of his adventurous journey from the Baltic Sea to the distant court at Palermo. If this is true, then he is to blame for the strikingly disproportionate share of Estonia's and Finland's descriptions. But who would dare to call him a scapegoat?

UPON THE GREAT BRIDGE

We don't even know what language was spoken at Palermo's royal court. The man's gaze wandered across the shelves of scrolls and weapons hung on the stone walls, through the ornate window trellis to the verdant gardens and blazing-white city walls below. It was early morning. The day would be hot and cloudless. Sharp-toothed strips of sunlight cascaded through the arched windows, spilling over the black-and-white tiles. He rested his palm on one of the dark ones. It was already scorching. For a

SIXTEEN YEARS WITH AL-IDRISI

moment, he imagined a grove of birch trees rustling in crisp morning air. He took a deep breath through his nose like someone who was drowning.

But the curtains were already being pushed open and yesterday's servants entered. They took quiet, stiff steps, bowing without pausing, their eyes trained to the ground and their ears pricked. Then, al-Idrisi appeared in the doorway, as slim as a pillar, dressed in a white silk tunic, and the red curtains collapsed together behind him as if by their own will. He nodded to the man (who had risen to his feet), pointed to a small bench, and took a seat across from him. A servant knelt on the rug, spread a writing skin over a small table, and dipped a quill in ink. The interpreter knelt next to the scribe, his arms crossed in his loose sleeves as if he felt cold. A third servant stood watchfully, holding more scrolls on a tin platter. Or was it tarnished silver? Apart from the interpreter, they all wore Eastern turbans, al-Idrisi's the largest and most brilliantly white. The scent of precious oils wafted from his clothing, as if he carried flowers from faraway gardens within its folds. His bony cheeks and pale face looked like a mask adorned by a thin strip of midnight-black beard. Wordlessly, he extended a hand, and the standing servant placed yesterday's scroll in his palm. His long, sensitive fingers did not rush to open it. A fly droned near the ceiling, and somewhere in the far reaches of the castle's courtyard, a load of firewood clattered against the stone pavement.

Idrisi: 'You named the land of Danemarch yesterday, did you not?'
He did not look away from the man as the interpreter arduously pondered the question.

Man: Danemarch, Denmark, yes.

Idrisi: 'Fine. Now, tell me: if you travel north from Danemarch, then what land do you enter?'

Man: Norway.

Interpreter: 'Norbaaga.'

Idrisi: Write that down!
As the interpreter looked on, four or five dashes and dots appeared. The rough scratching of quill-tip on skin echoed throughout the hall, muffling the dull and distant roar of the sea.

Idrisi: 'How does one travel from Danemarch to this land?'

Man: 'You must cross a strait. It isn't wide, perhaps a half-day's sail.'

SILVERWHITE

Idrisi strove to imagine the most distant continent's shore on the Twilight Ocean, the source of such exquisite furs. Some of the names heard long ago appeared to match those he was being told today. The interpreter monotonously conveyed the man's reply: about a half-day's journey, and the current is weak. That also sounded familiar.

Idrisi: 'You named Zweda. What more can you tell me about that land?'
The man faltered in his answer and Idrisi was too impatient to wait for the interpretation. He marched briskly to the shelves, drew his belt tighter around the caftan's embroidered lappets, climbed to the top rung of the ladder, and began searching through old scrolls. The words Zweda and Sigtuna spun around in his head: perhaps they were one? Sigtuna was also on the shore of a slow-moving current, was it not? He dropped a dusty scroll onto the silver platter the servant was straining to hold high above his head, and loudly asked: what lies to the east of Sigtuna?

* * *

All roads lead to Anhel.

Anhel is on the coast. It is the closest harbour for sailing to the Baltic islands. It is a three-days' sail from those shores. Kaleweny, on the other hand, is not on the sea. Did it really signify Revala, then? Perhaps Kaleweny/Revala did not yet exist in al-Idrisi's day? The possibility is invalid, as archaeology tells a different story.

What if Revala/Tallinn was not known by the name Kaleweny, but rather some other inland place? The possibility is not invalid, though it is extremely unlikely. Only sixty years later, chronicles recorded the name Kolyvan in reference to what is now Tallinn. Geographical reality and al-Idrisi's text contradict each other. An error was made. Only al-Idrisi alone can be to blame.

Why? The informant is not of unknown proportions to us. He has acquired an outline and colours, and started to speak. He was well acquainted with the eastern shore of the Baltic, and with Estonia best of all. True, he may have failed to mention that Tallinn was on the shore. It would have been odd, but the possibility is there. Yet, given all his fine knowledge, he could not in any way have stressed the opposite: that Tallinn was far from the shore.

Nevertheless, that is precisely what al-Idrisi recorded in his notes. What's more, when he had finished editing the final manuscript and began drawing its illustrative maps, he hid Kaleweny/Tallinn just as far inland as Novgorod. The informant's Kaleweny could only have been on the coast.

SIXTEEN YEARS WITH AL-IDRISI

Out of all his Estonian toponyms, al-Idrisi can name only one harbour: Anhel. It appears five times in the text: more frequently than any other place in Estlanda, and more frequently than any other toponym in the entire fourth section. Do we dare point out his error? To be fair, it was already done long ago: plainly and without wasting a single word. When the year 1154 was recognised as the beginning of Tallinn's written history, it simultaneously recognised the need to relocate inland Kaleweny to its proper place on the coast. 'Relocate' is a good word. We can lift the trees, bushes, hill, fortress, and entire landscape (along with the name, of course), and rest them all gently in a suitable coastal environment. Which is exactly what was done.

However, we could also simply change the labels and leave the landscape in place. The question is what al-Idrisi did. Perhaps he just temporarily exchanged Anhel's and Kaleweny's labels? Perhaps therein lies the reason for Anhel's frequency in being recorded?

Reconstruction is the only way to understand the error.

As soon as the mutton has gone the route of all meat along with the baked apples, chestnuts, and stewed asparagus, all in thoughtful silence, and the beakers have been emptied, the fruits bitten, hands rinsed of grease in warm vinegar water and sprayed with rose oil, as soon as the lute players have given way to the scholars, and the trellised shutters been drawn against the blistering sun, we enter the reconstructed castle hall as well. You may like it, or you may not. In any case, let it be noted that the sections of the previous dialogue that appear in quotes were taken from Tuulio's monograph on al-Idrisi (Tuulio 1934). Reconstruction is one method for exploring seafaring history.

What interests us is how the names Kaleweny and Anhel could have been switched. There must be an underlying thought process that was logical in terms of al-Idrisi's theory, but contradictory to reality.

Idrisi: What land lies east of Sigtuna?

Man: If you sail east from Sigtuna, then you reach Turku or Uusikaupunki. They are in a land known in your language as Finnmark.

Idrisi: We shall speak of cities later. What land lies east of Finnmark?

Man: That would be Tavastland, Häme. To the east and also to the north.

Idrisi: And if you continue sailing east from there?

Man: Well, then that is The People's Land. (*Smiles innocently*)

SILVERWHITE

Idrisi (*frowning*): Excuse me?

Man: What I mean is that is the land where I'm from. I could go straight from Revala home to Hanila.

Idrisi: We will speak of that later. Are there no other names in your land?

Man (*insulted*): Of course, why wouldn't there be? Revala is called Kaleweny and the Land is called Estlanda; you should know that yourself. But Hanila's just Hanila.

Idrisi: We will write Estlanda, Anhel, and Kaleweny. And if you sail east from Estlanda?

Man: First, there's the Votians, then the Izhorians, then …

Idrisi: And they are already of Eastern Roman faith?

Man: As if! They still light fires for their sacrifices and their dead.

Idrisi: Then they are Madzhus. And now, explain to me how you reach your Hanila. Precisely and in detail.

Man: First, I sail close to the shore until the Porkkala cape. From there, I cross the gulf to Naissaar and Revala …

Idrisi: Women's Island? That's interesting. We will note it.

Man: You can already spot the city hill from the other side of the island. It rises from a great, towering cliff. There is no other like it anywhere along our sea. It is so steep that none can get close.

Idrisi: Who can't get close?

Man: The Swedes, obviously.

Idrisi: Are there buildings, too?

Man: Sure, why wouldn't there be?

Idrisi: Churches?

Man: There's a church and a chapel, too. In Saha.

Idrisi: And craftsmen?

Man: The most famed silversmiths in the world.

Idrisi: Is it a king's city?

Man: No, definitely not. No one has yet overpowered Revala.

342

SIXTEEN YEARS WITH AL-IDRISI

Idrisi: Revala? What was that other name?

Man: Hanila? It's a six-day journey from there by foot.

Idrisi: But you mentioned Kaleweny?

Man: Yes, our sailors were once called Kalevs.

Idrisi: And in what direction must you travel?

Man: The highway crosses the Tõdva ford, so to the southeast at first.

Idrisi: Away from the sea?

Man: You could say that. Southeast is the only direction leading away from the sea for us.

Idrisi: And there are no cities between the two?

Man: Of course there are. There's quite a large fortress near Pala. My mother told me there are caves there where—

Idrisi: We will discuss the caves tomorrow. Right now, let us record distances.

Man: You could get there from Kaleweny in two days, and if you journey on from Pala to the mouth of the Pärnu River, then that can certainly be done in three days, and it's not very far from there to Hanila.

Idrisi: So, we will write that it is 100 miles from the mouth of the Pärnu River to Pala?

Man: I guess so. And if you take the shoreline route going southeast from Hanila, then it's only half that distance to the mouth of the Pärnu.

Idrisi: Ergo 50 miles. That means you arrive on the sixth day. Is it a large city?

Man: It's no city. It's a small fortress. Our farmland is much less fertile than in Revala, there's no grain to sell, but we do keep livestock. Our herds are rather large.

Idrisi: Now, let us speak of directions. Kaleweny is to the east of Revala?

Man: Not Revala, but Hanila.

Idrisi: Now, I fail to understand you. You said that you journey southeast by foot?

Man: Southeast at first, then south.

SILVERWHITE

Idrisi: Let us make this all quite clear. What is the westernmost place in Estlanda?

Man: If you leave out the islands, then Hanila, I suppose.

Idrisi: That is what I needed to know. And now, distances. If you sail to Estlanda from the west, then how many days is it?

Man: What Revala men do is first cross the Back Bay from Mälaren to Ulvila. That's two to three days with your average wind.

Idrisi: So, 100 miles?

Man: I'd say even more.

Idrisi: And how do they travel?

Man: We've got our trade friends there, I suppose, but otherwise you go along the shore as usual, nose to nose. From Ulvila, you go east around the headland, and from Porkkala to the south, so the route is twice the distance.

Idrisi: Then we will note that it is another 200 miles from Ulvila.

Man: Should be about right. And if you take the easternmost place on the shore of Lake Ladoga, Madsuna, then it'd be another one and a half times more to get there than from Ulvila.

Idrisi: Ergo, 300 miles, and that is what we will write. Now, let us speak of the islands in your ocean. How do you sail to them?

Man: You can get there from any harbour, but the shortest route is from Kaleweny, of course. Not more than three days.

Idrisi: Now, you are certainly mistaken. You wished to say from Hanila?

Man: You can sail from Hanila, too, or from Uusikaupunki, or from Ulvila if you please.

Idrisi: In any case, let's review the information. Anhel is Estlanda's westernmost place?

Man: Correct.

Idrisi: And if you continue on that route, then it is a six-day journey from Anhel to Kaleweny?

Man: That's right, if you go on foot.

SIXTEEN YEARS WITH AL-IDRISI

Idrisi: And it is a great, flourishing city right on the crown of a hill that is impossible to scale? They have no king?

Man: Correct.

Idrisi: From there to Ulvila is 200 miles, and 300 miles to Madsuna?

Man: Correct.

Idrisi: And it is a three-day journey to the Amazonians' islands?

* * *

We could continue knitting these stockings ad infinitum. The unknown is endless; the known is finite, plain, and simple. Even so, it feels as if we are closer to the truth if we place in al-Idrisi's mouth the question: which is the westernmost place in Estlanda? And if, in connection with another topic, al-Idrisi was informed that Revala/Kaleweny was Estlanda's first harbour along the classic old ship route, then switching labels on the map became fatally inevitable: the westernmost place in Estlanda had to be Anhel! Let us imagine a straight line – the shore of the Twilight Ocean. It did have a handful of bays and coves, but this didn't change the overall picture: the coast began near Norway/Finnmark and disappeared into the misty tundra of Pechora, pressing all the 'lands' and 'regions' close to one another like beads.

The Hanila man continued on his travels, and al-Idrisi was left alone with his notes. They needed to be systematised, dovetailed with the western, Scandinavian section and the eastern, Russian section. Al-Idrisi corrected Tallinn to Anhel and Anhel to Tallinn. The discrepancies disappeared, and that is the text that survived to our day.

What did al-Idrisi's manuscript look like before it was edited? What was the man from Hanila able to tell him about Tallinn? Using modern names, his account might have been as follows: It is 200 miles from Tallinn to the city of Ulvila. Tallinn is a handsome, noteworthy, flourishing 'city' that is part of the land Estlanda. Among Estlanda's other cities is Anhel, a small city or, rather, a large fortress. It is populated by farmers whose earnings are small, but who have large herds nevertheless. It is a six-day journey from the city of Tallinn to this place in the southeast. Travelling along the shore from Anhel, it is 50 miles to the mouth of the Pärnu River. […] It is 300 miles from Tallinn to the city of Madsuna. […] Tallinn is a great, flourishing city on the crown of an unscalable hill, whose people

345

SILVERWHITE

have fortified themselves against the Swedes' attacks. This city is not under the rule of any king.

To reach [the Amazonians' islands] by the shortest route, one must depart from Tallinn, and the length is three days. One may also journey from the cities of Uusikaupunki or Ulvila [i.e. the distances are similar, but it is still a shorter journey from Tallinn].

* * *

This is not Tallinn's oldest mention. If correct, it would be the oldest description of early Revala. All in all, it contains nothing new, and certainly nothing that contradicts Tuulio's findings. Tallgren summarised his brother's conclusions with regard to Tallinn in a single sentence (Tallgren-Tuulio 1930: 139): 'Based on the data al-Idrisi presents, it isn't possible to identify Qlwry as Kaleweny.'

The concerns of Tuulio's study were mainly linguistic and toponymic. He proved that Qlwry represented one of ancient Tallinn's name forms. Tallinn ended up away from the coastline, though this is of secondary importance from the standpoint of linguistic analysis. What matters more is proving that the man from Hanila and al-Idrisi genuinely used Tallinn's name.

In this travelogue, however, we are interested first and foremost in the actual landscape and its agreement with al-Idrisi's text and map, where actual-Tallinn acquired the shape and location of Anhel. But what modern place is then hidden behind the Anhel toponym? Switched labels on maps is one of the most common mistakes in ancient cartography, and all researchers account for it. Already in 1852, the Polish academic Joachim Lelewel determined that Anhel should be regarded as Tallinn (Lelewel 1852: III, 175–81). Konrad Miller, whom Arabists can thank for publishing all of al-Idrisi's texts and maps in a single work, arrived at the same conclusion in 1926–7: that Anhel should be geographically (not toponymically) equated with Tallinn (Miller 1926: I, 78; II, 147–8).

BLUE-EYED AMAZONIANS

Al-Idrisi's maps depict eight islands across from southwest Finland and Estonia. Populating the two larger ones are the Madzhus, who practise cremation; this therefore prevents us from identifying them as the Åland Islands, which had been Christianised by that time. Al-Idrisi was as

346

SIXTEEN YEARS WITH AL-IDRISI

accurate as one could expect in terms of religion. However, he does call the Madzhus' territory the 'Amazonian islands', sparking suspicions that the whole tale was borrowed from Greek myths.

Or does the name Amazonian reflect some local tradition, characteristic of the Baltic Sea? Are there any grounds for that hypothesis? Strangely, there are. For reasons that are not entirely clear, there developed an entire system of symbols that was focused at the mouth of the Gulf of Finland and connected, in one way or another, to Mother, Daughter, Mother-in-Law, Woman. Between Estonia and Finland are Naissaar ('Women's Island') and Tütarsaared (the 'Daughter Islands'). Near Iru Fortress, Revala's predecessor, was a famous boulder with magical properties called Iru ämm ('Iru's Mother-in-Law'), and to this day residents of Tallinn drink the fabled tears of Linda, widowed wife of the hero Old Kalev, from Lake Ülemiste. The Catholic Church dedicated its conquests on the eastern Baltic shore to the Virgin Mary. Adam of Bremen wrote of a land called *Terra feminarum*, Women's Land, placing it north of Aestland. *Historia Norvegiae*, written fourteen years after al-Idrisi's work, recognises a *Virginum terra*, Virgins' Land. Even Kvenland is derived from the name 'Women's Land'. Brother Nikulás, as we may recall, wrote of two Kvenlands, and based on their location, one should equate them to Estonia.

Or is this all just international folklore? Chukchi myths speak of an island far beyond the horizon of the Arctic Sea, populated only by women. These women become pregnant in the waves of the sea and give birth to girls. The legend was recorded by Colonel Friedrich Plenisner, who accompanied Vitus Bering on his 1741 expedition (Helmersen 1876). An identical legend was recorded on the western shore of the Arctic Sea (Hennig 1961: II, 462): 'finally, they floated to shore somewhere between Greenland and Biarma, where, according to their account, they encountered unusually large people and discovered a land of virgins where the women become pregnant by drinking water …' However, the last lines are from the year 1200 and described the discovery of ancient Svalbard in 1194! Six centuries and 6,000 kilometres of Arctic Sea constitute an almost cosmic abyss. If the bridge of folklore traverses that abyss, then it can only encourage one to show caution in topics of international migration.

Nevertheless, al-Idrisi's description of the Amazonians' islands leaves a pragmatic, almost humdrum impression. Their location is a three-day journey from Revala (= Anhel), and they are apparently only slightly farther from Uusikaupunki and Ulvila. Every spring, the men set sail. Al-

SILVERWHITE

Idrisi asserts it is a long-standing tradition. We can assume distortion of the causal association and perceive a work-related rhythm dependent on certain seasons. Tacitus employed metaphors from classical mythology to make pragmatic comparisons. The same could be expected from al-Idrisi. Does the Baltic Sea, and the Gulf of Finland more narrowly, offer actual supporting details for a comparison with the Amazonians?

North Estonia, as we are now well aware, was the world's northernmost grain producer for some time. It evolved into a unique distribution of labour between the opposite shores of the Gulf of Finland; one still practised during the first third of the 20th century under the title *sepra* [*sõbra*] *kaubandus* – 'friendly trade'. The academic Kustaa Vilkuna described it as follows (1963: 208–10):

> For Tammio fishermen (an average of 25 boats), the work year culminated with spring and autumn fishing. As soon as the season's catch was stored and the herring salted, the men joined together on a joyride to Virumaa's coastal villages on the opposite shore. The first crossing was made around St John's Day, the second in October. [...] Every Tammio family had old *seprad* [friends] who on the set date would arrive from inland with wagonfuls of grain to trade for fish. [...] Back in their home village, the women would keep a sharp eye on the wind and weather and calculate when the sailors might return from Estonia. When the time drew near, then most of the village's youth, children included, would gather on the island's southwest shore to await the men – they called this practice *vironmiesten vahtiminen* ['watching for the Viru-men'].

From the same author, describing Haapasaari Island: 'The two high points of the year for the leading fishermen were the two-week *kevätviro* [spring-Estonia] and *syysviro* [autumn-Estonia] [...] Children would make turf "wind ovens" with the mouths facing Estonia "to summon a fair wind so they [the sailors] would make it back from Estonia"'.

Widespread 'friendly trade' and age-old tradition tie these accounts from the turn of the last millennium to al-Idrisi's Amazonians. The given notion of *sõber*, 'friend', was also familiar on the shore of Häädemeeste, though in a much more refined sense than even in medieval Tallinn. The word has Baltic roots and made its home in the Finnic languages long ago, albeit in a more social sense than today's definition: the Lithuanian *sêbra* means 'group, society', and the Vepsian *sebr* 'cooperation'. The word underwent a phonemic transformation in written Finnish and is now *seura*: 'club, association'. In northern Finland, the word is used for members of

348

SIXTEEN YEARS WITH AL-IDRISI

joint fishing or hunting trips. Based on these details and others, Vilkuna concluded that the early days of friendly coastal trade date to prehistory and 'apparently developed at the time when a permanent settlement of Finnish fishermen developed on the Gulf of Finland's northern coast, which was once also Estonians' fishing waters'. Thus, what we have is a genuinely ancient phenomenon that appears to reflect in coastal Finns' ethnogenesis. With its numerous islands and labyrinths of channels, fish teemed more plentifully on the Finnish shore than on the Viru coast, attracting scores of fishermen during the big fishing seasons. While these trips were only seasonal at first, they gradually led to the development of permanent settlements. The fishermen maintained ties with their ancient motherland, from which they continued to import grain and, in the early years of the shore's colonisation, no doubt women as well: fishing was done by men, and the shanties on the Finnish coast originally made up 'womanless villages'. Fishing was not the only walk of life that separated men and women by season. Even today, the Komi's forest villages are emptied of their men twice a year: hunters vanish into the woods for four months as winter approaches, and the summer fishing and hunting season is only slightly shorter. Seasonal seal hunting was also a common practice along the Baltic Sea.

The widespread Nordic legend of Amazonian 'islands' can most likely be boiled down to the seasonal distribution of labour. Al-Idrisi also seems to allude to this: he positions the 'women's island' such that it lies northwest of Revala (Anhel), more or less in the area of actual Naissaar. The island's Estonian name may be a remnant of the significantly more widespread ancient 'Women's Land'. Al-Idrisi placed the 'men's island' close to today's Ulvila. The Oxford manuscript calls the channel separating it from the mainland 'the turning place'. Could this be a reference to Porkkala or the Hanko headland, where the ancient ship route took an almost 90-degree turn to the west (coming from Tallinn)? The Turku region and its later fortress town traded actively with the regions of Estonia, Livonia, and Latvia, starting from the 7th century. Consequently, archaeological findings do not disagree with al-Idrisi. On the contrary, he was correct when he characterised the biannual seasonal voyages as an ancient, long-established tradition.

IX

TALLINN IN THE HAZE OF LEGENDS

CONTRASTS

Early Revala, as we attempted to depict it through al-Idrisi, finds an irreconcilable opponent in Henry of Livonia. The chronicler leaves no room for doubt that Tallinn was built on vacant ground by the Danes. He describes King Valdemar II's landing in the summer of 1219 as follows:

> They all brought their army to the province of Reval and encamped at Lyndanise, which had once been a fort of the people of Reval. They destroyed the old fort and began to build another new one. The people of Reval and Harrien [Harju County] gathered a great army against them and they sent their elders to the king with deceptively peaceful words. The king believed them, not knowing their guile. He gave them gifts and the bishops baptized them, sending them back joyfully. (XXIII)

An abandoned fort, an empty harbour, unsettled environs … It is a bleak image that has since travelled through many hefty books.

Revala only received its city charter in 1248. However, the pope's legate William of Modena already called Tallinn a city in 1237: *civitas Revaliensis* (Höhlbaum 1874: 66–7). A bloody struggle broke out between the Teutonic Order and the pope's supporters. Listed among the victims are both merchants and 'Tallinn citizens' – and this in 1233 (Hildebrand 1887, doc. 15). Tallinn was a city eleven years before its charter and its residents were citizens fifteen years earlier? Where did the city begin?

* * *

An excursion into the history of a yet non-existent early town is time-consuming, troublesome, and – considering the pavement that's been pulled up – back-breaking. We can forgo it, sip gin and orange juice at the

SILVERWHITE

bar, and wait until the tenth chapter to appear back in the harbour where we'll be awaited by those same ships and those same sailors, ready to pull up anchor and continue telling tales of travel.

* * *

Where is the beginning? Mustamäe. In what is now one of Tallinn's sleeper districts, schoolteacher Arthur Spreckelsen unearthed a grey stone arrowhead and a tool there, both about 4,000 years old. The ancient hunter-fisher's home was a beach that bordered today's Rahumäe Cemetery. To its west was the Kadaka shelf, to the south sandy dunes that extended from the hills of Nõmme, and to the north a gentle rise separating the prehistoric site from Kopli Bay. The coast snaked through what are now dense blocks of occupation-era apartment blocks (Spreckelsen 1927: 103–4).

A farmer walked the beach after a storm. Waves had eroded the sandy bank. He crouched, collected coins, and sold them to Tallinn goldsmith Clementz. They had been minted in Rome between 139 and 161 CE. Coins discovered in the area of Õismäe and on Naissaar had found their way there from Volga Bulgaria, Tashkent, Samarkand, Bukhara, Baghdad, Badakhshan, and Antioch. Christian Martin Frähn dated them to 718–968 (Frähn 1848). The oldest Lund, London, and German coins were minted in the early 10th century (Kruse 1859: 25).

The oldest site of governance was Iru Fortress on the steep bank of the Hirve River. It was built at around 500 CE, burned down twice, and was finally abandoned as a defensive structure in about 1000. Did Revala out-compete it?

Tõnismägi Hill was higher than it is today. Together with the giant Toompea cliff and Kalarand's Suspipea Headland, it formed Tallinn's central rise. The landscape west of Toompea was similar to what it was at the beginning of our millennium; only Lake Harku might still have been a sea bay (Sulev Künnapuu to the author, 4.3.1974, and his sketch of the Härjapea estuary in the 11th–13th centuries, MS). The land east of Toompea was strikingly different, however. The nearly 300-metre-wide Härjapea river estuary emptied into Revala Cove, where today's financial district towers, and jellyfish floated in the lobby of Viru Hotel. The land was 2.4–2.7 metres lower. Century-old oaks rustled on the Kopli Peninsula. Remnants of that ancient forest include place names like Tamme ('Oak') headland, Tamme hill, Tammemäe creek, as well as

TALLINN IN THE HAZE OF LEGENDS

one sonnet, one primeval oak in Vanasadama Park, and Tallinn citizens' tradition of environmental protection, which reaches back to the late 13th century.

Archaeologists studied Toompea in 1952 and Town Hall Square one year later. According to the summary drafted by the academic Harri Moora, Tallinn was 'an Estonian settlement by the 11th century at the latest, not just a fortress' (Moora 1955: 84). Strange that Henry of Livonia didn't notice. Perhaps Revala was still just a small seasonal harbour? Then, he may have been right. If Henry of Livonia was right, then we must ask: to what *kihelkond* ('parish') did the harbour at Revala belong? If Henry of Livonia was not right, then we must ask: is there any information that speaks of Revala as a large, permanently settled, inter-*kihelkond*, perhaps even inter-county, harbour, just like its peers in other areas along the Baltic? And what did *kihelkond* really mean?

SHIPS CARRY BONDS

Coastal defence required monitoring, weapons, armed men, and many people's determined labour to build a fortress. These tasks were to be fairly divided between villages and families, depending on the number of able hands and the distance between their homesteads. For these and other reasons, Estonia's first major administrative units formed: *kihelkonnad*.

The Saaremaa schoolteacher and passionate local historian Jean Baptiste Holzmayer wrote these words in 1867 (Holzmayer 1867: 60): 'I have recently developed a very acceptable etymology for the word *kihelkond*. It comes from the word *kihlama*, "to swear by oath"'. Over a century has passed since then, and Estonian historians have added a great deal of new information to his correct explanation (Moora and Ligi 1970: 54). Our understanding of the ancient *kihelkond* has grown more detailed and accurate. Making any kind of a decision required consensus, 'societal agreement'. Every time a wide-reaching resolution was made, people had to gather to bless it for all. However, it is one thing to come to an agreement in words, but another, much more difficult thing to ensure that it is applied. In the 18th century, Saaremaa farmers had a tradition of sitting down and sampling porridge made from fresh rye after they finished harvesting it together. The meal celebrated the shared labour so it would be repeated again the next year. They called the food *leppepuder*, 'deal porridge' (Holzmayer 1872: 108). The deal itself or the meal accompanying it was called *kihlad*, 'oaths'. Even in recent Estonian wedding tradition, the word

353

SILVERWHITE

kihlad was used for gifts that the groom brought his future spouse and her parents. A pig head was prized among ritualistic meals. Contributions to the feast were brought in *vakkad* – bowls made of birch bark or wooden shingles. Some of the food was sacrificed to ancestors, in the hope of their support. The *vakk* was also venerated with sacred light and the feast itself was referred to as a *vakusepidu*, 'bowl party'. Over time, the voluntary gifts became mandatory tributes.

People were bound by a shared agreement and a mutual oath to fulfil it: a *kihl*. Parties to this agreement formed a *kihelkond* led by an elder. The *kihelkond* offered its members guarantees, protected their rights, and distributed communal fields. These fields became hereditary prior to Estonia's conquest and could even be sold, as Henry of Livonia reported: in 1184, a grey-haired man named Meinhard purchased a plot in Üksküla (Uexküll) for a church to be built (I). *Kihelkond* law was absolute; outside the community, their lives were for fate to decide. If a member was captured, then wealth was pooled for his ransom; if he was killed, then revenge was taken. The core of the defensive organisation comprised elders, relatives, and others bound by a common oath. There, the seeds of property division lay dormant. The concept of manor and lord existed prior to Estonia's conquest and meant a farmstead that had broken off from the community. Some were ten, twenty, or even eighty times larger than the average and required a corresponding labour force, free or slaves.

For greater defence, the *kihelkonnad* would unite into counties. A county meeting, *kärajad*, would be called to discuss complicated legal cases, receive or dispatch emissaries, and conclude peace treaties. If consensus could not be achieved, then the issue could be resolved by casting lots. When the final decision had been made, the longest stick was symbolically cut in half. Hostages were used as collateral to guarantee peace treaties. It was an international practice that dated back to the days of the Roman Empire or even further, and the original meaning of *kihl* was 'hostage'. The hostage's status should not be underestimated. Usually, hostages were chosen from among a leader's closest relatives and were treated with respect. The modern exchange of diplomatic missions is a distant echo of that common-law tradition. Despite their comfort, hostages would pay with their lives if a treaty was violated (Vilkuna 1964).

Kihelkond elders had their hands full in dealing with harbours. When negotiating with foreign merchants, the interests of one's people had to be defended while simultaneously showing respect for the strangers and Baltic legal traditions. A harbour had to guarantee profits for locals and

TALLINN IN THE HAZE OF LEGENDS

visitors alike. Foreigners had no rights in principle, though they could use their status as guests, which placed them on the same level as *kihelkond* members. The foreigner was expected to grant these same rights to members of a *kihelkond* who visited their home harbour in turn. They were 'friends'.

Harbour law applied to a limited area in small *kihelkond* locations. In the roadstead, it extended as far as where a ship's anchor could still touch the seabed, and on shore it covered a strip no more than eight *süld* (2.13 metres) wide: enough for bringing goods to shore, and no more.

Large harbours enjoyed greater independence from their *kihelkond*. They could be shared by several *kihelkonnad* and sometimes even multiple counties, as a result of which they required county-enforced laws, permanent administration, defence, and a justice system. One record from 1254 mentions the boundary of a harbour on Hiiumaa, *finis portus Sarwo*, despite the fact that the island was nearly uninhabited at that time. Delineating a harbour's exact borders was essential, as foreign merchants required wood for burning, timber for construction, and pastures for the purebred horses and livestock they had brought to sell: in short, they needed fields, which were shared property of the *kihelkond*. Conflicts were resolved in favour of the harbour, probably under pressure from the neighbouring *kihelkonnad* and counties. In one letter to the Lübeck city council, Revala's council claimed that the Paljassaar Islands, Aegna, and Naissaar belonged to the city on the basis of prehistoric law (LUB X, 537; cf. Niitemaa 1955: 43).

Ships carried *kihelkond* law and its traditions to sea with them. The following custom was recorded in Saaremaa: When heading out to sea, the men would take along a 'smoking bag', also called a 'witch's bag', the contents of which would be smoked if necessary. Sacrificial smoke was believed to cleanse and bring good fortune. 'Peace pipe' and 'incense' come from the same root. The contents of a Saaremaa smoking bag included snake heads and woodchips cut from every village family's doorway and from every boat (Holzmayer 1872: 111). The sacrificial smoke uniting the village was believed to protect the men at sea and in foreign lands.

Seafaring was a community undertaking from its very inception, and later out of necessity. Land became hereditary and even sellable, but ships still required shared labour. They no longer belonged to the community, but to a narrower group of seafarers. Revala's medieval records still remember their name: *ossemesz*, i.e. *osamees/osanik*, 'shareholder' (Johansen and Mühlen 1973: 134). Other terms were apparently used in the city

355

SILVERWHITE

as well. Near Vanaturu ('Old Market') in 1453 was a place called *zaell* (Johansen 1951: 49). It has been interpreted as *sálhus*, and is believed to have been an inn for foreign merchants, perhaps even in the days before Estonia's conquest. Estonian has a range of similar words that all share the compound *selts* – a group of people with common interests. Ruhnu seal hunters would form three *selse* to set out on the ice, and Revala records from around the 1600s name multiple sites termed *selsyweliat*, *selsze*, and *selschoppe* (Johansen and Mühlen 1973: 394). *Selts* is also directly related to the word *kild* when used in the sense of 'a group of friends' or 'a group connected by a common voyage or trade expedition'. Many shoots sprang from this linguistic stalk, some of which are still in use today. The historian Voldemar Miller believes that *killa* could be of Baltic-Finnic origin, and that it might even have been lent to Germanic languages, where it took the form of *gild*, 'guild' (1972: 179). Linguistics offers analogies: *kihl* in the sense of 'hostage' entered the Baltic-Finnic languages as a Germanic loan before it was used to refer to an administrative unit, and was then lent from Karelian back into the Scandinavian languages in its modernised form (Vilkuna 1964). And *kild*, *killa* is indeed a much older word than *gild*, *gildi*, 'guild'.

MYSTERIOUS KILD-MEN

Nevertheless, the problem is somewhat more complicated. Tracing *kild/guild* into the more distant past forces us to pause to consider men's associations, the concept of which is well familiar from the accounts of 19th-century explorers. A boy entering puberty is pronounced a man and dedicates himself to the secrets of life and death. The initiation involved difficult trials for the young man, in which he was meant to prove his maturity, bravery, and skills. The ritual culminated in a rite of passage, the participant's transition from one world to the next, the young man's symbolic death and rebirth as a man. The details of these traditions would lead us too far astray. Let us just note that young men preparing for these trials would be separated from the community and live together for a short time while focusing on exercise and mental improvement. They would set out on occasional masked jaunts through the village, which Estonian culture has preserved (in my own opinion) as *mardipäev* (Martinmas) traditions (also *mardijooks*, 'Martin's run'). In one of the oldest forms of national theatre, young men would dress in beggarly rags and go from door to door singing, playing instruments, and dancing until they were

TALLINN IN THE HAZE OF LEGENDS

let in and given offerings, not unlike the Celtic Samhain, the ancestor of today's commercialised Halloween. Participants were of a certain age group and were both feared and frightening. Their roving was wild and frenzied:

> I'll bang on doors of the sleeping,
> shake my member at them …

The impunity accorded such chanted threats was ensured by the notion that the youth embodied a connection with ancestors' war parties; that ancestors' ecstatic force was bubbling within them. A mask signifies identification. Similar initiation ceremonies are common throughout the world.

Outer aspects of this tradition are also similar to those of the oldest guilds, of our hypothetical *killad*. And what was a medieval guild like? Guilds were professional organisations that united craftsmen and merchants engaged in similar activities. Members were obliged to support one another. Guilds provided a collective spirit through social gatherings held at special locations called guildhalls, guild houses, or simply 'guilds'. Scandinavia's first guilds developed relatively late. This is important to keep in mind. Comparing Norwegian guild constitutions with those of Danish guilds shows that the latter served as examples for the former.

However, the word *kild*/guild appears on runestones carved even earlier. A runestone in Bjälbo, Östergotland, mentions a guild of warriors. Frisian *kild*-men, *frisia kiltar*, erected two runestones in Sigtuna in memory of their fellows, and the memorial to the merchants' guild dates to the 11th century.

Scandinavian myths speak of the *kild*/guild's foreign origins. South Germans already dealt with such associations (*per gildonia*) in Charlemagne's capitulary from the year 779, the sixteenth chapter of which bans or restricts their activities. Another document from 888 carries the same negative attitude and uses the words *geldam*, *geldoniae*. Guilds were to strictly limit themselves to religious education and upbringing. Feasts and celebrations were forbidden.

The etymology of 'guild' has not found a satisfactory resolution, as Miller points out, though the history of the word itself stretches somewhat further back in time. Old High German *gelt* meant 'sacrifice', Old English *geld* a 'sacrificial ritual', and in every Germanic language it possessed a parallel meaning of 'payment, tribute, revenue'. Linguists have attempted to determine which definition is older: ritual or legal.

357

SILVERWHITE

Setting up this problem seems artificial. Both ideas were too interwoven in the past to enable us to pragmatically differentiate between the two. More important is to recognise that in either definition, *kild*/guild was associated with a ritual feast, to which participants were required to bring offerings. From this perspective, the word must have originally meant 'a sacrificial association'. Skaldic poetry used the word in as early as the 10th century, also in reference to a feast or festivities (Cahen 1921: 64–5). Holding a large feast meant treating the dead; in other words, an All Souls celebration that was meant – among other things or primarily – to strengthen ties with one's ancestors. Church history tells of bishops' crusades against the remembrance of ancestors in guilds, which was thought to be a 'pagan custom'. Thus, the *kild*/guild is undoubtedly a pre-Christian social organisation. When later developing into urban medieval professional organisations, guilds maintained traditions that had roots in ancient beliefs – ones that church and state only tolerated with the greatest reluctance and open hostility.

But returning to our main question: was the initiation rite an original form of Martinmas customs? And men's associations an original form of the *kild*/guild? At our current level of knowledge, these questions remain unanswered. Men's associations took root in an era of clanships, where members were of common familial origin. Although *kild*/guild members were not connected by common ancestry, ancestral worship still played a central part in their shared 'sacrificial' festivities. Let us emphasise once more: they did not necessarily spring from the same source. What we are interested in here is the borrowing mechanism. The *kild* form may have spread and adapted very early on, as society recognised a similar organisation from long before – one with (perhaps) a different story of origin, but still similar in appearance.

The oldest preserved Baltic document regarding *kild*-men / foreign merchants is of Frisian origins. Who are the Frisians? Nowadays, the small nation of about 400,000 people live primarily in the northern Netherlands. Still, the Frisians' role in history cannot be measured by their current population. Long ago, they inhabited the entire North Sea coast between the Meuse and the Weser. The Frisians were famous seafarers who left a deep imprint on the history of navigation and shipbuilding. One apparent sign of their cultural contacts with the Baltic Finns is the word *kirik*, 'church', which Paul Ariste believes was borrowed from Frisian (Paul Ariste to the author 1.3.1973). Similarities in the names of some constellations are greater between Estonian and Frisian than

TALLINN IN THE HAZE OF LEGENDS

between Estonian and Swedish. This could be pure coincidence, of course. However, the dialectics of cultural contacts know no one-sided effects. Contact is always mutual. The giving and taking isn't necessarily equal and probably corresponds to the level of culture, population size, intensity of trade routes, and importance of hubs – however, it is still two-sided. It's a law of nature, too axiomatic for us to waste any more words on. *Kild* is older than 'guild'. Was it borrowed from Frisian? Or did the Baltic Finns lend it to Frisian, i.e. *kild > kiltar*?

The most probable possibility seems to be the third, which joins with the Silverwhite waterway: the Volga Turkish *kil* meant 'guest house' (Collinder), an inn meant for foreign merchants, something akin to a caravanserai. Earlier, we discussed shore rights, market peace, and foreign merchants who were forced to tack between freedoms and limitations. Captains constituted a world of their own; a society that was mobile but closed off to strangers, connected by ships, joined together by the ritual significance of shared meals on board, united by mutual guarantees, and bound by the legal traditions of their destination: those boarding in a Volga-Bulgar *kil*-house were *killalised*, *külalised*, guests. Let's not rush to shrug off the similarities.

According to wonderful Near Eastern custom, a rug would be spread out before a foreign guest and they would sit upon it. The rug is sacred, the person seated on the rug untouchable and even protected from blood feuds so long as they were there. As we will see in a short while, rugs also played an important part in ancient Estonian society. They are mentioned as being one of the most prized 13th-century spoils of war (*waipas*, Henry of Livonia, XXVII), and even in 17th-century records the Estonian word *vaip* was used to differentiate yellow-and-white or red-and-black textiles from bedsheets, *katted* (Johansen and Mühlen 1973: 394). Perhaps there is an Eastern air to this fact as well? Could ancient Estonian artistic textiles have taken inspiration from the Volga-Bulgars' *kil*-houses?

We receive unexpected help in designing this bridge from none other than Heinrich Göseken himself. Let us be immediately clear: Göseken does not prove anything. He simply gives evidence that this type of question is justified. On page 416 of the German–Latin–Estonian dictionary Göseken published in Tallinn in 1660, he gives us the Estonian word *killum* in the sense of 'a colourful rug'. Turkish Tatar languages – Uzbek, for example – use that same word with the same meaning today. There is no doubt that the word *killum* is of Eastern origin. More important, however, is the temporal factor: the word was not lent to Estonia through Hansa

SILVERWHITE

merchants or pious Bible translators, as was the case with the Hebrew word for 'ostrich' (*ya'anah* > *jaanilind*). Ariste dedicated an entire mini-study to the *killum/kilum* question, which found that the word only made its way into German in the 19th century (1972: 5)!

Let us put our facts in order and attempt to find a causal connection.

Foreign merchants formed a clearly delineated economic and nautical entity, whose members provided guarantees for one another and whose leader was tasked with conducting sacrificial rituals. Foreign merchants arriving in Volga Bulgaria boarded in *kil*-houses. In *kil*-houses, they enjoyed Eastern hospitality on a rug called a *killum/kilum*. Rugs played a more significant role in ancient Estonian society than we could imagine today. The *killum*-rug was borrowed into Estonian earlier than it was into German. The direction of lending was from East to West, not vice-versa. And, finally, the Estonian word *kild* is much older than the word *gild*.

Consequently, we could see in the Estonian word *küla* ('village'; *külaline*, 'visitor'; *'külas'*, 'visiting', etc.) a Silverwhite waterway-era organisation that formed among seafarers and that was named after a destination, Volga Bulgaria: the Volga-Turkic *kil*-house. This was a very common convention in language and seafaring. The *kild* form took root in Estonia because, for one reason or another (men's associations?), society recognised similar organisations from an earlier era. Baltic Finns could thus have been mediators in conveying both the activity and the phenomenon from the banks of the Volga to northern Europe. The Church's hostile attitude towards the 'pagan' *kild*/guild becomes more than understandable against this backdrop. And, finally, we find an answer to the question why Estonian used the ancient and more meaningful word *kild* parallel to the medieval word *gild*/guild.

There are striking commonalities between the rules and customs of a *kild*/guild and a *kihelkond*. A *kild*-member's obligations included bailing out a captured brother or helping to free him, standing guard, defending one's country, and putting out fires. Mandatory feasts played a central role, ideally increasing concord among members. Particularly strict punishments were dealt out for quarrels and fights at the table. We could see in this a fragment of ancient Estonian wedding customs, if we wish.

The fact that we know nothing about the founding of Tallinn's guilds makes the question even more intriguing. Estonia's first guilds were protected by the patron saints Canut and Olaf, both of Scandinavian background. This does not, however, speak to the guilds being imported by Scandinavians. Revala's seafarers formed friendly associations (*sõpruskonnad*

360

TALLINN IN THE HAZE OF LEGENDS

or *seltskonnad*) with their foreign trade partners. Both parties were to share equal legal foundations for communication and mutual obligations. In the Early Middle Ages, patron saints were marks of inclusion, a cosmopolitan language, a practical necessity that protected and provided guarantees for captains in foreign harbours, beyond their compatriots' defensive reach. In modern parlance, a guild was a passport, visa, and consulate; its patron saint, the coat of arms on the passport's cover and on the stamp inside. Time was international, and common activity perhaps an even stronger unifier than common language.

However, a gap yawns between the *kihelkond* and the medieval city. Let us try to fit the hypothetical early Revala with its no-less-hypothetical *kild* rights into that space. Although the bridge spans emptiness, it is firmly supported on both banks. *Kild* rights preserved all the administrative trappings of *kihelkond* rights, defence and justice included. The position of *kihelkond* elder was hereditary, and was apparently so in *kild* rights as well. *Kild* elders constituted the early town's administrative apparatus, which sooner or later would be called a city council (*raad*). This may incite a degree of protest in terms of legal history, but the question lies in the name. Perhaps we should say *selts* instead of *raad*, perhaps *kild* suffices. Either way, researchers believe that the Revala council was 'very likely' already functioning in the year 1230, long before Lübeck's charter (Johansen and Mühlen 1973: 60–1). *Killad*/guilds' leading role was preserved in Revala's self-government during the Early Middle Ages. The council was composed of twenty-four or twenty-six members, four or five of whom would take turns handling urgent affairs. In the event of two or more members' deaths, the council would fill the vacancies by a vote. Members submitted candidates to the mayor; all had to belong to the Great Guild. Also called the 'Children's Guild' (i.e. its members being symbolic descendants of the 'founders' or 'grandfathers' or 'ancestors'), it was first mentioned in writing only in 1363, when the distinction of status had come quite far. There are no surviving sources on the guild's earlier days, but the leading role of wealthy captains and merchants is beyond a doubt. After the new members were chosen, a summary of the important decisions – the *bursprake* – was ceremoniously read from the window of the Town Hall, after which the town executioner would chop a board in half in front of the door to symbolise finality. A similar custom can be found in Estonian tradition. On the second Sunday after St Michael's Day (29 September), two mayors were named and would take turns running

SILVERWHITE

affairs for the next year. By 1273 at the latest, the Revala *raad* was given the right to conduct justice in the city and the surrounding area.

Consequently, *kild*/guild elders maintained a leading role in governing the city and its common lands. Over time, the guild acquired a new, specifically status-centric nature that no longer concerns the history of exploration.

The *kihelkond* before the *kild*, the *kild* before the guild, the guild before the council, and the council before the city? In this way, the city concept appears to dissolve into something ambiguous and undefined. Again, it's a question of terminology. Let us stick to the entirely realistic conclusion that ancient Revala was ticking along nicely even before the Lübeck charter. It was no longer a village community, but neither was it an early medieval city just yet. In this travelogue, we've begun calling the transition its early town period. The Finnish urban historian Vilho Niitemaa believes that the Canut, Olaf, and Holy Body guilds were present and active in pre-conquest Revala, even though they weren't mentioned in historical records until 1326, 1341, and 1460, respectively (Niitemaa 1972; Miller 1972: 179). This is a new position.

Strange that Henry of Livonia didn't notice any of it.

CROSS AND COINS

In 1186, a former monk of the Augustinian Order in Segeberg named Meinhard was ordained bishop of the Livonians. He was given the title *Livoniae gentis episcopus*. By that time, he had spent two years as a missionary in Üksk\u00fcla and had built a church there. However, the new position brought conflict and Meinhard was forced to flee for his life. Henry of Livonia told the tale, albeit forty years later. He was still able to relate several colourful events and even a handful of quotes. Still, he was terse and vague about the bishop's escape: 'After taking counsel with his men, he proposed to go into Esthonia in order to go on to Gothland with the merchants who were wintering in that place' (I). And that was it.

Even so, many intriguing aspects can be peeled away. Merchant ships had already departed the mouth of the Daugava River. Meinhard had no hope of finding protection there. He was left with Estonia. But even in Estonia where the Goths (*mercatores de Gothlandia*) were wintering, he would have to wait for the next navigation season to make the crossing to Visby. He would be forced to reside in Estonia for some time. Therefore, his destination must have been under the protection of certain laws, as

362

TALLINN IN THE HAZE OF LEGENDS

he planned to claim the right to asylum, using today's terms. He was confident that these rights and his asylum status would be respected, as he set out immediately. The Livonians had no doubts, either: 'The Livonians, in the meantime, prepared to kill him on the road, but he was forewarned by Anno of Treiden and advised to go back. Much perplexed and unable to get out of the country, he, therefore, went back to Uexküll [Üksküla]' (I, 11). The right to asylum until shipping resumed naturally could not have been granted by the Goths (who were enjoying visiting rights), but only by a local administration. 'Winterers': it sounds naive. A travelogue needs landscapes, people, smoke rising from warm hearths. Let us imagine winter dwellings that offer shelter, covered markets where goods were stored, hayfields supplying the foreign traders' horses, and stables where the animals were kept for the winter; and also a blacksmith's forge, money chests, scales, cobblers, men who boiled tar and tarred ships and melted seal blubber, riggers and sailmakers; and, naturally, this landscape includes a chapel or a church, without which wintering in the ascetic early Christian environment of the North was unthinkable. Would this fit into the framework of village and *kihelkond* law? However, wintering first and foremost meant an opportunity to trade profitably in the cold season. Revala, primarily or alone, met these conditions in the 12th century and was thus Meinhard's likely destination.

Thirty-two years had passed since al-Idrisi's *Geography* was finished. Wintering merchants, legal relationships, and the whole complex harbour landscape are in obvious contradiction to Kaleweny, which 'is desolate in winter because people flee to caves'. It's hardly likely that Meinhard rushed into any cave. What did he hope to find? A merchant church.

The best surviving account of a merchant church describes St Peter's in Novgorod, which was built by Germans in a storage yard in around 1190 (Johansen 1963: 83–135). It was a strange institution – a storehouse in the most naturalistic sense of the word. Goods were stored in the sanctuary and the cellar; wine barrels were kept near the altar; tin and copper had to be pushed close to the wall, probably to keep the arches from collapsing. There was a measuring tape for cloth, weights and scales, a smaller silver scale, and the most important relic from St Peter's chest, which contained a constitution, privileges, and coins. Two guards were stationed in the church overnight. The door was locked from outside. The key remained with the foreign traders' leader, who was not allowed to leave the yard until the morning. The guards could not be the merchants'

SILVERWHITE

brothers or servants, weren't allowed to open windows or light candles, and, of course, were not allowed to have any guests in the church.

Even then, trade secrets played a very important role: no stranger had permission to enter the church in the daytime, either. If one managed to reach the steps, then the merchant on guard was required to pay a one-mark fine; if he entered the church itself, then the fine was ten marks.

When the merchants set sail, the keys were sealed and entrusted to the Novgorod archbishop and abbot of St George's Monastery. The coins were taken to Visby and locked in St Peter's chest in St Mary's Church. There were four keys to this chest: one in Visby, one in Lübeck, one in Soest, and one in Dortmund. An elder was chosen for every trade expedition: the *oldermann*, who picked two to four advisors in turn. An *oldermann* served as judge in disputes between merchants. Monetary fines and fees were generally paid to the merchant church. Thus, the church and its patron saint were the yard's conceptual owners and highest judges. The church and yard formed an indivisible whole; they were identical. It was a consecrated warehouse with religious immunity. Bringing trade beneath the protective shield of holy sites is an extremely ancient phenomenon that stretches back into time immemorial and that was even mentioned in the New Testament: 'and Jesus went into the temple, and began to cast out them that sold and bought in the temple, and overthrew the tables of the moneychangers, and the seats of them that sold doves' (Mark 11:15).

Thus, we can see the merchant church as a link between the ancient mercantile tradition and medieval trade. Time-wise, this would fit the early town. Lund's prehistory, which we discussed earlier, is evidence of the importance of ancient holy sites in a town's development. All Swedish bishoprics were originally places of worship or trade, unlike cities that developed during the Middle Ages (Herteig 1963: 82). What's more, Sigtuna's and Visby's churches were built upon prehistoric 'pagan' burial sites. According to one runestone, Goths had their own St Olaf's Church in Novgorod as early as 1080–90. Sails and rigging were stored above the arches of Tallinn's St Olaf's Church in the Middle Ages, preserving its earlier merchant-church qualities. On the other hand, the history of merchant churches allows us to trace how they intertwined with the *killad*/guilds and municipal government, as a result of which they became places for councils and courts, such as Tallinn's Church of the Holy Spirit. Merchant churches have therefore been seen as a crystallisation nucleus for future cities: the central focus of markets, buildings, street networks, forges, and workshops. Yet behind this smooth wording lies history's

364

TALLINN IN THE HAZE OF LEGENDS

perennial question: what is the cause, what is the effect? The merchant church is a mark of its hinterland's socio-economic development and a catalyst for that development at best, but it is not the cause. In fact, even the metaphor 'crystallisation nucleus' evokes entirely false associations. A nucleus may also be a foreign body, as infinitesimal as a grain of sand in an oyster, but it has still been brought in from outside.

Merchant churches' patron saints, which have not been studied extensively, allow us to determine the direction of trade and mercantile relationships, as well as their age, based on the 'mother church'. Consequently, they also shed light on shipping routes and seafaring more generally.

Folklore, the bottomless well of great truths and great errors, adds a shade of colour. Saha Chapel is believed to be older than Tallinn, but the same age as Revala's Canut Guild. Legend has it that on the site of the chapel grew a mighty oak that was chopped down and used in its construction. Most religious sites in Estonia are ringed by a dim nimbus of ancient beliefs woven with early Christianity. The local story of Kalana Chapel in Reigi goes as follows:

> Hundreds of years ago, a ship was lost at sea, couldn't find land anywhere and had been wandering the waters for nearly six months already. Then, the sailors vowed that wherever they should land, they'd build a chapel out of the oak timber they were hauling that time. The chapel had two bells at first; later, the second was taken to Soru Chapel. The bell was buried under a corner of the church during the war. Right now, nobody knows which corner the bell was buried under. The chapel also had two metal fish figurines (a cod and a herring), and fishermen always got big hauls while it was there. Then, some Finns stole the figurines and took them to Finland. But a local fellow sailed a little two-masted flat-bottomed boat to Finland one summer and brought those fish figurines back. Later, some Courlanders stole them and took them to Courland, where they've been ever since. There've been tons of fish off Courland since that time, but the fish have disappeared from this shore.
>
> There's been a cemetery by Kalana Chapel, too. Köön Suureps was the very last to be buried there. They used oxen to bring him because the horses wouldn't be hitched when it was time for his burial. Afterwards when everybody went home, those horses came out of the woods and went home and stuck their heads through the gate. They say Köön'd been a big witch in his day and the horses always chased him.

SILVERWHITE

The felled oak at Saha: isn't that a memory of the *hiis*, a sacred grove? Six stones upon which sacrifices were performed just a stone's throw from the chapel, one of them sat right in the churchyard! Saha was a source of great contention throughout the Middle Ages as to which diocese it belonged to. The ancient worship site was believed to heal and prevent ear afflictions. It fame endured to the late 19th century, attracting Estonians seeking help from as far as Järva County (Winkler 1900: 12–13). One line written in the 13th century appears to confirm the legend: *ubi fuit ecclesia et cimiterium adhuc est*, the parchment laconically informs us – here was a church and next to it is a cemetery (Johansen 1933: 583). Saha was owned by the wealthy Estonian feudal lord Hildelemb. In this case, 'wealthy' means he had twelve times the assets of an average peasant. He was somewhat overshadowed by the lord of Jõelähtme, as Albern de Osilia's lands were twice the size of his. And both seem downright miserly next to Albern de Kokael aka Part Koila, who, with his 89 *adramaad* (712–1,068 arable hectares), was the richest vassal of Valdemar II in north Estonia. It is probably pertinent to stress that the finer details of Christianity were the last thing that concerned a man such as Hildelemb. The new religion was seafarers' cosmopolitan tongue and the developing feudal aristocracy's plinth.

Indirect information supports the possibility of merchant churches, and thus also *killad*/guilds, in pre-conquest Revala. Not long after Revala's purported alliance (1018–35) with Cnut the Great, Johannes Hiltinus, abbot of Goseck Monastery on the Saale River, was ordained bishop of a mission covering the area of the 'Baltic islands' Findia and Hestia – Finland and Estonia. This first bishop was stationed in the merchant city Birka. From there, Johannes Hiltinus moved his bishopric to Linköping. A century and a half later, the bishop of Linköping arrived in Estonia as the Swedish king's emissary. Why, if not to represent the Church's continuity? The beginning of missionary work in Estonia, the first chapel or a stone cross, would be found among artefacts from Hiltinus's lifetime. He was an active man and the oldest parts of the Jomala Church on the Åland Islands were most likely built in his day (Dreijer 1960: 277).

We know a little more about Estonia's next bishop: Fulco, also written Folquinus, a monk from the monastery at La Celle, French by some accounts and Estonian by others (Edgar Rajandi to the author, 20.12.1973). The archbishop of Lund – head of the Church for all Northwest Europe – ordained Fulco as the bishop of Estonia in 1165 and thus gave him the title *Estonum episcopus*. He did not travel to Estonia

TALLINN IN THE HAZE OF LEGENDS

alone. Fulco was accompanied by an Estonian monk named Nicholaus, who had been trained in Trondheim (Nidaros) – Norway's capital at the time and the location of a cathedral that was built on the burial site of Olav II. Over the period 1169–78, Fulco and Nicholaus resided in Estonia and Finland. The post was, of course, for life. Fulco continued to bear the title Bishop of Estonia while living along the Aura River (near al-Idrisi's Aboa) in Finland. Could the Icelandic monk we quoted earlier have been referring to shared church authority with his 'both Finlands'? How are we to judge the results of Fulco's and Nicholaus's 'shepherding'? The archaeologist Ella Kivikoski (1955: 168) gives a very exact answer: 'Häme has no known burial mounds with pagan characteristics that can be dated to any later than the mid-12th-century, and I believe one can rightly say that Christianity came to dominate all of settled Western Finland at about the same time.'

Does Kivikoski's conclusion affect north Estonia, and if so, to what extent? Christianity's main concepts entered the Estonian language prior to Estonia's conquest, such as the word for 'church' (*kirik*) from Frisian. Some Christian names were also modified into Estonian prior to conquest and betray a Danish tradition, as the onomastician Edgar Rajandi has determined. Finally, let us also note that the church feast of the dragon-killing saint Margaret the Virgin was celebrated as *mareta-* or *karusepäev* in Estonia and southwest Finland on 13 July, contrary to other holy calendars (20 July in Western Christianity and 17 July in Eastern Orthodoxy). This could only have become a canonised anomaly if both areas still formed a single church province, which is to say before the 13th century.

Still, all these arguments are hitting a wall. The city, early town, and great harbour require forests for timber, wood for burning, pastures, and quarries. Together, they form its common land. Without common land, there is no city, and not even an early town.

Fifteen-year-old Danish king Erik V Klipping delineated the boundaries of pastures for the Revala lower town and fortress on 10 August 1265, which happened to be a Monday. The next Thursday, his widowed mother approved a commission composed of Tallinn citizens, the viceregent, and vassals, who were to draw and mark the borders of Tallinn's common land, as well as define the conditions of its use 'so that no injustice is done to the fortress or the city' (LUB I, 388–9). However, the limits of this area were already set by Erik IV Ploughpenny in 1248: with Lübeck law having been confirmed, he set punishments to be applied within the city's boundaries, *intra terminos civitatis* (LUB I, 199).

367

SILVERWHITE

Common land before the city, just like the council before the city charter?

True, the boy-king Erik Klipping plainly referred to his great-grandfather Valdemar II, who conducted the invasion of Revala. But the record is so ambiguous in regard to the common land's confirmation and delineation by Valdemar II that it remains unclear whether the king was confirming existing borders or drawing new ones or both. Was he expanding the pre-conquest common land, without which the early town and its harbour could not survive? There is no surviving delineation act of this type from Valdemar II's reign, which proves nothing. However, there is also no reference to one in any later surviving records, which is quite unusual in a medieval city, where tradition and continuity tended to be fetishised even in much more trifling areas.

One extremely noteworthy archival discovery does appear to support the possibility of a pre-conquest Revala common land. As it turns out, 'Estonian law', which historical documents refer to as *ius estonicum*, was applied in tandem with Lübeck law in medieval Tallinn (Mühlen 1969: 630–54). In certain cases, it limited inheritance rights. On the basis of *ius estonicum*, fishermen's houses on the shore and some other properties could only pass down to the city, i.e. the council, which saw itself as the legal successor of ancient common law (or hypothetical *kild* law?).

What do we know about ancient boundaries?

TWENTY-SEVEN SHEETS OF PARCHMENT

In 1929, a hefty manuscript labelled A 41 arrived at the Danish National Archives in Copenhagen, concluding an adventure-filled seven-year journey. The book had changed hands repeatedly, at least once falling into those of a dangerous owner who cut it open. The book's separated parts were later reunited between familiar covers once again. Those covers were wooden and leather-bound, and protected 153 pages of parchment, measuring 12.5 x 7 centimetres. The initial capitals were painted in red, blue, and gold; the lines of text were strictly and handsomely spaced – every element looked as if a master had held the paintbrush and goose quill. The most important section is known as the Danish Census Book – *Liber Census Daniae* (*LCD*). Contained in the work were several other texts, including our navigational guide from south Sweden to Revala.

Earlier researchers faced a mystery. What was the meaning of these long lists of names that lacked a single explanatory note? What, for example,

368

TALLINN IN THE HAZE OF LEGENDS

was the line written on the right side of the forty-fifth parchment supposed to convey: *Reppel 8 – Conradus non a rege par Jeeleth kyl Repel.* In the latter third of the 19th century, one researcher concluded that it must have been a workbook for beginner scribes – a collection of practice words to copy down. The truth unfolded gradually and dramatically.

Researchers of Baltic-Finnic languages soon realised that the *LCD* was an extraordinary document: the oldest record of Estonian linguistic history and a priceless resource to explain the history of north Estonian settlement. Specifically, one section of the *LCD*, albeit quite short, was dedicated entirely to Estonia. It contained 490 Estonian toponyms, 421 of which have been preserved to this day. Their decoding was the feat of Paul Johansen (1901–65), a Dane and long-time director of the Tallinn City Archive. Estonian place names fit on 27 parchments in the *LCD*. Their study and analysis take up 1,008 pages.

First, it was determined that the toponyms were recorded by monks who baptised North Estonians in the years 1219–20. The order of villages reveals the missionaries' direction of travel: hence, ancient highways that snaked from village to village, some of which still exist today. The manner in which the toponyms were recorded tells us the monks' ethnicities. Some lines begin on the coast and end on the coast. Consequently, the missionary arrived from Tallinn by ship. Taking navigational conditions into account allows the period to be narrowed down to summer months. Another black-cloaked man crossed bogs and swamps as if they didn't exist. Wouldn't that imply winter travel? '*Reppel 8 – Conradus non a rege par Jeeleth kyl Repel.*'

Letter by letter, word by word, a majestic, contradictory, and tragic picture was revealed to the researcher. At least 31 people crossed over 2,000 kilometres and visited 478 villages. And they each spoke Estonian, not to mention practised the rare art of writing in the language.

The lists, a report of completed missionary work, was handed over to the predecessor of Tallinn's Toomkirik (Dome Church, St Mary's Cathedral). But then, bloody years erupted. Danish rule was replaced by that of the Teutonic Order for a time, and only after the Treaty of Stensby (1238) was concluded were the Danes able to govern again. The knights' supporters were exiled. The Danes began restoring former property ownerships. To audit the titles, lists were composed along with the names of villages and the *kihelkond* and diocese to which each belonged.

SILVERWHITE

REVALAMAA

Revala County was composed of three *kihelkond*. The westernmost, which partially overlaps the area around today's city of Keila, was called Vomentakae. Ocrielae was the central *kihelkond*. The easternmost was likewise named Revala. Vomentakae is believed to have meant õhmatagus, 'beyond the swamp-island'. Ocrielae only appears once in Estonian history – on that parchment in the Danish census. Till today, its meaning is unclear. Paul Johansen believes it was misspelled.

Within Revala County was the Revala *kihelkond*, and within that was the village of Revala, 8 *adramaad* (40–48 hectares) in area, now called Rebala. The use of 'Revala' in three separate meanings is noteworthy, but not especially unusual. The same name can be found in Viru County and toponyms also coincide elsewhere, naturally. Even so, it adds another question to the agenda: did the village name expand to the *kihelkond* and project from there to the county, or vice versa? For Aethicus Ister's *Rifarrica* and the Scandinavian sagas' *Refala/Rafala* must have referred to a substantial and extensive place. Furthermore, they are evidence of the name's great age.

Every *kihelkond* once had its own defensive system and, for the most part, a central fortress. Johansen claimed that Tallinn was located in Ocrielae, stressing that there isn't a shred of doubt that it wasn't (1933: 187). How did he back this up? The fortress in Revala *kihelkond* was Jägala, 2.7 hectares large. 'Vomentakae' is believed to have been an insult and its people cowardly, the *kihelkond* having lacked its own fortress. We are left with Ocrielae, and Lindanise fits in well.

However, some aspects of the *kihelkond* borders arouse a bit of suspicion – could it really have been so simple? Johansen deduced that Ocrielae's eastern border ran north from Lake Maardu and met the shore near Muuga. East of the line was the Revala *kihelkond*, west of it Ocrielae. This is a reconstruction. The missionary's list gives no basis for it. All reasons given are indirect. The researcher judged the 'unsettled and unfertile' area between Iru and Lake Maardu to be most suitable for a border owing to its landscape. He presented other arguments as well.

But it is precisely here, just outside Tallinn, that the missionary's documented journey could provide the greatest accuracy and be trusted more than speculations. The parchment gives us an uninterrupted sequence of toponyms, according to which Iru, Nehatu, and Proosa (all three on the eastern bank of the Pirita River) belonged to the Revala *kihelkond*, the 'capital' of which was the fortress at Jägala. Väo and Lagedi on the river's

TALLINN IN THE HAZE OF LEGENDS

western bank, however, belonged to the Ocrielae *kihelkond*. What could be simpler than using a river as the border between two administrative units. Still, this was not the case. After visiting Lagedi, the missionary crossed to the eastern bank and visited the villages of Uusiküla (now destroyed), Saha, and Kärsi. They also belonged to the Ocrielae *kihelkond*.

If the logic of landscape did not contribute to forming the *kihelkond*'s boundary, if the boundary formed contrary to landscape's logic, then there can be only one explanation: it did not form, but was deliberately drawn.

The Revala *kihelkond* jutted out to the west. Harju County, which did not extend to the sea at that time, also jutted out towards Tallinn as a narrow peninsula, and penetrated deep into the Ocrielae *kihelkond*. The neighbours had their claims on Revala's harbour. Consequently, the harbour at Revala did not belong to any one *kihelkond*. It should rather be seen as an inter-*kihelkond* or even inter-county location.

AN EMPTY BEACH?

This contradicts earlier interpretations of the Danish Census Book (*LCD*), which regarded Revala as the Ocrielae *kihelkond*'s harbour. Johansen reasoned that the surrounding villages' prosperity was directly contingent on their use of a favourable harbour location. This prosperity declined steeply when village communities lost their shared harbour after Estonia's conquest. Consequently, it must not have been a large harbour, but small; similar to neighbouring villages' shared fishing beaches. Johansen judged villages' prosperity by their available hectares of arable land.

The logic is sound, but his conclusions are based on a prejudiced approach. The *LCD* does not confirm them. When the *LCD* was composed, Tallinn's twenty-seven neighbouring villages consisted of an average 48 hectares of farmland each. Two to three hundred years later, the average increased in ten villages, decreased in nine, and remained the same in eight. The lack of notable change means there must not have been any changes in the relationships between Revala and its environs, either.

'A mere glance at population-density maps shows that the coastline was uninhabited almost everywhere,' Johansen claimed, giving the following examples to prove his point: Laiduse (Laoküla near Paldiski) lies 1.5 kilometres from the shore, Viama (extinct) 1 kilometre, Lillevere (Rootsi-Kallavere) 2 kilometres, Randu (Viru-Nigula's Rannaküla) 2 kilometres, Muldilippe (Lüganuse's Muldova) 1 kilometre, Satse 1 kilometre, Ontika 0.5 kilometre, and Utria 1 kilometre.

SILVERWHITE

It is quite obvious that Johansen had a very narrow definition of the shore. Neither did he take into account the ancient waterline, which extended much further inland seven or eight centuries ago. Additionally, fishing played a relatively insignificant role in Revala and Viru counties, which had long since focused on grain production. Fishing and seafaring are not twins – on the contrary, fishing was a despised profession that was left to slaves and servants.

Far stranger is the emptiness that yawned around the supposed early Revala. The missionaries' baptismal expeditions began in the village of Sõrve, 15 kilometres from Tallinn, and the vacuum's half-circle extended from there along the villages of Vatsla, Hüüru, Pääsküla, Mõigu, and Iru at an average radius of 8–12 kilometres from the harbour at Revala. Within this arc, and on the Viimsi Peninsula as well, the *LCD* does not list a single toponym. It is missing the large, and consequently ancient, villages of Rannamõisa, Ilmandu, and Harku to the west. Viimsi to Revala's east and Järveküla to its south were added later, after the mission was complete (Johansen 1933: 64). Johansen explained the apparent desolation around Revala with hypothetical missing reports.

A simpler explanation, and one requiring less violence, would be that Revala's environs had already been Christianised prior to Estonia's conquest, and thus formed the early town's common land. What does the story behind Tallinn's name say about that?

HOW OLD IS TALLINN? MANY NAMES FOR A GOOD CHILD, ONLY ONE FOR AN ADULT

Estonia's capital has been called by nineteen names throughout history. The better-known are Tallinn, Reval, Lindanise, and Kaleweny/Kolyvan; lesser-known include Finns' Keso, Russians' Ledenets, and Latvians' Danepils. Even if we are unable or unwilling to derive anything else from this fact, the plurality of names speaks to the location's international renown; to its central location on many nations' nautical routes. There is also a sceptical side to this conclusion: if no single name form became dominant, then we should view Tallinn as more a stopping point than a centre; as an empty windswept shore instead of a fortress town.

Nevertheless, this flip side is false. And were there really so many names? Let's take a closer look.

The youngest name is, of course, Tallinn – Estonians' Waterloo, an honest admission of grievous defeat. Tartu, Otepää, and Viljandi also fell

TALLINN IN THE HAZE OF LEGENDS

into enemy hands but preserved their traditional names. Was the defeat in 1219 so severe that it even erased the settlement's earlier name from memory? Or was it, conversely, irreconcilability that forced the adoption of a new name – one that mocked and gave prominence to the fact that a foreign flag flew above the ancient fortress? Can we conclude that the fall of this early town was more extraordinary to those who witnessed it than the defeats at Tartu, Otepää, or Viljandi?

Tallinn's ancient name form was *Tanlinna*. The first syllable derived from the Old High German word for Denmark, which was also lent to Estonian: *Taani*. Ergo, 'Tallinn' meant 'Denmark Town' and sprang from a very specific event. It casts no light on the more distant past.

Latvians' *Danepils* and the *Kalevala*'s *Tanikanlinna* are Tallinn's Latvian and Finnish equivalents, respectively. We can leave these aside as well. An unexpected mystery stems from the name *Ledenets*, however. Researchers believe it is a Russian adaptation of Lindanise. And 'Lindanise' itself? Could it be one of the several drifting names, which adds weight to the empty-beach argument? No, not at all, because Lindanise and Tallinn are not overlapping concepts. As the linguist Eduard Roos has explained (1963b: 605–12), Henry of Livonia's *Lyndanise* is a compound word. *Lynd* had the same sense as *linn* ('city'), joined to a form of the word *ase*, which is extremely common in Estonian toponymy and signified a place of permanent settlement. The first half of the word was still used in 17th-century religious texts and also appeared in the names of two fortresses: Agelind and Somelind (which today would translate to 'Daybreak City' and 'Finnish City'). True, the *lynd* compound is at the end of the latter examples and at the beginning of Lyndanise. This also appears to confirm, however, that Revala's fortress was no longer the city's singular feature and consequently not its only toponym – that another, more important notion and name had developed, encompassing the fortress town itself. Therefore, Lindanasen > Lyndanise, which should be interpreted as the part of the Toompea clifftop that was used as a permanent settlement. If we wish to defend our hypothesis of early Revala, then we must treat the harbour, fortress, and city as separate entities. Every fortress-town component had its own name, and Lindanise/Lyndanise referred to the upper defensive structure.

It appears that al-Idrisi's *Kaleweny* and the Russian *Kolyvan* did not originally refer to the city itself, however, but to its most prominent social stratum: the captains and sailors. Estonia hosts dozens of toponyms that shed light on the social relationships of their day. Sixty-three place

373

names spread across the entire territory include the word *kuningas*, 'king'. Linguist Valdek Pall notes that 'the word "king" likely functioned as an anthroponym. The "king's land" and the "king's village" would have meant "land or a village that belongs to the rich". "King" is thought to have evolved into an anthroponym from these place-designating compound words' (Pall 1969: 92–3). The notion of 'king' as a rich man or leader took root in the Estonian language before the Common Era. Seafaring played an important role in the development of an aristocratic class, and this social stratification developed more rapidly in harbour settlements. Captains also had to create a term to refer to their own village in a way that highlighted maritime activity and power in addition to overall prosperity. The word that united all these notions was *kali* – a technical ship term, synonymous with heroic might, and seemingly the source of the name Kalev. It appears in two different versions in Estonian folk songs – sometimes as a name, othertimes as an adjective. The first written mention of Kalevipoeg was made by Mikael Agricola in 1551, and of Kalev by Heinrich Stahl in his bilingual German and Estonian book of sermons, *Leyen Spiegel* (1641). Even in the latter, it carries a dual meaning: the young shepherd David kills the great *kali*-powerful Goliath with his sling (*se suhre kalliwehje Goliathi*) (Stahl 1641: 305). Thus, the word originally had a much more sober and pragmatic meaning that was still common knowledge in Stahl's day. Like the title 'king', the Estonian *Kalev*, *Kalevine* (Kalev-like), *Kalevipoiss* (Kalev-boy), and *Kalevipoeg* (Kalev-son) could have referred to a specific person or a *kild* of people, their field of activity, or strength. Al-Idrisi's *Kaleweny* and the Russian chronicles' *Kolyvan* derive from the genitive form *kalevan* – *kalev*-like, belonging to *kalev* – much like the toponyms Kuningamägi ('King's Hill'), Kuninganiit ('King's Meadow'), or Kuningamets ('King's Forest'). Here, the toponym was formed from a societal feature of its residents. Similar toponyms around Estonia include, for example, place names with the root word *ori* ('slave'): Orjaku harbour, for one, which in 1254 was written as Oryocko. Analogous Danish slave-related toponyms include Trællebjerg, Trælleborg, Trællerup, Sweden's Trelleborg, etc.

Kalev and *Kalevine* are common around the Baltic Sea as terms of status or profession. Contrary hypotheses have been put forward, claiming the name form 'Kaleweny' was only used in Russia and was conveyed to al-Idrisi by a Russian merchant. This conflicts with the facts and with al-Idrisi's own account: he had no Russian informant when composing the section on Russian geography. The result of this leaves no doubt: al-Idrisi did not even know the Russian name for Novgorod. Furthermore, Turku/

TALLINN IN THE HAZE OF LEGENDS

Åbo borrowed its Finnish name from travelling Novgorod merchants: *torg* (market) > Turku. Al-Idrisi was only familiar with the Scandinavian name. On the other hand, Kalev was prevalent in Scandinavian myths as *kylfingjar/ylfinger* and, with one exception, was always used in reference to a social group. This exception is 'king' Gylfa in the *Poetic Edda*, which is etymologised from the name Kalev (Gordeyev 1973: 67):

> Swift keels lie
> hard by the land,
> Mast-ring harts
> and mighty yards,
> Wealth of shields
> and well-planed oars;
> The king's [*gylfa*] fair host,
> the Ylfings haughty
> ('Helgakvitha Hundingsbana', I, 51)

Estonian folklore compliments the Kalevs by attributing to them the qualities of city- and harbour-builders. Ultimately, we must consider that the Kaleweny name form hasn't left the slightest trace in Revala toponymy. However, the adjective *kalevivägine/kalijas* has been profusely preserved in Estonian dialects near Tallinn (Saareste 1951: 89–101). This would make the greatest sense if the name-giver was a social group that had lost its ruling status, or at least its title, after Tallinn's conquest in 1219.

Thus, the Tallinn name forms we have so far examined reflect a specific historical event, one component of the city of Lindanise (the fortress), and the profession of the city of Kaleweny's highest class of people. Different names for different subjects. In our next discussion, we will proceed from the following logic: a natural harbour was the most substantial prerequisite for urban development. Out of all the components that formed the city, it was the oldest. Most Estonian harbours acquired unique names even if they were located near villages. Consequently, we should seek the name of ancient Tallinn's harbour, and we will likely find it among the city's oldest name forms.

The oldest of all known Tallinn toponyms is Revala, which the Irish priest Feirgil recorded as *Rifarrica* around the year 750 and which was written as *Rafala* in the 11th-century Scandinavian *Njáls saga*. Context makes it clear that the given toponym is tied to Estonia: Gunnar and Kolskegg sailed to Saaremaa and *Rafala*. Although the *Njáls saga* was written only in the 13th century, the events it describes took place several centuries earlier.

SILVERWHITE

When etymologising became a social game for scholars in the 19th century, an array of witty and unscientific interpretations of 'Revala' were made. The most innocent might have been the German *Rehfall* ('deer's fall') and the most persistent the Swedish *räfvell* ('sandbank'). Estonian etymological proposals were also made: *rava* ('reef'). Alexander Heinrich Neus, who has been called the first Estonian folklorist, tied *rava* to the Khanty word *rep* ('cliff'), the latter being a reference to the Riphean Mountains, which is what ancient Greek geographers called the Urals, and thus connected the roots of Tallinn's oldest name to the purplish fog of proto-Finno-Ugric. This patriotic etymology found no support. The borrowing of both 'Revala' and 'Lindanise' from Scandinavian languages was regarded as more probable, and even provable. Professor Per Wieselgren (1900–89) also adopted this position in 1947.

An orderly parting in Revala's orthography stands out when the Danish Census Book and Henry of Livonia's chronicle are juxtaposed. In the former, Tallinn's county is *Repel*, its *kihelkond Repel*, and a village near Jõelähtme that still survives today *Reppel* (Rebala). Henry of Livonia writes that the county is *Revele/Revelis*. Perhaps this is coincidence? Luckily, we can use Viru County to check. Near Rakvere, Kadrina, and Haljala was a *kihelkond* of the same name. The Danish Census Book writes *Repel*, Henry of Livonia (conjugated) *Revelensis*. Coincidence is ruled out. The departures are orderly here and just as orderly elsewhere.

The linguist Gustav Must arranged Revala's various pronunciations chronologically (Must 1951: 303–11). He was curious as to how the toponym's mutation could be explained: Rifarrica, Refala, Refaland, Refalir, then Repel or Reppel, and, finally, in Latin sources, Revel or Reuel. Linguistics tell us that Scandinavians lacked the sound similar to the Estonian *b* and the letter to signify it. Therefore, it was replaced with the closest sound when necessary: an audible spirant that in Irish and Norse writing was represented by the letter *f*.

Germans encountered ancient Tallinn's name via the Scandinavians. When the spirant later turned into a *v* in Germanic languages, then the toponym also changed: *Refala* > *Revala*. This is likewise how Henry of Livonia and European history came to know it.

Lastly and latest, the name of Rebala village was recorded. Now, the Low German ear picked up the breathy Estonian sound as a *p* and it was written in the Middle Low German record as *Reppel*. Rebala's first syllable is indeed short. Otherwise, the village would have been pronounced

TALLINN IN THE HAZE OF LEGENDS

'Reebel' or 'Reepel'. Thus, Rebala was the original Estonian-language form for county, *kihelkond*, and the village of Revala.

When researching the roots of 'Rebala', one linguist concluded that it comes from the common Estonian word *rebu* ('yolk'). From signifying an egg-yolk yellow to reddish-brown, the word was applied in taboo form to a familiar animal: *rebane*, 'fox'. In folk songs, the Alutaguse dialect, and certain Finnish dialects, *repo* is still used to signify both 'yolk' and 'fox'. *Rebu* also appeared in the toponym *Urkereppen*, recorded in Tallinn in 1432. The site was on the slope of Toompea, adjacent to today's Rataskaevu Street, and in modern Estonian would mean 'fox den' (Johansen 1951: 39). Toponyms derived from animal names are common in Estonia: take for example Karula (*karu*, 'bear'), Hargla (*härg*, 'ox'), Sigula (*siga*, 'pig'), Kirbu (*kirp*, 'flea'), Lutika (*lutikas*, 'bedbug'), Varese (*vares*, 'crow'), Kurgja (*kurg*, 'crane'), etc. 'Rebase' can be found applied to a hill, a village, farmsteads, and – derived from the latter – surnames. Recorded historical residents of Rebala village include Olaf Reppele (1391) and Andres Rebja (1688). Colours are also popular components of Estonian toponyms, such as Mustametsa ('Black Forest'), Sinisaare ('Blue Island'), Punamäe ('Red Hill'), Halliparra ('Greybeard'), and Rohe ('Green'). Any attempt to disentangle *rebu* and *rebane* would be for naught, as the word comprised both meanings. Either way, it is an ancient Finno-Ugric name.

This incursion into the mysterious world of phonemic mutation would be completely unbearable in any honest travelogue, were it not to lead us back to the subject of navigational routes. It's clear that 'Revala' could only have spread through the civilised world as 'Rifarrica' in the 750s, Feirgil's day. Likewise, Refala and Refaland were its inevitable forms in the Early Viking Age. Let us be pedantic and repeat the conclusion more precisely: Refala and Refaland survived in texts that were written relatively late, the *Njál's saga* only by the 13th century. The chronology of phonemic mutations now confirms that the name forms are regular, but had to have formed long before being recorded; that they must have been in use during the Early Viking Age, prior to the notions of Lindanise and Kaleweny. Rebu > Revala must have been linked to the harbour by the 8th century at the latest, as one otherwise couldn't explain the toponym's use in Feirgil's *Cosmography* and its broadening to refer to the 'island' or, rather, county. The development had already happened by the Viking Age: Scandinavian sources refer to both Refala and Refaland, *Revalamaa*. In Finnish myths, migratory birds fly 'south' to winter in Repola (Toivonen 1924: 103). These myths are believed to be ancient.

377

SILVERWHITE

And what about today's Rebala, demanding heritage protection? Is there any connection between Jõelähtme's Rebala village and the harbour at Revala? The question is made even more intriguing by the multiple meanings of 'Revala', which historically could be a village, *kihelkond*, and county simultaneously. Estonia has many place names that are variations of Rebu/Rebase. Should we assume it was simple coincidence and see Jõelähtme's Rebala as a random namesake that helped us to recreate the toponym's ancient phonetic mutations?

Whereas we just put the reader's patience to the test once more and took them for a bumpy ride along the Ocrielae/Revala *kihelkond* border, now is the time to apologise and point to this wearisome marathon's destination. It is Revala's plurality of meanings: simultaneously a village, *kihelkond*, and county.

Out of the 490 Estonian toponyms in the Danish Census Book, 130 end in the suffix *-la*, just like Rebala village. These 130 villages are larger than the others, ergo significantly older. We could therefore also assume that Rebala village is ancient enough for its name form to have broadened to the entire *kihelkond*. The population grew denser and more land was cleared. The convergence of sea routes at the Bay of Tallinn accelerated development in the western part of the *kihelkond*. Just as Mustjala and Sõrve broke off from Saaremaa's Kihelkonna *kihelkond*, so the western part broke off from Revala *kihelkond*. We know from the Danish Census Book that Revala County comprised three *kihelkonnad*: Vomentakae, Ocrielae, and Revala. Within the last-mentioned were today's Rebala and the prehistoric Revala, nearly two millennia old and lending its name to the *kihelkond*. When the western section broke off from the mother *kihelkond* under the name Ocrielae, the Revala name still applied as its county. Three stages, three epochs, three Revalas: first village, then *kihelkond*, and finally county.

But Ocrielae? What does that mean? There isn't a single ancient toponym of Baltic-Finnic origin in Tallinn's vicinity. Ocrielae is the only *kihelkond* name in the Danish Census Book that hasn't found an Estonian-language equivalent or even a satisfying explanation. A misspelling has been suggested. Some have believed its correct form could have been Otriälä, or even Heckeral (Johansen 1933: 187, 296). The argument isn't convincing. The book gave us nine *kihelkond* names. Eight have been decoded (with various degrees of difficulty) and were recorded in approximately similar forms later; some still exist today. *Kihelkond* is not comparable to small villages or farmsteads, whose names can fade

378

TALLINN IN THE HAZE OF LEGENDS

away with the destruction of a village or a family's murder. And even if it was the case of an incorrect spelling, then the correct form would still appear in a later document. Alas, Ocrielae is surrounded by a total void. It was written thirty-two times in the Danish Census Book and afterwards disappeared without a trace, leaving not a memory or the tiniest clue in writing, myths, or poetry. It is too strange to be true.

The key to Ocrielae's mystery lies in the variations of meaning we defined in Rebu > Revala. Again, these can range from 'fox' to 'reddish yellow' to 'yolk yellow'. Medieval Latin's equivalent is the word *ochre*: a colour 'varying from light yellow to brown or red'. 'Ocrielae' would therefore be a Latin translation of 'Revala', to which the diminutive suffix *-la* has been added in the possessive, i.e. 'Little Revala', '(belonging to) Little Revala'.

This interpretation offers us far-reaching conclusions. Firstly, it confirms the *Revala* < *rebu* etymology. Any attempt to derive Tallinn's ancient name from Germanic languages falls away. Revala's parallel Latin form renders them baseless. The name's Baltic origins are indisputable. More important, however, is the bouquet of conclusions we can draw from the coexistence of so many Revalas. A county, a *kihelkond*, and a namesake 'Little Revala', Ocrielae. Why 'little'? Little in relation to what? In relation to the rest. We can see in the Latin 'Little Revala' an early town that broke away from its mother *kihelkond*, as well as its Christianised environs that over time became its common land.

Christian missionaries did not comb through Tallinn's near vicinity because there was no need to. That is what Henry of Livonia failed to mention: the early town's environs were already Christianised. Only under these conditions could the Latin name be formed and adopted by the Danes upon their arrival. The Danish Census Book's aims were sober and practical. A name that had no validity or broad usage would have been pointless to record. The Latin translation of 'Little Revala' could only have spread and shaped tradition if organised by the pre-conquest Church. Have we stumbled onto Bishop Fulco's tracks? Where was Revala's first chapel or church built, and when? Let's refrain from granting it too much importance. In Christianised Kyiv, baptised noblemen were buried along with their horses before the altar – a vivid example of the complex interweaving of traditions. This obviously also applies to early Revala. And thus the reason for Ocrielae's sudden disappearance becomes clear: it vanished together with the Christianised town's exceptionality; after the conversion of north Estonia was complete. It was not in colloquial use

379

SILVERWHITE

but was found only in church language, in pre-conquest church letters, perhaps on seals. In 1250, Saaremaa's elders used a stamp that bore a coat of arms and the words *Osilianorum munimentum* (LUB I, 369; LUB III, reg. 321; Johansen 1933: 748).

And where did Tallinn get its coats of arms? For extraordinarily, just like everything in the odd city, it had two. Valdemar II gave it a large coat of arms, kin to those of Denmark and England, the oldest surviving reproduction of which dates to 1277. More significant here is the small coat of arms: a white Greek cross on a red background. The town magistrate used the large coat of arms. The small coat of arms was used by the *killad/* guilds, which were active before the council, which formed before the city, which was born before the city charter. Only once the *killad/* guilds had merged with the city's council and administration did the coat of arms with the Greek cross become the lesser of the two. Is it the same age as the guilds? '*Ocrieliarum munimentum*'?

SOULS OF EARLY REVALA

Growing in Saaremaa's Viidumäe environmental protection area is a little flower unpoetically named the alpine butterwort. It can be found nowhere else in Estonia. Suurisoo Marsh is home to the endemic Saaremaa yellow rattle. The plants are remnants of ancient climatic periods, once widespread but now relics. When environmental conditions became unfavourable, many species died out or were confined to a narrow area. However, they give us living pictures of an erstwhile landscape, and that is precisely the charm of relics.

Relics can be surprisingly tenacious in tradition. Three buttons on the sleeve of a blazer are a relic of knighthood: they were for attaching gauntlets. Some Tallinn relics hint at seafaring's leading role in the pre-conquest town. Saha Chapel stood on the property of Estonian feudal lord Hildelemb. Living in nearby Puiatu was his neighbour Vililemb, a lower vassal with five *adramaad* (25–30 hectares) (Johansen 1933: 835; Rajandi 1966: 224). We know more about his heir. Queen Margaret I feudalised Puiatu to 'blessed' or 'memorable' Vililemb's son Johannes in 1274. '*Pater suus bonae memoriae*' applied to deserving noblemen, and particularly clergymen. The beginning of the document is much more significant: it turns out that Vililemb and his son lived in a nobleman's house in the Toompea fortress: *una curia in castro Revaliensi*. Both Puiatu and the house

TALLINN IN THE HAZE OF LEGENDS

belonged to the family prior to Danish rule, *prius a regno Dacie* (LUB III, 439 a).

Medieval Toompea was a community for rural noblemen, a city within the city, and it was crowded. Some couldn't find room and were forced to settle in the district that is now Lai Street. For Estonians, no matter how 'memorable', living on Toompea would have been more than unusual under the Danes. The Puiatu family's residence must have been inside the fortress before the Danish conquest. Thus, the full truth of that era begins to come into focus through Henry of Livonia's half-truth: 'They [...] began to build another new [fort].'

Studies of Toompea's soil have shown that the castle, *castrum minor*, was built (in 1227) upon a steep natural escarpment. The slope had to be filled. There was even a limestone quarry beneath Toomkirik, Dome Church. That also had to be filled. Such gruelling extra labour could only be the consequence of a lack of space in the fortress, as well as one other thing: its residents' legal status, which had to be respected.

West of Tallinn lived a nobleman named Leemet aka Clemens Esto, a great feudal lord compared with Vililemb and the owner of 34 *adramaad* (170–204 hectares). After the Teutonic Order was driven out, the Danes confiscated the Teutonic vassals' lands in Sõrve, Vääna, Liikva, and Vatsla, returning them 'to the Estonian Leemet'. This adds some support to Vililemb's pre-conquest dwelling on Toompea. The Battle of Lindanise (1221) and the Saaremaa Estonians' attempts to capture Revala (1223) apparently didn't diminish the standing of Vililemb, Clemens Esto, or others like them in the eyes of the Danes. With a certain degree of risk, we could regard Vililemb and his entourage as those early town elders who greeted Valdemar II with words 'demanding peace', and without any great risk we can assume there was contention between the early town town and the surrounding lands.

Literature and folklore have given us a somewhat romanticised vision of Estonians' 'paganism'. This treatment exaggerates Estonians' religiosity, as the passionate denial of one system of belief assumes the equally passionate approval of another. A thriving society knows neither ideological vacuums nor blind fanaticism. Let us believe the Germans who, according to Henry of Livonia, sent word to Riga that the Estonians 'would never hereafter accept the Christian faith so long as a boy a year old or a cubit high remained in the land' (XXVI). Those are words worthy of an antique tragedy, though they were spoken at a time when – by Henry of Livonia's own accounts – Christianisation equated to the killing of men

381

SILVERWHITE

and boys, the torching of villages, the enslavement of women and girls, great looting, and petty theft. Christianity could take root in the early town because it was agreeable to the upper crust of society and beneficial to captains. The lack of religious compulsion must have created a situation in which Christianity existed side by side with ancient beliefs for quite some time, and was even infected by it.

Revala's relics corroborate this theory. The urban elite's speech and customs preserved *kihelkond* traditions and particularities; Martinmas and St Catherine's Day were (and are) still celebrated with the songs and door-to-door roving mentioned before (Kohl 1841: II, 263–4). Most consequential, however, was the preservation of All Souls' Day, which held a major role in ancient Estonian beliefs. It is based upon the notion that the souls of the dead visit their former homes and hearths at that time. They were addressed as if they were alive, and families refrained from making too much noise or speaking loudly. A meal was set for them and the 'men of old' were asked to be kind and help their descendants in all their tasks and difficulties. It was a late-autumn time for quiet visits, soft conversations, and telling riddles. This account was recorded in the 19th century (Holzmayer 1872: 82–3):

> [The master of the house] ... calls his guests to enter by name, friends and family who were deceased, and ushers them into a room where two nice candles burn instead of pine splinters and the table is set with the family's finest foods. The man sits down at the table and insists his guests eat, calling them out by name, waving a white cloth each time and encouraging them constantly: 'Eat, drink, eat, drink!' He finishes his plate and then allows his guests a little more time to satisfy their hunger, occasionally splashing a little beer onto the floor, all for the spirits.
>
> When everyone has had their fill, the man takes a pine splinter, chops it in half on the doorstep, and asks his guests to leave after first humbly asking for protection and shelter for the animals [...].
>
> Dead souls are fed because they worked for the good of the living and toiled for them.

To fulfil this 'pagan' tradition, medieval Revala's city council would gather for a souls' feast every year on 2 November. The food was bountiful, the accounting pedantic, and the reason for the feast marked in the ledger was *Hinkepeve* ('Souls' Day') (LUB III, 1346, reg. 1631; Johansen 1951: 105–6).

The conclusions are more telling than the fact. All Souls' Day could have been imported, as it was inseparable from cemeteries, relatives,

382

TALLINN IN THE HAZE OF LEGENDS

generations, permanent settlements, and permanent places of residence. Renegades could not have brought it from the country to the city, much less to Town Hall. All Souls' Day accounting has been preserved from the year 1370 onward, its roots no doubt reaching into the days of the early town. It is the only way the ancient custom could have triumphed and thrived as official city tradition in the wake of the bloody crusades. In other words, it is a relic of fortress-town administration and adds hefty weight to the theory that the council regarded itself as legal successor of the pre-conquest government apparatus, just as one can conclude from the simultaneous validity of *ius estonicum* and Lübeck law.

EARLY REVALA LIVING

Early Revala's advantage was its natural harbour on an ancient sea route. Its nucleus from which it grew was a marketplace next to the harbour. If the early town hypothesis is true, then we should also be able to find relics in the market's organisation and the harbour itself.

The oldest surviving information from the early 14th century confirms that several ancient workshops were collected around Revala's market. The most famous of these were the jewellers, then called *ettekenmekers*, who ran the silver forges. Working alongside them were the *missincmekers*, who made knife hilts and belt buckles. The tinsmiths, *tyngeters*, produced buttons, jewellery, and pitchers. These craftsmen assembled on a street of their own, named *Tynnepatten* ('Tin Pot'). Hatters, *hodemekers*, made felt hats for peasants; *pistemakers* crafted saddles, belts, and metal ornaments and jewellery for straps. We can very likely regard their predecessors as the authors of the Revala region's rich archaeological discoveries.

One craftsman whose field provides us with a wealth of inferences was called a *sulverberner*, 'silver burner'. Germans quickly monopolised goldsmithing in medieval Revala. This was not a legal consequence, however, and the names of some Estonian goldsmiths are also known. The reason for the profession's concentration in the hands of newcomers lies primarily in the fact that, unlike silver, gold did not play a noteworthy role in traditional Estonian craftsmanship. Thus, the latter were referred to as 'silver burners' (also *argentifaber* in Latin) to draw a distinction between them and the German goldsmiths. Peasant jewellery was also made of silver, but it counted among the *ettekenmaker*'s (jeweller's) duties. Although the professions seemingly overlapped, they all had different names. So, what was the difference?

383

SILVERWHITE

Western merchants also recognised 'silver burners' in Novgorod, where they were responsible for checking silver's purity and weight. Later, this developed into minting. The analogy stands, of course. Thus, silver burners might have constituted the most important profession in the early town, given silver's prominence in ancient society. Silver was simultaneously a currency and a commodity, a conclusion which is supported by coin scales discovered in Estonia along with weights in both Arab and Scandinavian units. Silver bars and jewellery were similarly used for payment, broken up if necessary. Determining the metal's purity was an essential part of the transaction, otherwise there would be no use for the silver scales. On the basis of analogy, we could see the early town's silver burners as an earlier stage of the city mint. These responsibilities were later passed to the Revala council as legal successor to the early town's administration. The profession's title lasted longer than the profession itself, and thus the Estonian word for 'money' in the era of medieval Reval remained *ra* (Johansen and Mühlen 1973: 383). Silver purity was marked using a stamp that bore Reval's lesser coat of arms, that of its earlier *killad*/guilds (Vende 1967: 13).

We should ask ourselves if relics could have endured with such vitality even in the medieval city. A decree made by the Reval council in 1551 leaves no doubt. The document specifies what a *pistemaker* could and could not sell. Among the latter were '*kuddershens, helmede und snekenkoppe* [...] *der undeutschen dracht*' (Johansen and Mühlen 1973: 389), i.e. beads and snake heads used in Estonian dress. The first two items in the German text – *kuddershens* and *helmede* – are themselves Estonian words for beads, and the latter is a literal translation from Estonian. *Ussipea* or 'snake head' is a term used for cowrie shells which were imported from tropical India; they were used in many countries as currency, and also as jewellery along the Volga and in areas of Baltic-Finnic settlement. *Kuddershens* (*kudrused*) are small (silver) beads. Tradition is so conservative that it ties 16th-century fashion demands to archaeological discoveries. Some of the items listed were imported goods; thus, trade ties were also passed down from generation to generation. In light of this, we should appreciate the documenting of ethnic Estonian silver burners as late as the 14th century: Melenthewe (1336), Mathias (1343–7), and Conemann (1391) (Friedenthal 1931: 58–60). Prehistoric Estonian silver worship persisted and still evoked fascination in the 18th century. August Wilhelm Hupel (1774: II, 42–3) wrote this account of Estonia and Livonia: 'One could collect an immense amount of silver from both governorates, as most

384

TALLINN IN THE HAZE OF LEGENDS

craftsmen will only have their coffee served in a silver pitcher and our peasants wear a great deal of silver necklaces and brooches.'

In addition to guaranteeing pure silver and accurate scales, the market was required to maintain order and security. A relict of this survived in medieval Revala, also. By common law, the Revala market bailiff was required to be Estonian (Niitemaa 1952: 86–102). In European cities, the market bailiff was one of the highest city officials; in Revala, it was the opposite: owing to the Estonian requirement, the position's social status dropped lower and lower as antagonism increased. The conclusion is quite telling: abandoning western European tradition turned out to be much easier than changing local common law.

Craftsmanship is tied to sale, and thus ancient workshops were arranged around Tallinn's central market: today's Town Hall Square. It was the most favourable site for merchandising. It's unthinkable to assume that Estonians only seized the location in the Middle Ages, during an intense competitive struggle with the privileged gentry. The only satisfactory explanation for Estonians setting up shop around the marketplace is that their presence there was a relic from the days of the early town.

Linguistics offers support for this theory. A presence on the edge of the market meant permanent settlement. The Estonian word *maja* ('house, building') appears in Henry of Livonia's chronicle in 1211, though in a different meaning from that of today: 'They [...] captured all the children and the girls, and gathered the horses and flocks at the village of Lambit [Lembit], where there was their *maia* [maja], that is, their assembly place' (XV). The word originally meant a place for meetings or councils, and the definition still persisted in local dialects as a military term in the 16th century. Even so, *maja* referred to a dwelling in medieval Revala – in German as well. Another Estonian architectural term recorded in historical documents is *nurka/norke* (*nurk*, 'corner'). This shouldn't be interpreted as the corner of a room but as the intersection between protruding logs in a log structure (Johansen and Mühlen 1973: 381; Hupel 1795: 161). Archaeologists have discovered an ancient well stacked from thin limestone slabs beneath Town Hall Square. This adds stone walls to the fortress-town landscape. Henry of Livonia mentions stone walls when describing the defeat of the fort on Muhu Island in 1227: 'the hill was high and icy, the wall was on top of the hill, and it was made of stones as smooth as ice' (XXX). Winding between the facades and stone walls was a *tänav* ('street') that became a road outside the town's boundaries. There is no reason to believe the word was adopted

SILVERWHITE

in a later period: Hupel (1795) already recognised the *tänav* as both a shepherd's path and a city street.

Let us add a few mercantile concepts that can be dated somewhat more precisely. The part of Revala's population that spoke German used the word *Kauf* in its Estonian sense. In official German, the word meant 'buying' or a 'purchase'. In Estonian, however, it (*kaup*) primarily referred to 'merchandise'. There was a saying in Revala: *der Kauf ist* [...] *geschlossen* (Hupel 1795: 107), which is a literal translation of the idiom *kaup on koos* ('the goods are together', i.e. 'the deal is done') from the era of 'friendly trade'. There is no such expression in official German.

The Estonian *kaubasõber* ('trade friend') was also lent to German as *sõbber*, in addition to several units of measurement, the oldest of which is Henry of Livonia's *talentum Livonicum* (VIII) – the Livonian pound, which circulated throughout the entire Baltic Sea area. Also, there was *külimit* (a grain measurement equal to about 14.8 litres), *kimp* ('bunch'), and even the Silverwhite waterway-era Arab exchange quantity *nogate*, used in the meaning of a coin or money (Niitemaa 1952: 25; Neus 1849: 4; Hupel 1795: 69; 1 Livonian pound = 20 pounds, since the 18th century 8.19 kilograms; Johansen and Mühlen 1973: 381). One medieval Revala job title adds a geographical dimension. Eastern trade was handled by so-called Russian agents – ethnic Estonians who, in addition to Estonian and German, spoke Russian and Swedish fluently. Their surnames betray the profession: Peter *Vallmysraea* (1473, a combination of *valmis* ('ready') and *raha* ('money')), Peter *Waytay* (1557, *vahetaja* ('exchanger')) (Johansen and Mühlen 1973: 150). The Estonian surname Tolk, which is still prevalent today, is at least just as historical. The name comes from *tõlk*, 'interpreter', also known in Low German, Swedish, and Danish as *tolk*. The word has a fascinating biography. It is preserved in modern Russian with the meaning 'mind' or 'comprehension', but was used in archaic speech to mean 'negotiations', 'discussions', 'explanations', and even 'military camp'. The direction of lending isn't clear, though its area of dissemination is astounding.

A LONG LEG TAKES A LONG STEP

Toponymic relics are meagre.

Karja Street's current name and location have most likely been the same since the days of early Revala. It formed as an ancient herding route and established a firm name that carried over into foreign-language historical

TALLINN IN THE HAZE OF LEGENDS

documents: *Kariestrate* in 1366, and later translated into Low German as *Veestrate* and into Latin as *platea pecorum* (Kivi 1972: 47; Johansen 1951: 39).

Translated names also lead us to the oldest form of Harju Street: Smiths' Street (*platea fabrorum*) in records starting from 1339, and later also Tinsmiths' Street (*tinngeterstrate*). The Estonian language did not preserve these name forms. Why? A street lined with forges should be one of the city's most prominent, given smiths' significant role in the early town.

Smiths' Street originally overlapped with today's Niguliste Street, which led from the Old Market. We can assume that the blacksmiths' district was limited to that section at first, but as the town grew, the street expanded along Harju Road, which is now Harju Street. This must have happened prior to Estonia's conquest. Why? No road exiting Tallinn was named after its source, but rather after its furthest destination: Viru Street, Narva Road, Tartu Road later as well. Was Harju Street an exception? A street and a road leading nowhere?

No, because recall from the Danish Census Book that Tallinn was not located in Harju County prior to its conquest, but Revala County. Harju County was indeed the road's furthest destination. Naming the street (road) 'Harju' with post-conquest borders in place would have been absurd – therefore, the tradition must predate the Danes' arrival and have remained a relic after Revala and Harju counties were merged.

Throughout the centuries, only two routes have led up to the fort on Toompea from the Revala lower city: Pikk jalg ('Long Leg') and Lühike jalg ('Short Leg'). Wittily, Revala was called a 'limping city'. In Latin, they were *longus mons* ('mountain', 1342) and *brevis mons* (1353). The Estonian names are ancient, as we can tell from their archaic form; today's ordinary Tallinner wouldn't notice. Estonians understand subconsciously that 'Long Leg' is the longer route and 'Short Leg' shorter. However, the Estonian language offers no basis to equate 'leg' with 'road'. So what is their meaning? The Estonian word for 'leg' appears in the strangest associations: you can lift a leg, go a leg ('by foot'), pull a leg ('drag'), give a leg ('kick'), let a leg ('scram'), and even pull a leg on your back ('get going'). The words *jalamaid*, *jalapealt*, and *jalalt* express swiftness. A leg may be light, hot, busy, slanted, or open – but colloquial speech has not retained any memory of a 'long' or 'short' leg. In certain dialects, the word does have parallel meanings of 'going' or 'stepping'. As one old saying goes, 'The road is poor and the leg short' – here, it clearly refers to one's steps, i.e. slow pace. Consequently, the archaic meanings of Long and Short Leg must be

SILVERWHITE

the exact opposite of what we now presume! 'A long leg takes long steps': Pikk jalg was the faster and more commonly used route from the town to the fort, while Lühike jalg was slower and probably only used in case of emergency – not even a road, but a winding, treacherous, craggy path. In the event of an unexpected threat, the residents of Smiths' / Harju Street would have been too far from the end of Pikk jalg and would therefore climb straight up the steep hillside to seek safety in the fort. The difficult route required short steps, 'a short leg'. And from this, can we conclude that the danger must have approached from the sea?

Let's slide the mystery behind Neitsitorn, 'Maiden's Tower', between these thoughts. There were several ways that defensive towers along the city wall could be named: Munkadetagune (Behind the Monks) and Saunatagune (Behind the Sauna) were obviously named in relation to city structures; gate towers were named after the road (Harju) or the craftsmen gathered along it (Smiths' aka Harju Gate); and often a tower would be named after its builder or chief military officer. At the intersection of Harju and Müürivahe streets was a yard belonging to an Estonian named Asso, who lent his name to that tower (Assauwe, 1413) (Zobel 1966: 40, 48).

Neitsitorn (built in the 1370s), the Kiek in de Kök artillery tower's northern neighbour, was known in German as *Mägdeturm*. *Magd* does mean 'servant girl', earlier 'maiden', but the tower's association with the word is folk etymology. In Revala, a girl was called a *piike* (Estonian *piiga*), which is absent from official German. The linguistic differences were remarkable on the whole. In Germany, a city gate was *Thor*; but in Revala, *Pforte*. This itself arouses suspicion in regard to the tower's translated title – suspicion that is deepened by an even older name: *Meghede* tower, in a document dated 1373. It isn't even a distortion of the Estonian *mägi*, 'hill / mountain', but the clearly recognisable genitive plural (*mägede*), wonderfully justified by the tower's location. It was built at the intersection of two hills – in a place where a slim knoll now known as the Danish King's Garden joined the great fort hill. Further evidence is found in a document dated five years earlier that mentions *porta karie*, Karja Gate (1368), as well as the usage of Estonian names for Asso Tower in official documents recorded four decades later.

Even so, it is odd. Located on the slope of the hill was Kitsetorn ('Goat Tower', demolished c. 1533), as well as Tallitorn (Lambs' Tower) in the Danish King's Garden, not to mention the Lühike jalg gate defences, all of which were connected to the hill in one way or another. So why was the topographical feature only applied to one tower out of multiple options?

TALLINN IN THE HAZE OF LEGENDS

This forces us to consider the connection of Hills' Tower to the Mägiste clan, descendants of the Maekiuses, who were mentioned in the Danish Census Book (Johansen 1933: 50): several of the clan's members were active in Tallinn and, as we shall soon see, also left their name in seafaring history. This, of course, is only of secondary importance in the given association. Let us simply settle for the conclusion that several indigenous toponyms were preserved on the fort's eastern slope.

If we now take a step from toponymy into folklore and move from the Lühike jalg gate towards Meghede Tower, then the Danish King's Garden could also slide onto the chessboard of history as the true location of Valdemar II's tent in the year 1219. Little urban folklore has been collected in Tallinn, and even less has been researched. All the more valuable is every insignificant legend relating to place, as their longevity is significantly greater than historical records and architectural monuments. The area that is currently known as the Danish King's Garden seems a suitable place for the king's tent for several reasons. Earlier, we discussed that the locations of the Toompea castle and Toomkirik point to a dire lack of space in the fort. For security reasons, and certainly those of protocol, the Danish king couldn't have set up his tent in the lower town or before it. Yet it's as if the king's garden was created for that very purpose. It hadn't yet merged with Toompea in the 13th century but constituted a narrow series of steep hilltops – the former coastal bank that extended from the citadel as a peninsula. The impenetrable king's garden provided a perfect vista of the lower town and harbour, though the entire military camp naturally could not fit on the narrow crest. Judging by Henry of Livonia's account, the king's main tent was separate from his army: 'All of the Danes gathered with the king' (XXIII), though this took place only once the Battle of Lindanise was concluding. The chronicler also implies that Bishop Theoderich had set up his own tent elsewhere, probably next to a lower-town chapel, cemetery, or (even a) church, where he was killed.

If we imagine the king and his men galloping down Pikk jalg to the rescue, then another of Tallinn's legends acquires a dangerously vivid basis of proof: the origin of Dannebrog, the Danish national flag. Every Danish elementary textbook teaches that the flag was miraculous, a divine gift received outside Tallinn in 1219, descending from the sky into their hands at the decisive moment when the Harju Estonians and their allies were about to gain victory. The legend is widely known but wasn't recorded until the 16th century. The Dannebrog's white Greek cross on a red background echoes the lesser Revala coat of arms, which derived from

SILVERWHITE

that of the *killad*/guilds. Can this poetic tale provide clues regarding the actual power relationships in the early town? A painting depicting the Battle of Lindanise was recently discovered in a rural church in Northern Denmark, created just a few dozen years after the event. Although it doesn't add anything to our topographic knowledge of early Revala, it is worth mentioning as the oldest known artwork depicting Estonia.

ALL ROADS LEAD TO THE HARBOUR

Revala's boatmen and seamen represent the supposed early town's most fascinating relic. The former profession was apparently a product of the adoption of cogs. Their vessels were lighters (*lihterid*), flat-bottomed utility boats used to transport cargo to and from ships in the harbour. The Estonian name comes from the German *leichtern*, 'to lighten'. As we know, medieval harbours were quite basic. Deepening the seabed at Revala only came onto the agenda in the 17th century. Still, cogs sat deep in the water already in the 13th century. Their rigging didn't allow for fine manoeuvres. Steering a ship into harbour with its sails up could end in misfortune, damaging other docked vessels or the pier itself. At the same time, a cog with reefed sails was too heavy to row or tow. Lightening a vessel at a fore-pier was crucial and, in some harbours, mandatory (Heinsius 1956: 195). It was the boatmen's task to ferry cargo back and forth to and from shore. Several specialised boat types were used for this at Revala: the *bordinger*, *mündrik* (which also lent its name to the boatmen), *praam* ('ferry'), and *lootsik* ('skiff'). Their prototype was the ancient Baltic ship, and they were, for the most part, also fit for individual seafaring. Even with its low draught, the boatman's boat could not beach directly on the shore. Cargo was unloaded once again, this time onto a special cart with wheels large enough to be pushed right up to the side of a boat in shallow water.

Every Revala boatman was required to have two larger vessels and a rowboat. Workers were employed to man them. Thus, the boatman was a small businessman, sometimes even not-so-small. Boatmen had the right to conduct trade, though this right also obligated them to pay a citizen tax. Their by-laws, the Old Norse *skraa*, did not limit the number of boatmen in Revala. Evidence of their prosperity was the steep registration fee, twelve Riga marks in silver, plus a ten-mark feast for their colleagues, who were to be treated with two barrels of beer, two sheep, two pigs, and a cornucopia of other dishes – like an ancient Estonian wedding

TALLINN IN THE HAZE OF LEGENDS

feast. We can tell from Hinrik Münrik's 1527 testament that the men could lead a life of opulence. Hinrik left behind 276 marks in silver coins in addition to 'one ship in harbour with a tonnage of 40 *lastid* [80 tonnes] and everything accompanying it: anchor, sail and rigging, 3 boatman's vessels, 1 ship boat'. Among his numerous debtors was a Gotland manor lord, apparently his trading partner. Another Estonian, Luder Münrik, sent his son to apprentice for ten years with the affluent merchant Pauell Techler in Lübeck (in 1552) (Niitemaa 1952: 225–6; Johansen and Mühlen 1973: 166).

Boatmen were tasked with both unloading and loading large ships. However, the latter would only be performed after the owner displayed his cargo's 'release ticket'. In other words, boatmen had to verify exports. They were additionally required to set up seamarks, i.e. attend to shipping safety. And, finally, they had to keep pirates away from Revala's navigational routes, even monitoring pirate ships or entering into battle, thereby allowing us to regard them as a coast guard or sea police.

Strangely, there were several city officials whose duties overlapped with those of the boatmen. The harbour had a customs house, and customs officials verified exports. Boatmen took care of harbour security, but it was also the duty of the harbour bailiff. Why the duplicate authority? How could the simple unloading of cogs entail so many sundry obligations? Or should we ask instead: which brought what? Which was the primary task, which were the later ones?

In about the 1390s, *bursprake* confirmed boatmen's privileged status on the bay of Tallinn and in its harbour, including their sole right to receive ships arriving from Bruges (LUB IV, 1516, C34). However, their first by-laws, the *skraa*, were only ratified in 1506. For nearly three centuries after Estonia's conquest, boatmen's rights and activities were protected by tradition alone. This is incredibly curious, especially given that medieval city laws even regulated the length of sleeves on dresses and the number of wedding dances. Consequently, boatmen's common law must have been astonishingly effective, but also detailed enough to successfully compete with written law for hundreds of years. This also explains a certain restriction that applied to medieval Revala's Germans: the boatman's profession was off-limits.

The more exceptional the common-law provision, the sturdier the tradition upon which it could have functioned and endured. The medieval boatman's distant predecessor was the defender of ancient Revala's sovereignty, the protector of the early town's coasts and coastal waters, and his Estonian heritage was an inherent requirement for the position. The warship and land defence constituted an indivisible whole. This is also implied by the ancient word *malev*, 'which we should probably interpret as *maaleeva* – military fleet and harbour' (Vilkuna 1964). The job's defensive function began to fade after Estonia's conquest, but the replacement of ancient ships with Hanseatic cogs gave boatmen a new task. Thus, the relic entwined with another function, finding support and justification, remaining surprisingly resilient, and, as we shall soon see, evolving into a springboard for the coveted captain's status.

Pre-conquest Tallinn, aka budding Revala, was able to guarantee uninterrupted development for ancient Estonian seafaring even in the oppressive framework of the Middle Ages.

X

SON'S LAND AND MOTHER'S LAND

REVERSE PLANKING

The appearance of the first cogs on the Baltic Sea was a more momentous event in terms of its consequences than any naval battle in that era. For as long as foreign merchants were able to sail from the Baltic Sea onto the Neva River and from there down the Silverwhite waterway to Volga Bulgaria, the clamorous meeting place of the Orient and Occident, so long was the use of deep-draught ships unthinkable. Such trade expeditions could only be accomplished in light Viking-era cargo vessels. The situation changed when Novgorod's power extended to the shores of Lake Ladoga. Free trade, which sometimes relied upon ancient law, othertimes on the law of the fist, adapted to the defence and limitations of state authority, the international waterway became Novgorod's trade route, and the young capital quickly shoved distant Volga Bulgaria aside.

Novgorod was connected to the Baltic Sea via the Neva, Lake Ladoga, and the turbulent Volkhov (Olhav). Merchandise was transferred to Novgorodians' flat-bottomed boats at a pier or in the estuary, from whence they navigated the frothy waters to Novgorod's market. Cargo transfer was a new phenomenon on the Baltic Sea that nullified the ancient ship's former advantages and raised the modern-sounding question of transportation costs. The first historical record to analyse the drop of cost prices after large-tonnage cogs were put into use was written only in 1304, though there is no doubt that financial considerations were behind the rapid spread of the new class of ship from the very beginning (Heinsius 1956: 109–11). The quick, light ancient ship lost its purpose. Large, heavy ships came to dominate, though they could sail no further than rivers' estuaries.

The oldest depiction of a cog can be found on a Lübeck seal from 1226. On closer inspection, it's clear that both this image and several others show clinker planking incorrectly, with the lower plank covering the edge of the one above it. In shipbuilding, this is as absurd as installing a roof's shingles from the top down; from the ridge to the eaves. Reverse planking would make a vessel less watertight and less seaworthy. With correct planking, each edge acts as a miniature keel that reduces roll as well as drift in a side wind. Images confirm that reverse planking was common among the first cogs. Carpenters lacked shipbuilding experience. This isn't unexpected if you consider that Germans only acquired their first Baltic harbour with the refounding of Lübeck in 1159. A shipbuilding tradition isn't born overnight.

However, the cog's depiction reveals a much more consequential innovation. Reverse planking meant that the first pair of planks was not installed next to the keel, but just below the gunwales. The ship was built from the top down. Only one condition could make this possible: the ribs were assembled prior to planking and had been done so in a way that kept them steady, as only then could the upper row of planks be installed.

The innovation was revolutionary. It was the birth of a ship's frame, to which an outer hull was attached. The strength of the load-bearing frame determined the vessel's tonnage. It sealed the Viking ship's fate. As you recall, the Viking ship's load-bearing element was its outer hull, to which the ribs were added last.

Shipbuilding repeated the logic of natural evolution. Researchers have seen this as following the model of Central European house construction: first the erection of a frame, which held the inner and outer walls. Perhaps there is a degree of truth to this. It would explain why shipbuilders ignored Northern nautical experience and installed upside-down planking, which inevitably reduced the cogs' longevity and gave them weak bows. The innovation wasn't a sudden explosion but emerged as the fruit of lengthy, error-riddled experimentation that ultimately produced a new type of vessel.

The cog was born on the Baltic Sea. Other ship types were born on other seas, adapted to the local needs and natural conditions, wind directions, wave heights, and available building materials.

Papal authority kept a vigilant eye on seafaring's rapid development. These cutthroats – according to the clerics in Rome who discussed the matter with rising approval – these captains who, in the name of avarice, are prepared to sell their fellow man to the gentiles as eunuchs: they could become the Church's ally one day! And so the Church, which had hitherto looked upon captains with scepticism, took the first friendly step: it went

SON'S LAND AND MOTHER'S LAND

to war against shore rights and appointed St Christopher as the captains' patron saint.

It was at this time that both al-Idrisi and Henry of Livonia set their sights upon the world.

AL-IDRISI AND HENRY

Only twenty-two years divide al-Idrisi's death from Henry's birth. They were nearly contemporaries and, in the greater scheme of history, they were: the 12th and 13th centuries share a common denominator for Arabs and Estonians alike.

The Baltic-Finnic and Arab worlds met for the first time during the era of the Silverwhite waterway. The contact was not one of equals. Careful Arab hands had translated, transcribed, supplemented, and reproduced classical cultural heritage. Estonia was in the age of the *Kalevala*, dominated by the might of seafarers and the *kalid*, gripped by the passion of waterway exploration, immersed in ship-worship. Estonian folklore has preserved the song of an anonymous captain titled 'Knower of the Winds', a folkloric precursor to wind theory:

> … I sure know the winds this way,
> winds this way, seas to sail,
> to look against the days,
> to romp along the days,
> to stop along the edge,
> to sail to Turkey-land.
> I sail to the red sea,
> I sweep to the flaxen sea,
> my own sea's the righter,
> Sweden sea's the yellower,
> Finnish sea's the bigger.

Manuscripts have preserved al-Masudi's wind theory, written at the height of the Silverwhite waterway in the year 956 (Krachkovsky 1957: 177):

> When the sun's distance from the equator increases as it leans north, the air in northern latitudes warms and southern air cools. From this, we can assume that the southern air contracts and requires less space while the northern air expands and requires more space, for there is no emptiness in the world. And from this, we can assume that for those who live in northern areas, the winds blow mostly from the

SILVERWHITE

north because it is moving southward from them, and that the wind is nothing more than the movement of air and its waves.

Different cultural levels, different climates, different languages and ethnicities – and yet, they influenced each other's folklore at the dawn of history. What a colossal bridge! And on top of that, al-Masudi's astonishing pearl that is carelessly cast before the reader: 'for there is no emptiness in the world'. It would be 687 years before Torricelli himself arrived at that conclusion, and a little over 700 years before experiments with the famous Magdeburg hemispheres were undertaken in a Tallinn gymnasium. One axis of Estonian life was the Baltic Sea, another was the Gulf of Finland. The Arabs expressed this more poetically: a journey was the world's fifth pillar. Both were guided by the North Star prior to the compass, sometimes high at its zenith, othertimes low on the horizon. As one academic has stated, 'Estonian folk astronomy can nearly be compared to that of the ancient Greeks and the Arabs in terms of its numerous constellations and many mythological tales created about them' (Prüller 1968: 54). One cannot fault him for the noble exaggeration. And yet, when Ulugh Beg (1394–1449) calculated the latitude of his observatory at Samarkand, he placed it at 39° 37′, which was the highest degree of accuracy that could be achieved by pre-optics astronomy. Samarkand's actual latitude is 39° 38′ 50″, and the man who managed to correct Ulugh Beg by less than two minutes four centuries later was the young Otto Wilhelm von Struwe, son of Friedrich Georg Wilhelm von Struwe, the first director of Tartu's Old Observatory.

These are random crumbs that add more colour than significance.

The second time that Estonian and Arab fates intertwined was in the time of al-Idrisi and Henry of Livonia – as well as in their respective work.

* * *

A mild Arab with highly refined intelligence, tolerant of all but intolerance, a poet when the moment was opportune, an amateur botanist, a scientist by profession. The globe floats in szpace like 'a yolk within an egg', he wrote with no hesitation, as religious dogmas and spiritual bliss were what concerned him the very least. Nevertheless, the world of the Koran was just as flat as that of the Bible, and equally as cruel towards those who showed it any doubt. Al-Idrisi was enthralled by the real world.

Henry of Livonia, on the contrary, was enthralled by the ideal world as he understood it. To achieve it, the real world had to be destroyed. His

SON'S LAND AND MOTHER'S LAND

surroundings were to be destroyed by fire and sword, repressed within himself with self-denial and asceticism. Al-Idrisi was enthralled by the world's diversity, Henry by the uniformity of the world to come. The world was to become the Kingdom of God ruled by one will, one law, one goal. The subjugation of the self to this conception was unavoidable, the dissolution of the self and its assimilation into the system the final objective.

Al-Idrisi and Henry were opposites; they were also adversaries.

The crusades lasted for over two centuries, ending when the opposites blended. The Orient adopted the Occident's dour intolerance, the Occident was infected with the Orient's grinning curiosity. Jihad and the Renaissance are Henry's and al-Idrisi's offspring.

The crusades' indeterminable front joined the fates of the Estonians and Arabs. They could hardly have been aware of it, but their common enemy certainly was. As Henry of Livonia wrote of 1198 (III):

> In the presence of the king an opinion was asked for as to whether the goods of the pilgrims to Livonia were to be placed under the protection of the pope, as is the case of those who journey to Jerusalem. It was answered, indeed, that they were included under the protection of the pope, who, in enjoining the Livonian pilgrimage for the plenary remission of sins, made it equal with that to Jerusalem.

It constituted a single front to him and those who shared his mindset, and that front's victories and losses were also one:

> At that same time, indeed, the Christians from the land of Jerusalem had taken Damietta, a city of Egypt. They lived in it and the church of God had victory and triumphs over the pagans everywhere throughout the world, though with us it was not for long. For in the year immediately following [1221] the Oeselians came with a great army after Easter and besieged the Danes at Reval. They fought with them for fourteen days and, lighting many fires, they hoped to take them in that fashion. (XXIV)

One of the chronicle's most celebratory pages is dedicated to the Fourth Lateran Council in 1215. It was attended by four hundred patriarchs, cardinals, and bishops, as well as eight hundred abbots. The council was led by Pope Innocent III. The Church's highest body decided to join the two extreme wings of the crusades – the shores of the Mediterranean and the Baltic – under a single title. Henry of Livonia stressed the importance

SILVERWHITE

of the event with the use of direct speech, which he otherwise only rarely employed:

> The Bishop [Albert] spoke: 'Holy Father,' he said, 'as you have not ceased to cherish the Holy Land of Jerusalem, the country of the Son, with your Holiness' care, so also you ought not abandon Livonia, the land of the Mother, which has hitherto been among the pagans and far from your consolation and is now again desolate.' [...] The supreme pontiff replied and said, 'We shall always be careful to help with the paternal solicitude of our zeal the land of the Mother even as the land of the Son.' (XIX)

The land of the Son and the land of the Mother, Palestine and the Baltic lands ... Poetic metaphor? Or should we seek a true context to it?

Geographical knowledge had spread very inconsistently. In the 'learned master and scholastic' Adam of Bremen's work (1076), Aestland's closest neighbours to the east are the *Scutis* and *Turcis* along the Sea of Azov. Even the unparalleled al-Masudi believed the Baltic extended to the gates of Volga Bulgaria! Europeans' conceptions were an entire magnitude hazier. What does that mean, 'an entire magnitude'? Over time, Europeans became familiar with Arab geography but were literally miles off in converting the Arab mile. It had far-reaching consequences – and outcomes. With clever measurements and calculations, the Arabs defined a degree of longitude: 56⅔ miles. We now know that the length of an Arab mile was 4,000 black cubits, or 0.4932–0.4933 metres, which made the length of a mile 1,972.9–1,973.2 metres and the length of a degree of longitude 111 kilometres (rounded). Only in the 19th century, with the help of Friedrich Georg Wilhelm von Struwe (1793–1864), was a degree of longitude resolved at a final 110,938 metres. The Arabs' results turned out to be astoundingly accurate a full century before al-Masudi! Europeans relied upon the Arabs' measurements throughout the entire Middle Ages, failing to realise what the 'black cubits' and Arab mile might actually mean. They were believed to be shorter. And, therefore, cautious Columbus dared to embark on his expedition ... He believed he was sailing to India, which the Arabs had described, mapped, and measured; to a land that he thought was half the distance closer, somewhere in the western Atlantic Ocean. The ships' paltry tonnage would never have sufficed if Columbus had known the actual distance of a voyage to India and attempted to stock rations accordingly. Blessed are the ignorant, though ignorance was significantly greater during the time of Innocent III.

SON'S LAND AND MOTHER'S LAND

Calling the Baltic the 'land of the Mother' cannot be ascribed to mere symbolism. We must realistically assess the Baltic lands' modest share of the European map, but also take into account Church tradition. The Virgin Mary was at the very apex of the Catholic hierarchy. She was equal to the Son, but more popular because she was given human characteristics. The lofty 'land of the Mother' title could only be given once in the entirety of Church history, so naturally it needed to be granted to a region connected to and inherently equal to the land of the Son. In geographical history, mistakes have sometimes played a fateful role and othertimes a fruitful one. Observed from Rome, the distant battlefields on the shores of the Baltic and the Mediterranean might have appeared ideally symmetrical, but the symmetry of the titles contains, first and foremost, an emphasis of equality – in other words, a judgement that was based on a false geographical conception. Ptolemy equated the distance between the Black Sea and the Baltic Sea to the width of the Sea of Azov on his map, and Ptolemy's maps were still the most accurate in that era. On pre-Ptolemy monastery maps, a short waterway ran east from Narva into the Saracen steppes.

Calling the Baltic the 'land of the Mother' stemmed from a post-Pytheas notion that the lands of northeast Europe bordered directly on the Near East. We can recognise the dulled sparkle of the Silverwhite waterway, sparking the imaginations of explorers and foreign merchants once again and leaving such a bizarre imprint on Church history – coupling the land of the Mother to the land of the Son; the Baltic lands to the scorching Arab East.

Al-Idrisi and Henry.

You really don't have to lie? Henry didn't lie. He was silent. 'The chronicler is trustworthy, all checked facts have proved true,' concluded one analysis (in 1963). However, the silenced facts cannot be checked.

Valdemar II's arrival in Reval did not elicit hostile activity. On the contrary, relations were pragmatic and adhered to protocol, with both sides exchanging gifts. 'The king believed them, not knowing their guile,' Henry of Livonia foretells (XXIII). Yet the king's arrival was anything but a simple military expedition. He disembarked in the company of the Slavic prince Wenceslaus, Archbishop Andrew of Lund, Bishop Theoderich of Estonia, Bishop Nicholas, the bishop of Roskilde, and the king's chancellor Petrus Jacob.

What could have been the reason for this diplomatic mission? Four months earlier, Rigans had violated the Revalamaa border. It was the

399

SILVERWHITE

first time this had happened, which is beyond strange. Archaeological discoveries confirm with total certainty that Revalamaa was a wealthy county. Horses, cattle, sheep, oxen, home-woven Estonian *waypad* ('rugs') that would adorn the residence of a French bishop in Avignon a few centuries later: the spoils would have been tempting. Loot was crucial to the crusaders. They were often starving. Provisions were delivered by sea. It was an economic war. But not once had they crossed the Revalamaa border. Until the 1219 military campaign, undertaken in the frigid cold of February, Revalamaa was veiled in a mysterious void. The county's name itself did not even exist for Henry of Livonia.

However, the crusaders needed only cross the border for Valdemar II to show up with his diplomatic mission four months later, upon which the elders of Revala addressed the king 'with deceptively peaceful words', as Henry of Livonia asserts. What held the crusaders back? The early town and its immunity were protected by the tradition of a 'sworn compact' with roots in time immemorial. When that immunity was violated, the Estonians turned to Valdemar II. This would explain several gaps and questionable details in Henry of Livonia's account. 'Sworn compacts' were surprisingly enduring; and the older, the stronger. The Goths did not violate their agreements with the Saaremaa Estonians even when the pope's legate incited them to crusade in 1226: 'The Gothlanders refused. The Danes did not hear the Word of God' (XXX). Thus, one can understand why the battle broke out only some time after Valdemar II arrived. The sea routes and coasts were under careful watch. His appearance could not have come as a surprise. Henry of Livonia himself acknowledged this when a storm blew the crusaders' cogs into Saaremaa waters: 'At the first light of dawn the sea opposite us looked black with their *pyratica*' (XIX). The only way to make sense of this is to assume that the fleet was assembled in a single day, despite the storm. It is easier to repel an overwhelming force as they come to shore, not giving them the opportunity to set up defences in an almost impenetrable citadel. This did not happen in Revala, as the elders saw Valdemar II not as an enemy, but as an ally. However, Revala's harbour was shared by several counties – evidence of this is the roads that converge upon it. It was the only access to the sea for the people of landlocked Harju County. Revala was crucial even to the distant Saaremaa Estonians. The Battle of Lindanise, which Henry of Livonia attributed to the treacherous deception of 'pagans', speaks to conflicts between the town and the counties. It was fought by '*Revelenses et Harionenses*'. The former should be interpreted as a *malev* military force

SON'S LAND AND MOTHER'S LAND

organised by the mother *kihelkond*. Based on what the Danish Census Book can tell us, the battle did not harm the relations between Revala noblemen and the Danish king.

And if that truly was the case, then why did Henry of Livonia remain silent? In fact, he did not: 'They all brought their army to the province of Reval and encamped at Lindanise, which had once been a fort of the people of Reval' (XXIII). So he described the arrival of Denmark's King Valdemar II (1202–41) in 1219. And one year later, Henry of Livonia himself was already in Tallinn: 'They went after this into the fort of the Danes with the priest and made the same statement to the venerable Archbishop Andrew of Lund' (XXIV). And that was it. Not one word more. Henry of Livonia was rarely so laconic. Could he have had a special reason for it?

Baptising was usually seen as a formal act of Roman jurisdiction. Yet the situation was different in Baltic lands: the pope intended to essentially maintain control over the conquered lands as *Terra Mariana*, the land of the Mother (Biezais 1967: 86). This could only be possible in areas that the Church had taken by sword and cross. Was Henry silent because Tallinn did not fulfil this condition?

Henry of Livonia, doggedly loyal to church–state ideology, recorded in his chronicle all that served the crusades' purposes, and left out all that cast a shadow upon them or put them into doubt. The principles of Christianity had been borrowed into the Estonian language already prior to the conquest, though to search for any sign of this from Henry of Livonia's account would be fruitless. We should have a greater appreciation for his ability to be silent in enrapturing and truthful ways. His chronicle wasn't unintentionally later added to the Riga archbishop's personal library. He stuck to his rules through and through, albeit revealingly: the absence of a source is a source in and of itself; one that evokes questions and enables conclusions. With his silence on Tallinn, did Henry also conceal the presence of the city's Christian congregation, and thus the lack of motive for a crusade? Conspiratorial silence?

That isn't the best expression. Early Tallinn wasn't alone in its fate, and the lips of historical records are sealed even more firmly on the others prior to conquest. Birka was the greatest trade hub in the North. It radiated upon every Baltic coast, and Iceland was the North's central historical archive that documented colourful details of even faraway Constantinople, Novgorod, and, naturally, Estonia. Nevertheless, Iceland's sources are silent on Birka.

SILVERWHITE

Helgö thrived and developed for ten centuries. The Nordic fortress town tied classical antiquity to the fall of the Carolingian dynasty and Caesar's Gallic campaign to Yaroslav's campaign to the borders of Tartu – and then it vanished, sinking into the ground for ten centuries, only to surface once again as archaeological relics that flipped cities' ancient history upside-down. Not one written source even mentions Helgö's name (Kumlien 1970: 10).

Al-Idrisi or Henry?

Al-Idrisi and Henry.

XI

THE FIRST CAPTAINS

SEA DEMONS

Does early Tallinn owe its early development to seafaring and the Silverwhite waterway? And did it grow smoothly into a medieval city? Preserving its continuity? If so, it should verify the continuity of Estonians' seafaring tradition. If the function was preserved, then the factor that shaped it should be preserved as well.

And so it is. The first Estonian captains step up, conveyed by historical records. Allow them to pass in order:

Oseleer (Saaremaa, 1360)
Salt transporter Kassowe (1369)
Captain Sundy and *schipman* Kurge (1402)
Skipper Surepee (1405)
Captain Pujalke (1429)
Skipper Otto Mekis (1437)
Salt transporter Clawes Ubies (1462)
Skipper Michael Sturmann, owner of a house in Travemünde (1477)
Skipper Ludeke Kaseke (1491)
Skipper Oleff Sesenkar (1492)
Skipper and large-scale merchant Hans Surenpe (1501)

The names grow more frequent in the 16th century, though it doesn't mean that the number of captains, ships, or voyages increased in correlation. By order of the Tallinn city council, captains were required to start getting their departure documents in order: *we segelen wil, de scal komen vor den rad* – whosoever wishes to sail, come before the council! Until then, names had only been recorded in the event of accidents, court cases, or inheritance issues. From here, the dates and names grow closer:

SILVERWHITE

Skippers Mattis Musta, Laurencz Witte (1539)

Captain Matis Kalle (1543)

Captains Jurgen Podder, Reinolt Vos (1556)

Captain and grain merchant Tonnies Berch Tallepoisz (1557)

Tallinn patrol ship captain Andreas Jumetack Kaubipooicke, of whose crew only one man spoke German, though he himself was not a captain (1559)

Captain Tonnies Kock and his crew Andreas Lifflender, Matz Ssuwe, Clawes Grote (1576)

Skipper Clemendt Sevenborgk and his crew Jurgen Meil, Andreas Olleloic-Stroschnider, Matz Hein, Jacob Kaharpe (August and September 1576)

Captains Andreas and Jacob Lifflender (1576)

Captain Lorentz Staal, son of a jeweller (*pistemaker*) (1577)

Hans von Kurküll, who wasn't a Lübeck nobleman, but rather Captain Ants Küti, who had moved to Lübeck from Küti village in Viru County (1579)

Simon Syseke and his crew Hieronimo Petersen, Jürgen Fohrman, Jacob Oloffson, Ewertt Pryn, and Erich Marcussen continued this list in the 17th century, and in our day it would be concluded by Max Krasna, an instructor of future captains and a man who has ploughed the waves of every ocean for the last quarter-century. Some of the names seem strange. Modern Estonian is to blame. Kaharpe would now be Käharpea ('Curlyhead'), Olleloic now Õlelõikaja ('Stalk-Cutter'), Surepee and Surenpe now Suurpea ('Bighead'), and as a toponym Suurupi: one 1566 Dutch *leeskartenboek* refers to the cape as *Surpe* < *Suurpää*, which appears to be eponymous with the home village of a famous Tallinn family. Captain Otto Mekis (1437) is none other than a distant descendant of the Mägiste family, familiar to us from the days of the Danish Census Book: Johannes, the grandson of Mäoküla's great vassal Pärt Mägiste aka Bertald de Maekius, rose to become a *Landrat* (a state authority) and several of his own descendants became Tallinn officials, though the seaman's two brothers were Tallinn blacksmiths (Johansen 1933: 881–2; Johansen and Mühlen 1973: 426–7; Moora and Ligi 1970: 88–9). Perhaps the list should begin instead with the year 1312: one in which a prosperous Tallinn merchant who claims his name is Lodewicus Rootslane ('Swede') swears peace before Stralsund city authorities. Perhaps the man was a Swede. Nevertheless, he curiously spoke in Estonian. Rootslane's only given name in the historical records is *Lodewicus Rozleyne de Revalia* (Johansen

404

THE FIRST CAPTAINS

1951: 77). Was the ethnic and social environment of Tallinn's seamen in the late 13th and early 14th centuries really so homogeneous that even foreigners adopted the Estonian language and culture?

One Swedish historian estimates that the German portion of the population in early medieval Tallinn was approximately equal to that in Stockholm – quite small. Tallinn's distance was one reason; another was its social atmosphere, which was unfavourable to migration in the 13th and 14th centuries, a time of otherwise vigorous colonisation. In this sense, Tallinn and Riga have little in common. Early Revala was an exception, and medieval Tallinn turned out to be just as unusual. Estonian and German words and phrases blended in both people's colloquial languages, records of which still survive to this day (Johansen and Mühlen 1973: 381).

* * *

The devil once promised to help [sailors]. The men had him man the pump, sticking the end of it through the hull and into the sea. The devil still hasn't finished the job. Once, the men hoisted the sail at night. The cook-boy also lent a hand, but grabbed the devil's tail in the dark. The others saw he was pulling and went to help. They yanked the tail right off him! The devil then said, 'May the devil take those sailors who want him!' and left. (Pärnu)

SALT TRANSPORTERS

Until recently, researchers believed the ancient Baltic ship became obsolete as early as the 12th century. This isn't true. Visual images depict numerous ships that still followed the ancient traditions in shape and construction in the 14th century (Olsen 1963: 34). The rapid, seaworthy, and easily manoeuvrable Baltic warship maintained its place – unlike the old cargo ship, which did indeed become obsolete. In 1207, King Sverre of Norway had his cogs chopped shorter and equipped with oarlocks, as they would otherwise be useless in a naval battle.

Further evidence of the cogs' inadequate manoeuvrability is the fact that after a battle with the Saaremaa Estonians at sea, the crusaders were forced to spend three weeks in Saaremaa's New Harbour and wait for a wind that would take them to Visby because of the 'daily storms at sea, sudden squalls, and contrary winds' (XIX, 6). A constant contrary wind coming from the west on the erratic Baltic Sea – especially in July – is unlikely. A touch of drama was added by the threat of the islanders' attacks,

SILVERWHITE

not to mention Gotland's taunting proximity, at a mere 80 nautical miles away. Yet despite all this, the fleet was immobile for a full three weeks. This speaks to one of two possibilities: damage sustained during the battle, which Henry of Livonia fails to mention with his usual discretion, or the cogs' rudimentary manoeuvrability – the need to wait for a single suitable wind out of many.

In 1893, the skilled captain Magnus Anderson sailed a replica Viking ship, based on the Gokstad discovery from Norway, to the World's Exhibition in Chicago, averaging a speed of 10–11 knots. This alone allows us to disprove the German historian Paul Heinsius's claim that 'In some cases, the Hanseatic ship was faster than even the Estonian *pyratica* on open sea' (Heinsius 1956: 168). Every technological innovation, no matter how outstanding, requires time to become viable and competitive. The ancient ship and the cog were used parallel to each other on the Baltic sea for two centuries, even as cargo ships. This is an important fact. When Westphalian and German merchants elbowed their way into the Baltic trade, they were originally sailing companions, accompanying foreign Norse merchants on the Baltic sea routes. They stayed in this role long after the founding of Lübeck, their competitiveness only increasing gradually (Christensen 1963: 166–72). Thus, the Lübeck/Hanseatic trade system did not bloom overnight or as a result of the Teutonic Order's organisational skills, but instead built itself up on the basis of local traditions and experiences.

It's often said that Tallinn was founded on salt. What facts lie behind the catchy saying? A sea route is a link between production and consumption; in other words, it reflects general bold economic lines. However, it does so differently from highways. Navigation has its own logic. The course of a sea route depends on the wind, sheltered harbours, and even pilotage, as a result of which it does not always coincide with the bold economic lines. Prior to the Hanseatic League, Gotland was a point of convergence for Baltic shipping routes. Visby's prosperity did not reflect Gotland's level of economic development, but navigational strategy. When these routes drifted away from Gotland in the early 14th century, Visby ceded its central functions to rapidly developing Tallinn and quickly became a humble provincial centre. The conflict between Lübeck and Viru County's coastal population played a role as well, but other major factors sealed the city's fate: Visby's rise had been a product of ancient navigational logic, so the cog prevailing over older vessel types made decline inevitable. The same would later be repeated with Tallinn: when shipping routes drew away in the 16th century, Visby's successor also declined to the level of its

THE FIRST CAPTAINS

surrounding lands and became another little provincial city in northern Europe – 'wild and desolate', as Balthasar Rüssow would complain. Yet there's still time before that.

The shipping route ended at the mouth of the Neva, where cargo was unloaded from the cogs and placed on river boats. What came next was apprehension. Conditions on the Volkhov and Lake Ladoga and in the Neva watershed were uncertain. A resolution passed by Hanseatic merchants in 1346 stipulated that no one was allowed to sell goods to Russia without visiting Tallinn, Pärnu, or Riga. From then on, loading cargo from cogs onto smaller vessels or carts could only be done in safe and protected harbours. Tallinn was the most favourable. The new rules required scores of captains familiar with local waters and increased the proportion of Estonians on the sea routes even further.

Global sea routes, to be specific, because the very first known record of an Estonian seaman sailing on the Atlantic dates to the 15th century: Captain Pujalke (1429), who received 100 *lastid* of salt in Baye, France. There is no information from earlier years because the lists of vessels that visited Tallinn contain only cogs (Johansen and Mühlen 1973: 74). Captain Pujalke (i.e. Puujalg, 'Wooden Leg') was also only included in the document because of his merchant status, not actual sailing ability. The poorer the merchant, the more those two overlapped. Nevertheless, Puujalg was on the fringes of being a wealthy merchant with his 100 *lastid* of salt and he heads our list of Estonian captains because of his destination: salt was transported to Tallinn from Lüneburg (on the lower Elbe), from Lisbon, and primarily from La Baye, whose harbour was at La Rochelle on the Bay of Biscay. It was brought in loose bulk, which allows us to draw certain conclusions about the dryness of the cargo hold. The quantities were impressive even by today's standards: up to 20,000 tonnes annually, two-thirds of which travelled on as transit cargo. Such large trade volumes pushed Tallinn up among the world's largest ports. In 1468, the city received 95 large seagoing vessels, 67 of which carried salt (Johansen and Mühlen 1973: 74; Niitemaa 1872: 356; Heinsius 1956: 190). Tallinn's salt transporters were commonly referred to as 'Baye-goers', taking regular trips back and forth to La Rochelle during the navigational season, which lasted from April to November. This is regarded as an outstanding achievement.

* * *

SILVERWHITE

Line-crossing ceremonies are carried out in several places: the Danish straits, Gibraltar, and at the equator. Do we know how old the ritual is? We do not, we cannot remember.

In Estonian waters, a man was pulled beneath the keel three times when he was to make peace with the sea or other sailors.

Classical authors spoke of the Pillars of Hercules in the North. The southern pillars were in the Strait of Gibraltar. And their northern counterparts? There are several possible sites, mainly the majestic Kullen cape between the Kattegat and Öresund, but also some on the Estonian coast. Perhaps we should regard the northern Pillars of Hercules as the mythical Baltic-Finnic pole?

The North also had its own Hellespont (Helle Strait, to translate the ancient Greek name for the Dardanelles): the Daugava River, known in Livonian as the Väina (Johansen 1964: 7). It's as if the North was a mirror image of the Mediterranean. And this could be true: many geographers assumed that the world was symmetrical in nature. The conception of antique authors? In that case, however, it would be difficult to explain how the notion became so ingrained in Estonian traditions that the Kullen cape was still compared to Gibraltar in the 20th century ...

* * *

Here, I'd like to write about an old seaman's tradition that today's sailors still practise with great care: a seaman's initiation, or rather the christening of a young sailor or ship's boy who's sailing a greater sea for the first time. It can only be done in one place: when passing Kulli Island [the Kullen cape], which is a craggy little island above Daneland. As soon as someone's ship that's carrying these young seamen who are sailing past Kulli Island for the first time approach it, as soon as the island can be spotted with the naked eye, then unusual activity begins aboard the vessel. All the sailors are brought up on deck, where they're lined up. [...] An incredibly long rope or line was pulled beneath the ship running from port to starboard. A young seaman was then tied to one end and pulled beneath the vessel. It was done three times. [...] After that, he was tied to another rope and pulled up to the tip of the mast three times using a block. Only when that was finished was the sailor's initiation all right and finished. [...] The coastal villagers have been doing those initiations since time immemorial and that's why they row up to ships passing Kulli Island to sell a liquor that's incredibly cheap but also incredibly strong. (Viljandi)

408

THE FIRST CAPTAINS

An 'old sea man' lived on Kullen's tall peak, and many Estonian sailors are buried in the tiny Arild Cemetery at its foot. One anchor-shaped iron cross, the sole surviving memory of an unknown Estonian ship, warns: *År 1849 förliste det Estniska skeppet Concordia* ... (Kurgo 1965: 136): 'In the year 1849, the Estonian ship Concordia sank before Arild. Its crew of seven men are buried here. You carry us through every storm to a safe and blissful harbour. And we are comforted by Your name.'

MASTS AND MASTERS

How large was a cog?

The first bridge over the Daugava River was built in Riga in the late 13th century. A 33-foot-wide lane was left open in the middle of the river by order of the city council, putting the cog's width at about eight metres. We know from Lübeck that channels were kept clear to a depth of 5–6 cubits, making its draught 3 metres. These facts don't give us the cog's tonnage just yet.

In 1241, Saaremaa Estonians were forced to conclude a peace treaty with the Teutonic Order representative Andreas von Velven. The Estonians promised to give the Germans half a sheaf of grain for every *adramaa* (5–6 hectares): altogether 125 *lastid* from 3,000 *adramaad*. The medieval *last* was equal to 1,933 kilograms or 3 square metres. The Teutonic master or the bishop was obliged to transport the tribute of grain at his own expense, a condition that was worded in detail: the tribute was to be taken away by cog. If the rulers could not provide their own vessels, then they were required to rent the ships (*naves*) and employ the captains (*gubernatores*). Thus, the Saaremaa Estonians' grain tribute – nearly 250 tonnes – was equal to the capacity of one cog.

Early in its career, the cog boasted the greatest tonnage of any vessel and was the king of the seas. They must have been constructed of strong materials. Archaeological discoveries confirm that planks were 5 centimetres thick, 45 centimetres wide, and had a 5-centimetre overlap. Shipbuilders therefore counted on the vessels being encased in ice in the winter. Its keel was straight, the stems at bow and stern nearly vertical, and its body rounded. It wasn't exactly handsome. Drawings show the ends of crosspieces extending through the planking. They are placed horizontally even though the gunwales rise at the bow and stern. We must assume that an equally horizontal deck was laid over the crosspieces. This would require support posts in turn, resting against the keel. Consequently, the cargo hold could have been divided into several spaces that would also need to be accessed while at sea, as the grain that was transported to La Baye in exchange for salt had to be turned over frequently.

In depictions of the cog the rising gunwales conceal the deck structures, but they were there. Henry took part in the Trafalgar of 1215. He wrote: 'When the flames of [the Saaremaa Estonians'] fire, which was taller than all of the ships, reached out towards us, we called the bishop from his cabin, where he was praying day and night' (XIX, 5). We shouldn't picture the cabin as being overly modern. An Italian document from that time states that the ceiling of a cabin was less than a metre and a half tall.

The helmsman's place was naturally on the roof of the stern cabin, from which he steered the ship with a long tiller. In the following century, shipbuilders starting building a second deck on the half-decks at both the bow and the stern, forcing the helmsman to move yet another storey higher. Railings were installed around the roof of the aft structure, and often also a taffrail with ornately carved embrasures. Thus, a small fortress rose around the helmsman — a *castellum* that resembled a bridge. Of course, the latter word has acquired another nautical meaning in every language. A second, slightly smaller 'castle' arose at the bow of the ship. It also had an important navigational role: depth was measured and anchor manoeuvres performed from the forward cabin. The rather clunky cog depended on an anchor more than its predecessor and its successor, and sounding depth remained a primary means for determining location until the Late Middle Ages. In the 15th century, one Italian quipped that vessels sailed the Baltic Sea more by plumb than by compass.

It was hard to steer the giant ship with a tiller. The innovative pintle-and-gudgeon rudder may seem elementary at first, but it required a vertical transom board or at least a sternpost for attaching the hinges. Ancient Baltic vessels were rounded at both ends — handsome indeed, but ruling out any possibility of a pintle-and gudgeon rudder type.

Even the Chukchi whaling boats have rounded sternposts. At the Uelen collective farm, these vessels were even equipped with first-class Swedish Electrolux outboard motors. The Chukchi shipbuilders masterfully resolved the dilemma of joining tradition with progress, still steering their speedy and almost soundless boats with an animal's long leg bone. They cut a hole in the bottom of the boat for the motor and build a watertight well around it.

However, the cog's vertical sternpost already contained possibilities for a new solution. The Elbląg (Elbing) city seal shows the first known

THE FIRST CAPTAINS

cog steered with a rudder tiller (Heinsius 1956: 119). Did this crucial invention happen on the Baltic Sea? It would be nice to believe so, and some researchers of Baltic navigation do indeed believe it. However, long-distance captain Buzurg ibn Shahriyar was born in a place called Ramhormoz on the shore of the Persian Gulf in the 10th century. In his prime, he sailed the seas; in his twilight, he wrote travelogues. One illustration in his India-themed work *Ajaib al-Hind* of 953 leaves no doubt that ships in the Orient were already steered with a pintle-and-gudgeon rudder. A four-armed anchor dangles from its bow. Of course, the history of technology offers ample examples of the simultaneous and independent emergence of important inventions at different latitudes if needs even just slightly overlapped. The argument is rather unconvincing in this case. Nearly three centuries separate the innovations. Could they have been unrelated? On the Mediterranean, the helm evolved into a fantastical gadget that attempted to connect several tillers at once. It was an innovative dead-end. Why did the Eastern invention find itself so at home on the Baltic Sea? Why was this the launching point for its triumph throughout Europe? Or have we perhaps improperly underestimated ancient cultural exchange?

The cog's principles of construction turned out to be capable of evolution. We know of a 15th-century equation for calculating the length of a mast and the area of a square sail. The mast's length equalled the width of four ships. Therefore, a vessel 6 metres wide would have a 24-metre mast, taller than a modern eight-storey apartment block. The width of a square sail was four-fifths of the mast height, its length half of that: a 6-metre-wide vessel's yard would be 20 metres wide and the sail's area up to 175 square metres. The sail's area could be increased even further with two or three additional sails attached to the bottom of the spar, increasing the area to 335 square metres. This was a great, majestic ship that had little in common with the plump vessels of Henry's day. The square sail could be lowered. Winches and blocks were used to reef the immense heap of canvas, sometimes with quite ingenious constructions. One 1429 seal shows a chain of beads attached to the tip of the mast. The halyard used to hoist the square sail would slide along them with little friction, perhaps constituting the first ball bearings in technological history. Moving rigging was once braided with linden bark, later with hemp or seal skin (following ancient vessels' example):

SILVERWHITE

Ma mies merime poiga,	I'm a man, a seaman's son,
purjetan punasta merda,	I sail the red sea,
oma merda oigejada,	our own sea the right sea,
Viru merda virgilista,	the Viru sea the furrowed one,
Rootsi merda ruugejada	the Swedish sea the rouge one,
sinisilla siilavalla,	beneath the blue jibs,
punasilla purjeella –	beneath the red sails –
sinised on servarihmad,	blue are the straps,
punased on purjeained.	red are the sails.

(Kuusalu)

Although this is a poetic document of which nothing more can be asked than what its structure produces, the precision of folk memory is astonishing. Although it may seem to every Estonian speaker that *sinised on servarihmad* was the bard's sacrifice of accuracy for alliteration, the metaphor is nautically flawless. *Punased on purjeained?* I'm not ashamed to say it is my favourite verse, though not merely because of the pleasant symbolism, alliteration, parallelism, or even the enchanting word *purjeained* (literally 'sail materials'). Those, too, of course, but we have now reached the point of our travelogue where we may place folkloric metaphors side by side with documentary witnesses, just as in court. And what do we find? The poetic texts are much more accurate than they may seem at first glance. They are facts, in a certain way: historical sources that have admittedly broken from reality in the collective memory and the works of generations of bards, and have fitted themselves into a poetic structure, but, when unravelled, the thread still leads us back to some tactile, pragmatic piece of information. We could compose the most accurate biography of a classic author based not on the dates of their birth and death, but on their written works. Earlier, I quoted 'The silk ships sailed past' in connection with the Silverwhite waterway, and I am inclined to believe it. I see no need, or even opportunity, to treat the line as an abstract metaphor; as the original bard's 'contrivance' (if we may employ such a notion). Silk is one of the most prized goods ever transported by ship – naturally, it sparked the bard's imagination. Silk ships did sail the Gulf of Finland in the days of the Silverwhite waterway, no earlier or later. And what of it? Estonian runic singing is believed to have developed around the middle of the 1st millennium CE. The masculine essence of silk ships later wore into a feminine symbol: wore, but was also preserved as a result. The later

THE FIRST CAPTAINS

development of runic singing outside realistic circumstances is unlikely given the aim of folk songs to depict and generalise what is standard. Returning to the last lines:

> blue are the side-straps,
> red is the sail canvas.

More often than not, a cog's sail was red or brown. The 'red sails' so gleefully exploited in Estonian folk songs contains a memory of the medieval cog. Later vessels' sails were striped or white.

Ornamentation was stitched onto a cog's sail, most often Tallinn's lesser coat of arms. A pennant bearing the colours of the ship's home city waved on the mast. Later, it was moved to the stern. The earliest details of Tallinn's nautical flag were recorded in the 17th century (Greiffenhagen 1927–8: 139). Its symbolism was different from that of our era: it marked the harbour. The medieval Tallinn nautical flag had six vertical stripes, alternating light blue and white. A war cog would carry a broom on the bowsprit. Although Henry of Livonia provided early descriptions of anchoring manoeuvres, the implement wasn't mentioned in Tallinn records until the year 1367 (LUB MX, 1367). Specifically, a Tallinn captain had acquired two anchors and their ropes as loot in a battle with pirates, but they had in turn been plundered from a Lübeck ship and the original owner demanded their return. Anchor chains, which had been ordinary components of classical vessels, weren't used again until the early 19th century.

Finally, we also possess rather credible data for calculating the cog's average speed. One Tallinn council member departed the city on 5 June 1418 and arrived in Lübeck on 11 June 11 – less than six days later. The vessel covered nearly 100 nautical miles in a day, giving it an average speed of 4.5 knots. Despite all its technological improvements, the cog still could not have competed with the ancient Baltic ship in that era.

Purjetin punasta merda,	I sail the red sea,
rohelistra Rootsi merda,	the green Swedish sea,
Idamerda irmusamba,	the East Sea more frightful,
Kagumerda karskeemba,	the Southeast Sea stiller,
Läntsmerda veevahusta,	the Western Sea frothy.
Kuha lendas alta koudi,	A fish flew beneath the sheet,

SILVERWHITE

vesi alta arjustimmeid.	water beneath the shrouds.
Kääd mul koudida kohendid,	My arms tightened the lines,
jalad mul aasida arisid,	my legs gripped the loop,
peial pistis pirkelida,	my thumb clutched the sprit,
elmad oitsid elmarida.	my coat flaps held the wheel.
Kui tunnen tuuled tulema,	When I feel the winds a'coming,
pahad ilmad pääle käima,	poor weather approaching,
keeran turja tuule poole,	I turn my back to the wind,
parra paha ilma poole,	side to the poor weather,
salgad sadude poole,	hair to the rains,
itse mina puljan purjeessa.	while I trim the sails.

(Haljala)

The cog's side planks were oak, though pine was also used above the waterline. This determined shipbuilding sites. The vessel's name (*koge* in Estonian) has been tied to many villages such as Koguva, Koggham, Kogg-kro, Kogg-stain, and Kogg-näs. Captains were forbidden to harvest mast-pines from Naissaar and Aegna in 1297, after which the islands were reserved for Tallinn citizens' use (LUB I, 566). The city council's resolution, which was sanctioned by the Danish king, thus laid a foundation for Estonia's tradition of environmental protection. Under particularly strict protection was a handsome [< holy?] oak grove on the Kopli Peninsula. During the Middle Ages, anyone who felled a tree there was obliged to plant two thickets in its place. To leave the grove untouched, the renter of the city's Kopli brickworks was required to ferry firewood over from Naissaar!

Visby's historical documents show the speedy rate at which Nordic shipbuilders mastered cog-building techniques. Entire teams of Estonian shipbuilders worked there alongside Estonian blacksmiths and 'tower builders'. The *Este* (Estonian) surname was very common in the city from 1485 to 1487: Andreas Este, Nils Este, Mattis Este, Anders Este, etc. (Johansen and Mühlen 1973: 423). This doesn't point to the emigration of Estonian shipbuilders so much as to the better preservation of Visby's archives.

Tarring began when the very first planks were installed. Birch tar mixed with sap, sulphur, and flaxseed oil was forced into the gaps as sealant.

THE FIRST CAPTAINS

TENDERS AND FIRE STARS

The history of exploration remembers the fate of Lieutenant Aleksei Chirikov, Vitus Bering's expedition companion, aboard the *St Paul* – a two-masted schooner of about 90 tonnes. Chirikov's scouting party landed near present-day Prince of Wales Island on the western coast of North America and never returned. The lieutenant dispatched a search party which did not return, either. Tempting smoke pillars rose from the island. Was it a signal? The men waited for days. It was a question of life or death, as Chirikov had no more boats and was running low on drinking water. They didn't step foot on dry land again for the rest of the expedition – land that was temptingly close and yet tragically distant for a vessel with such great draught. Men perished of hunger and thirst, albeit just in reach of rivers and forests.

As cogs were made more specialised, tenders became crucial. The Estonian language has preserved several names for utility vessels. Aside from the north Estonian *mündrik* there was also the *kaan*: a yardless two-masted ship (with a shorter mizzen). The latter was a very common auxiliary vessel in the Middle Ages: open-topped, flat-bottomed, of up to 10 tonnes, used in harbours and on the roadstead. Russian loanwords include *praam* < *parom* ('ferry') and *lodi* < *ladja* ('barge'), both of which were used primarily for transporting firewood. In Riga, the *praam* even became a unit for measuring its favoured commodity. Up to 200 barges arrived in Tartu annually in the 19th century, including exceptionally large vessels (a few dozen metres long of up to 190 tonnes) from Vasknarva, which lies at the headwaters of the Narva River on the northern shore of Lake Peipus.

In an 1822 copy of the Estonian-language newspaper *Marahwa Näddala-Leht*, Otto Wilhelm Masing informed that 'A few more *pornikud* [...] are running between Riga and Tallinn'. The linguist Ferdinand Johann Wiedemann wrote the word as *pording*: a ship type native to the Baltic Sea, recorded as *byrdingr* prior to Estonia's conquest, which also appeared in Hanseatic documents starting from the 14th century. However, the *pording* could no longer be classified as an auxiliary ship: with a crew of up to thirty men, it constituted an independent seagoing vessel. Equally respectable in age and importance was the Estonian *kutter*, Baltic *skuta*: a swift sailboat of up to 50 tonnes, later also used as a general term for smaller vessels. Henry of Livonia's *liburna* should apparently be interpreted as either the *skuta* or *byrdingr*. Liburnas were fast warships used by the Dalmatian tribes and adopted by the Romans (along with the name) in the 1st century.

415

SILVERWHITE

However, Henry of Livonia applies the classical name to a different type of ship: a cargo ship unlike the cog, which was also used by Estonians in the blockade of Riga and the battle at Sõrve. Can we perhaps associate Masing's *pornikud* with the ancient cargo vessel?

We already discussed the 'galley's little northern sister' in connection with the 'snake ships' or *snäkkja*. The Häädemeeste dialect has preserved the ancient name as *nääk*, and medieval dialect as *schnigge* and *nikk* (Saareste). The vessel was so large that it could be loaded from six horse-drawn wagons at once, making its length approximately 20 metres. The *nikk* carried the same size of crew as a cog in the late 13th century, and has been listed along with the largest ship types. It had space for twenty rowers in addition to a sail, making it the fastest vessel on the Baltic in the cog's era. In the 14th century, *nikid* were used to patrol and commandeer vessels in the Bay of Tallinn.

How much did a ship cost? In 1450, Tallinn's overall expenses totalled 2,595 silver Riga marks. One mark was equal to 208 grams of silver. That same year, Tallinn acquired a warship that cost 460 marks, i.e. 95.68 kilograms of silver. It constituted 18 per cent of the year's expenses. An additional 38 marks (1.5 per cent) was spent on harbour upkeep, and, together, the costs totalled twice the amount that the city managed to spend on its most important seamarks: churches (Johansen and Mühlen 1973: 68).

According to an old navigational guide:

> If one approaches from open sea and Riga's spires come into sight, three altogether, two of equal height and a third slightly shorter, then one should sail such that the lower spire remains between the taller ones, and maintain the course until one reaches the roadway, where the sea is either 8, 10, or 12 feet deep for anchoring.

One Estonian folk song describes Tallinn's spires somewhat more gaily. A ship is still far off and the foolish townsfolk falsely believe it is a rich Hanseatic vessel:

> thought it a dear cargo ship,
> thought it a big salt ship.

The spire of St Olaf's Church, on the other hand, stands out so prominently to seafarers that the men start to smell the scents of home and develop not-at-all pious yearnings:

THE FIRST CAPTAINS

It was Olaf's tower
a brother's whisking sauna,
where we go wedding on Saturdays,
washing sweat from our brows [...]

* * *

A captain was caught in a storm at sea and

> thought his own ship and all who were aboard it were surely doomed,
> fell down to his knees, called out to saint Niglas [Nicholas] and prayed,
> oh god Niglas, if you get me and this here ship, its goods, and all those
> who're aboard with me, out of this peril alive and well, then as soon as
> I step foot on land, I want to have a candle that looks just like this here
> mast made in your honour.
>
> When his son, who was kneeling behind him, heard this said, then
> he asked in a wretched-sounding voice: 'Oh-ho, dearest father – where
> are we going to get so much wax?'
>
> His father replied: 'Be still, my son. When we get to land, then
> we'll make the candle quite small.'

That nice anecdote was read from the pulpit of the Church of the Holy
Spirit in Tallinn on 1 November 1605 (Suits and Lepik 1932: 3).

* * *

The continuity of navigation, seamarks, and sea routes is even greater than
that of shipbuilding techniques. The Classical Era lighthouse in Boulogne-
sur-Mer was rebuilt by Charlemagne, and the famous Tower of Hercules
in Spain survived to see the advance of the crusaders. Valdemar II brought
the antique tradition to the Baltic Sea, granting permission for a seamark
to be erected at the perilous Falsterbo shoals near Malmö in 1225. During
the 19th century, Estonian *rannapapid* or *kallaspapid* were men tasked with
observing schools of fish along the coastline, though their ancient job was
to guard the shore and light fires (*tuletähed*, 'fire stars') to guide fellow
Estonians and warn others away.

Natural seamarks were passed down from generation to generation
like folklore:

> Every place on the sea has its own name, and those places can be found
> using family and *küli* marks. Family marks are points you can see from
> sea. In Leesi village, for example, there's a hayfield called Päälüs where

417

SILVERWHITE

a big birch grows right now; the village buildings lie between that birch and the sea. If somebody's far out at sea and that one birch can be seen over Mats's house, then people say: 'The birch is over Mats's house.' [...] *Küli* marks are other trees, coppices, points, and hilltops like those. (Kuusalu)

This passage is perhaps the most convincing evidence of the extent to which pre-map navigation was based on local experiences and conservative, self-enclosed, folklore-like oral tradition. We should see it as one reason why the Tallinn city council didn't take its first steps towards ensuring safe seaways until 1470: the Vahemadal (Middelgrund) shoal between Paljassaar and Naissaar was marked by two barrels, and a 'barrel tax' was levied on ships. It took two and a half centuries for the classical seamark tradition to migrate from the mouth of the Baltic Sea to the mouth of the Bay of Tallinn. The shoal's marking leads us to another conclusion: the ancient sea route from Tallinn to the Finnish shore, and from there following the Åland Islands to Swedish waters, had begun relinquishing its role to a direct route. Ships were more seaworthy, angular courses were bent into arcs, and the arcs straightened into lines in turn. Just thirty-five years after Vahemadal was marked, construction began on the lighthouse on Hiiumaa's Kõpu Peninsula. With immense effort, the structure was completed over the period 1514–26. Lighthouses on the Sõrve Peninsula and Ruhnu Island also began guiding sailors in the 17th century.

What was a medieval lighthouse like? The lighthouse at Kõpu is just a quarter shorter than the Pikk Hermann Tower in Tallinn. It was a mighty structure made of rubble and limestone, square at the base, and on the top a bonfire was lit on dark nights. The fire burned on a raised base and lent its name to the Estonian word for lighthouse – *tuletorn*, literally 'fire tower'. It consumed thousands of cubic metres of firewood in a year. The lighthouses at Sõrve and Ruhnu used imported Dutch coal in addition to wood. A warning fire burned from twilight till dawn from 1 August through 31 December, and was kept going all day long in a storm or foggy weather (Soom 1940: 206–7 and map). The Tallinn White Lighthouse (1773) was added in the next century, and the Tallinn Red Lighthouse (1835) in the century after that.

Still, we would be wrong to assume that the Tallinn city council paid no attention to seamarks in the era of folkloric navigation. Before its penultimate fire, the steeple of St Olaf's Church – Tallinn's 'honour and pride', as Léouzon Le Duc called it (in Estonian) in his French-language travelogue – was the tallest in the world. To frugal Tallinners, its height

THE FIRST CAPTAINS

was not of value for the purpose of records, but of practicality: St Olaf's Church was Tallinn Harbour's landmark; a point of navigation far out at open sea and even from the Finnish islands. This is not conjecture but a fact stated in the minutes of a Tallinn city council session on 18 June 1596 (calling the spire a seamark) and in an even earlier document from 1379, which speaks of the church as a storehouse for flotsam (Johansen 1951: 127–36). It's no wonder that St Olaf's was so popular in blasphemous Estonian folklore.

> Long, long, ago, Tallinn was tiny. It just wouldn't seem to grow, no matter what. Its people longed for their town to be great, famous, and booming with trade [...]. Suddenly, someone came up with a fantastic idea. He immediately declared it to the rest of the townspeople:
> 'Let's build a church that's bigger and taller than anywhere else. Then, ships at sea will spot our steeple and visit with their goods!'

* * *

And some Italians laughed that the Baltic was sailed by sounding?

> And places here in this sea have their own particular names according to depth. For instance, the water between the boulders near the shore is called *vesikivik*; a little further where the boulders don't stick out anymore and the sea's maybe 2–4 or 5 süldä [süld = 2.13 m] deep is called *rannu*; 5–15 is called *lau*; 15–25 is *penger*; 25–50 is *haud* ['grave']; and after you cross the haud, the water's a little shallower; around 20–30 süldä is a *laat* – first *maahaualaat* ['land-grave fair'], then *vahelaat* ['inter-fair'], then *maluhaualaat*; and now the water starts to get a bit deeper and you come to the *maluhaua* land-side where the water is 50 süllä deep, now there's *kesk-maluhaud* ['middle *maluhaud*']; next from here is *maluhaua*'s upper side – *onihaud*, 60 süldä deep [...]. The sea is deep onward to the west-northwest; only to the southwest does it get a little shallower, till [...] you get to the Malumadal shallows it's 4½ süldä of water; to the northwest lie Malusi Island and Vähämaa. [...] When you cross that *haud*, you come to the Kinturi Shallows. (Kuusalu)

RURAL HARBOURS

Wars and deaths tend to find their ways to the pages of historical records much more than mundane tasks or artistic work. A peddler wanders from

SILVERWHITE

village to village, a heavy pack perched on his back, and somewhere near Tallinn, the Grim Reaper catches up – his sole travelling companion. The weather was already warm, fresh birch leaves were sprouting like mouse ears, and on 27 May 1534 Tallinn's city officials recorded the late man's possessions. In addition to peddling tools, jewellery, a couple of books, and one harmonica, he carried compasses (Niitemaa 1952: 228). His death thus becomes part of the Silverwhite waterway chronicle, for what would cause a peddler to carry compasses in his satchel? A customer base that demanded them, of course. Zyrian hunters prowling the limitless primeval forests of Komiland already used compasses at that time, but in Estonia only sailors and sea fishermen required the tool called a *tuulekodarik* ('wind rosette') or *põhjakivi* ('north stone'). Three years after the peddler's death, a representative of the Tallinn city council complained at a landowners' council meeting (Estonian *maapäev*, German *Landtag*) that peasants were sailing all the way to Lübeck on their cutters and boats and selling fish there. The reason for his protest was not the coastal Estonians' seafaring, but the 'unusually' steep fish prices at Tallinn's markets as a consequence. Rural harbours were a thorn in Tallinn's side from the city's earliest days, as it wished to monopolise shipping. Nevertheless, it didn't declare them 'unlawful' until 1545 (Niitemaa 1952: 173, 223).

Seafaring abilities, shipbuilding, and the art of navigation existed and evolved in two contradictory environments: urban and rural. Loyalty to the sea remained steadfast in both: citizens perpetually adapting to systems of status, and coastal dwellers engaged in an equally perpetual struggle against them. This isn't to contrast the two, and is much less meant to highlight the rural over the urban. In its essence, the medieval city was already anti-feudal and in permanent conflict with the nobility. Tallinn is a classic example and the execution of Johann von Uexküll (1535) one culmination of the antipathy.

The rural seafarers' struggle centred around shipping and shore rights. It was an echo of Estonians' ancient struggle for liberty that reverberated throughout the entire 14th century and still boomed – sometimes louder, othertimes less so – in the 17th and 18th centuries.

In the kingdom of Denmark, the entire shoreline was declared property of the king. After north Estonia was annexed by Denmark, the law should also have applied here. It did not. It was beyond royal power to annul the peasants' shore rights, which became a front line along which unusual alliances of opposing forces were made. Cities and merchants were the most rigorous opponents of shore rights. Seaways were poorly marked

THE FIRST CAPTAINS

and shipwrecks frequent. There was nothing protecting merchants' lives if their vessel ran aground, not to mention their cargo. Coastal Estonians regarded such ships as their rightful property. First they would cut the shrouds, causing the mast to fall and rendering the vessel immobile. Any crew who attempted to defend the ship would be mercilessly tossed overboard. Native shore rights grew even more severe after Estonia's conquest: given that the profits of trade were disappearing into feudal lords' pockets, coastal Estonians lost any interest in guaranteeing shelter or asylum in certain places.

In 1287, a Lübeck merchant vessel sank off the coast of Viru County for unknown reasons. Coastal Estonians salvaged the flotsam and sold it to vassals and the Kärkna Monastery. Lübeck's city council ultimately found out and appealed to Queen Agnes of Denmark for the cargo's return. The dispute involved high-level politics. Strangely, Tallinn's citizens came to the Viru villagers' defence. Could the matter have reflected Tallinn's social climate and its attitude towards the rural coast? A raid of Viru County villages came away empty-handed, finding no peasants, vassals, or goods. A high-level conference of merchants from Lübeck, Riga, and Gotland was held in Tallinn on St John's Day 1288. It broke up in clamour, with even the bailiff showing sympathetic bias towards the coastal villagers. Unexpected and dramatic consequences ensued. Gotland's peasants had maintained their ancient ties with coastal Estonians, helped even by the Gotland Roman monastery's property in Viru. News of the German merchants' defeat flew to the island like lightning and inspired an uprising meant to overthrow Lübeck's hegemony once and for all (Niitemaa 1955: 267).

This balance of opposing forces on Estonia's coast lasted remarkably long. The reason was Old Livonia's relative independence, but especially the coastal Estonians' resistance. Archbishop Albert II Suerbeer's manifesto (1253) declared shore rights illegal, the punishment for which was exceptionally harsh in the light of the recent crusades: there would be no religious services in the *kihelkond* until the stolen goods were returned (LUB I, 566). Of course, the seafarers couldn't care less. The archbishop's judgement didn't coincide with that of the coastal dwellers, or even the vassals. The manifesto was to be read aloud four times a year in every coastal church starting from the date it was published and for all eternity: even to this day. This demonstrated the impotency of the central power structure; hence the Estonians' haughtiness. Fifteen years later, the Livonian Order and the Curonians concluded a treaty on salvaged goods, the stipulations

SILVERWHITE

of which probably also applied to Estonians. Whoever recovered the flotsam could claim one-third as their own; the rest was to be set aside for one year and one day while the legal owner or inheritor was identified. If none was found or the deadline passed, then the remaining portion was split between the finder and the local lord. This law was enforced in Old Livonia throughout the entire Middle Ages – in so far as enforcement was possible, naturally. Four centuries later, coastal villagers deftly stashed away a shipment of iron and copper from a Swedish ship that sank near Kunda, and a journey to the area by Tallinn's viceroy could do nothing about it (Soom 1940: 180). Coastal rights were tenacious. Coastal dwellers and authorities alike saw any attempt to curtail them as an exceptional and, even more importantly, one-off event that created no legal precedent. However, a new Estonian word did form: *vandiraiuja* ('shroud-cutter'), referring to coastal villagers who pillaged ships that had run aground; it was used as a derogatory term for western Estonian islanders more generally. The term triumphantly crossed the finishing line, i.e. it is still used today. What mattered most from a historical perspective was not the centuries-long struggle to protect shore rights, but a clearly perceptible sense of ownership of and responsibility for one's home coast and waters; a certain psychological sense that in no way fits with the pietistic Christian cliché of 'Estonians' seven hundred years of slavery'.

Ancient harbours' fates varied during the Middle Ages. Knight and wealthy vassal Helmoldus de Lodhe regarded Mahu Harbour – also called Kaupasaar ('Merchandise Island') after the ancient trade site at the mouth of its bay – as a favourable location with sufficient mercantile activity to found a city there. Only its name is preserved in historical records: *opidum et civitas Cogkele*, now Koila village. The nearby Kokaranna toponym appears to contain memories of cog traffic. Helmoldus de Lodhe founded another city at the ancient Laoküla harbour: *Lodenrodhe*. His utilisation of ancient Estonian nautical sites stands out in both cases. Hence, they must have still played a fairly active role in trade and seafaring in the late 13th century. Lodenrodhe was founded in 1296. The coastal Estonians were permitted free trade with Lübeck. Nevertheless, both ventures ultimately failed. In that era, geographical advantages and traditional trade ties were no longer enough for founding new urban environments. And to what extent can cities be 'founded' anyway? The establishment of a mint was more of a catalyst in Lund's history as well. Perhaps Lodhe's failed attempts can help us better understand early Revala.

422

THE FIRST CAPTAINS

A document penned by Sweden's King Albert in 1368 sheds some light on the focus of coastal Estonians' seafaring and the intensity of their mercantile voyages. It is a trade treaty granted to Saaremaa's Kihelkonna and Sõrve villagers, which treated them as Swedish subjects. The Estonians were allowed to trade all goods (except horses) throughout Sweden 'just like the citizens of our kingdom'. Horses were a strategic commodity even prior to Estonia's conquest, and were generally addressed in separate treaties even during the Middle Ages. For example, a 1420 document stresses that Finns had the right to sell horses to both Germans and Estonians, *den Esten, unsen luden*, but Estonians were only permitted to export draught horses (LUB V, 2424; cf. Niitemaa 1952: 355).

Cities were unable to monopolise trade owing to opposition from coastal villagers, and this led in turn to their deepening antipathy towards rural harbours. Vasknarva's bailiff issued a complaint to the Tallinn city council alleging that Estonians were sailing onto the Narva River and going from there onto Lake Peipus to trade with Russians. The Narva city council also sent the grand master of the Teutonic Order a letter of protest over Viru coastal villagers engaging in constant trade with Russia.

From this period comes also one of the few surviving documents that list the members of a rural harbour crew: Olaf Bake, Erick Bul, Erick Presser or Pres, Knuth Kauke, Las Peterboi, and Mathias Holm. Only the last-mentioned spoke German. The ship was captained by Andres Jummetack Kaubipoike (literally 'Merchant Boy'), a coastal villager from Juminda, Kuusalu *kihelkond*, whose compound surname speaks to a family seafaring tradition.

> The aforementioned Kolga Beach is a tiny peninsula and people call it Juminda Cape and that cape runs northwest-by-north; those deep sea troughs also run along the cape going from northwest-by-north to south-by-southeast. The seabed drops off suddenly when going up from the land, though it deepens gradually if you head to open water from the bay. [...] Between Leesi and Kiiu-Aabla villages, which is to say between Leethaud and Logapohja [deeps], there's a shallow part of the sea with holes or ruts that's called Hargu. [...] Juminda has Kolju grave, Kolju tip (the head of the cape), and Kolju heap – a pile of stone that was once about 5 sülda wide and 2–3 sülda tall, [...] though there still is a whole lot of them there to this day. Legend has it that an army (of sailors) was once there and they're the ones who piled up all those rocks, but when that was, nobody knows. (Kuusalu, Kolga)

SILVERWHITE

The reason why Juminda skipper Andres 'Kaubapoeg' appears before us lies in the Livonian War. Tsar Ivan IV conquered Narva in May 1558 and turned the city into Tallinn's rival overnight, attracting ships from near and far: Denmark, Lübeck, and even France and Scotland. Tallinn immediately began capturing western European ships as they sailed past, and Andres Kaubapoeg captained one of those vessels. We can thank the extraordinary circumstances for his name's preservation, as well as perhaps even a shortage of sailors in Tallinn (as 42 men crewed a seized vessel [Spreckelsen 1907: 106 jj]), though 'extraordinary circumstances' still do not explain the man's hereditary profession. Juminda entered nautical history as a place where cargo vessels were commandeered in 1422 and 1423. We do not know the relationships between pirates on the Baltic Sea, and it's not impossible that Andres Kaubapoeg had ties to Juminda piracy. He was recorded as the captain of a warship. The only difference between commandeering and pirating was that captains could produce a document sanctioning the act in the case of the former. Tallinn shipowners would hardly have entrusted a vessel to Andres Kaubapoeg if they regarded him as an ordinary coastal sailor. And what did such a surname really mean? Names have their own unique magic and reality.

A YOKEL MADE CAPTAIN

How did a sailor become a captain? What was the nature of an Estonian captain's work?

A Tallinn edict (*bursprake*) stipulated that if a merchant set off on a voyage, then *sole ... enen guden man to hus laten* – he had to leave a trustworthy person behind (LUB II, 982 and Bursprake B 8). It was easier to just stay home and entrust the sailing to a captain.

Captains were shipowners, but the larger and more expensive that vessels became, the more the captain could also be a stakeholder. A shipowner did not personally take part in sailing, but employed a captain and sailors for the job. Thus, captains also performed the role of shipowner and had to find themselves a new freighter in the destination harbour. Over time, this new legal form dominated and drove out ancient law. The development of mercantile correspondence has been used to explain this legal shift, albeit incorrectly, or at least insufficiently and overstating its significance. The real reason is that Lübeck's first citizens came from Rhineland and Westphalia and were not seafarers. The early medieval Hanseatic merchant turned out to be useless at sea, incapable

THE FIRST CAPTAINS

of steering the ship as a helmsman or hoisting the sails as a sailor. Thus, a system of dual ownership evolved. Captains were sole owners of smaller vessels. Merchants were forced to pool money and share costs to build a cog. Shipping law arose from the latter in turn. Lübeck merchants had to borrow their first nautical laws from Hamburg in 1199.

These details are highly significant, as they explain why the ancient Estonian seafaring tradition was able to survive uninterrupted and adapt to new vessel types on the Baltic Sea, particularly in Tallinn.

The boatman's profession was a springboard to the status of captain. Boatmen had a basic right to conduct trade so long as they paid taxes to the city. Yet as soon as one purchased a cargo ship, he had to give up his former job. The restriction may seem unjustified at first, especially given that Tallinn had no cap on the number of boatmen. However, the purpose lies in common law: it was impossible to merge ancient defence and security responsibilities with the captain's duties. One excluded the other. Boating was certainly a form of training to become a long-distance captain, but the two could not overlap. The boundaries between them were extremely generous. Boatmen's mercantile relationships extended as far as the Finnish coast, Lübeck, and Denmark. Here, we can deduce what the title 'captain' entailed: he was not involved in the loading or unloading of cargo or coastal sailing, and his voyages went further – beyond the Kullen cape.

The Estonian word for 'captain' was *kipper* (skipper), which comes from the Norse expression *schipher*, 'shipmaster'. Latin sources also call the position *nauta*, *dominus liburnae*, *capitaneus*, and, in Henry of Livonia's chronicle, simply *ductor navis*, 'ship leader'.

Skippers had personal contract with their crew. The men were not bound to the ship, and when a captain sold his vessel, they did not necessarily stay. Vestiges of ancient Estonian common law could also be found in nautical relationships. If a sailor committed a punishable act, then the fine was paid not to the captain, but to the entire crew (Heinsius 1956: 230). In the event of theft, the offender was cast out, forced to leave the ship, or marooned on a deserted island.

The captain, like the sailors, would engage in personal merchandising. This can also be seen as the survival of ancient Estonian customs and a simultaneous cause of innumerable conflicts and misunderstandings. Sailors' ancient trading privileges soon clashed with the city's rules and restrictions, which were enforced with increasing stringency and intrusion – and ignored just as consistently. Finally, the Tallinn city council began stipulating (1515–54) that only Tallinn citizens could charter shipping, and a captain '*egenen kopenschop willen nergen hen segelen*' – would not engage in independent trade voyages (Johansen and Mühlen 1973: 152–3). There was no question of wealthy merchants holding any monopoly, at least at that time.

Like all ordinary citizens and visiting merchants, a skipper was allowed to sell his wares at Tallinn's market for the first three days after (and including) his arrival. Thus, the skipper did not charter his vessel only to a merchant, but to himself and his whole crew. This was not speculation at his own initiative but a common sailor's bonus: if, for some reason, a seaman did not take advantage of the opportunity, i.e. he left a certain portion of the vessel's tonnage unused, then his crewmates would compensate him accordingly. Thus, the refined Hanseatic era intertwined with ancient Estonian customs. Quantities of personal cargo were regulated, of course. A sailor was allowed to bring a barrel of wine from La Rochelle, a quart of beer from Hamburg, and so forth. Consequently, sailors maintained the right to commonly use the cargo hold even though the vessel itself had meanwhile become the private property of the skipper or a group of merchants. During the era of cogs, relics of communal practices even echoed in sailors returning their portion of pay in the event of early disembarkation or inability to work because of sea sickness, and the money was evenly distributed between the captain and other crew members.

Seamen were paid more-or-less fixed wages for a voyage of predetermined length and accrued a fee for every additional week. The sum depended on profits. Percentages were settled between captain and crew. Sailors had the right to receive direct monetary compensation from the chartering merchants if the destination was changed during the voyage or if the cargo required special care, such as shovelling grain at sea. The wages were approximately equivalent to the profit made from personal merchandising.

The skipper's rights and obligations in relation to the chartering parties were stipulated in just as much detail. Once the anchor was raised, the freighter was required to pay the transportation fees in full, regardless of any change in destination port, shortening of the journey, cancellation of the voyage, or shipwreck. If the freighter cancelled before anchor was raised, then he still had to pay half the fee. If some cargo had to be jettisoned during a storm, the freighter bore the expense. The crew's personal goods for sale were only jettisoned after at least half of

426

THE FIRST CAPTAINS

the freighter's cargo had already been tossed overboard. Sailors retained privileges at sea that they were unable to keep on land.

Sailors were bound by common financial and legal interests, but especially by their incessant struggle with the city's wealthy merchants to guarantee that their rights were upheld. A memory of those distant times survives in modern Estonian: *antsakas*, *huntsakas*, and *untsantsakas* ('city slicker'), which were originally used in the sense of 'Hanseatic merchant' (*hansa* > *antsa*) (Saareste). The broad diffusion of the expression and the judgement it contains show that antagonism between seamen and Hanseatic merchants was typical in that era.

However, we still cannot conclude that the sailors themselves were free of societal conflicts. The equality of ancient *kalev* brothers inevitably had to retreat before the distinction of wealth. Skippers were aided by a *gubernator*, i.e. 'mate', and also a second mate starting from the 15th century. Unlike his immediate superior, the second mate slept in the forecastle with the rest of the crew. The mast became the ship's societal border post. Also inhabiting the forecastle were the boatswains, sailors, ship's carpenter, and cook – an important and respected member of the crew. Located in the aft were cabins for the captain, first mate, foreign merchants, and passengers.

To a certain extent, the legal status of an Estonian sailor abroad was guaranteed by the ship's home port and the captain even if the man had no citizen rights in Tallinn. If a sailor served on a Hanseatic ship, then he had the same rights and obligations as a citizen of Tallinn when in port towns. In the famous Flemish Hanseatic office in Bruges, Tallinners and Gotlanders formed their own 'eastern' third, and later had an office in Amsterdam (Johansen and Mühlen 1973: 74). Lübeck was one of the most common destinations apart from the salt ports, of course, and was the 'gateway city' through which the Estonian word *puju* spread into European shipping terminology as a word for coarse linen: *superior puik*. The strange relationship between chance and probability is embellished by the fact that Italian merchants purchased hand-woven Tallinn *vatmal* from none other than Bruges – a city that produced Europe's finest broadcloth (Johansen and Mühlen 1973: 393). A few other examples reveal sailors' close ties to Estonian craftsmen. In 1568, one merchant offered to sell 372 pairs of mittens that were knitted on Saaremaa. Artistic Estonian textiles appear to have enjoyed an uninterrupted presence on the international market.

We know from Henry of Livonia that the crusaders had a great appreciation for Estonian rugs: *tributa et waypas*, he notes when listing

427

SILVERWHITE

loot acquired from Tartu County, and the chronicler's use of the Estonian words seems to refer not only to the rugs' value, but their genuine quality. The latter is confirmed by our knowledge that the home of Archbishop Friedrich von Pernstein of Riga (d. 1341) in Avignon, France, was decorated with Estonian-made rugs, towels, and cloths – thus connecting ancient Estonian artisanship to medieval trade in its products. Captain Jurgen Podder (1556) took Reinolt Voss along on his voyage. Who might he have been? Nineteen years later, *pistelmacher* (craftsman) Tonnies Voss sails from Tallinn for trade. There's no doubt that the men are related or that the connection shows ties between sailors and craftsmen/merchants. Captain Lorentz Staal (1577), who appears as a blip in historical records a few years later, is again the son of a *pistelmacher*. Such associations were naturally illegal and attempts were made to conceal them, as a consequence of which it is difficult to reconstruct an objective picture of the true extent of relationships between sailors and Estonian craftsmen in Tallinn. In any case, they must have been much more extensive than the records reveal. An Estonian man named Pujalke (1429), who purchased 100 *last* of salt from La Baye, bequeathed 120 silver marks to three Tallinn churches – mostly to St Olaf's. That was more than a quarter of the cost of a warship in that era.

Our conception of seafaring Estonians would be incomplete if we failed to mention the goods that enticed men to sail. Grain constituted Estonia's wealth from the time before conquest to the modern era, making up nearly half of Tallinn's exports in 1586 (20,000 tonnes). Rye and, to a lesser extent, barley and oats were shipped to the Netherlands, Portugal, and Spain. 'Estonian grain' was an expression in Italy – *hartkorn*. Added to these were flax and hemp (also processed), wax and honey, seal blubber (also used to treat ship planks) and fat (which was still a delicacy in Linnaeus's day), squirrel and weasel hides (last recorded in 1391), livestock, timber, ash, tar, limestone (raw and ashlar), and, unexpectedly, Karksi beer. Salt from France, Germany, and Portugal held an unrivalled first place in terms of imports.

Earlier, we discussed 'friendly trade' on the Gulf of Finland. Viru Estonians exchanged freshly harvested autumn grain for Finland's spring herring. Primitive friendly trade could only last so long as the ancient relationships between producer and consumer endured, and such was the case on the Viru shore (Luts 1970: 288–314). But how are we to explain the friendly trade practised by wealthy Tallinn merchants Ficke and Kappenberg? Ficke's trade partners were peasant farmers, blacksmiths,

428

THE FIRST CAPTAINS

and millers. The compound surnames of Kappenberg's associates speak for themselves: Andrus Puseppe ('Carpenter'), Jan Woress Muresep ('Mason'), Jan Kassesepp (< Kasuksepp, i.e. 'Coatmaker') – altogether thirty-three men living in villages on Tallinn's furthest fringes and in Türi, Võhma, Suure-Jaani, and Karksi. During war, Kappenberg's friend Tytte Altema travelled to Tallinn and entrusted him with 190 silver marks – a fortune in those days – which the man faithfully later returned in three payments. Did this wealthy merchant with European accounting practices and international connections reawaken the ancient form of trade and its accompanying terminology? Ficke and Kappenberg more likely inherited the ancient mercantile associations through family connections, as the latter's surname appears in Tallinn's historical records as early as 1310 (Niitemaa 1952: 240–5).

* * *

> Men used to sail to Finland often, taking potatoes and eggs, bringing back herring. Matches and tobacco and cloth – those were contraband, you weren't allowed to openly return with them. Salt transport was how coastal folk earned a living in my childhood […]. As for Finland, men went to Helsinki the most, then also Porova, Loviisa, Hamina, Kotka, and Viipuri [Vyborg]. […] Those're all old Estonians there. Many migrated [to Finland], ran off […]. (Viru coast)

* * *

Even wealthier than Pujalke was a captain named Hans Surenpe ('Bighead'), who died before May 1511. He arrived in Tallinn from Turku, where he'd acquired citizen privileges in 1489. The surname and its variations appear quite frequently in Tallinn's records (Surenpe, 1369; Meynecke Surenpe, 1369; Peter Surenpe, 1407; Surepe, a captain who remained on the Latvian coast, 1405) (Johansen and Mühlen 1973: 422), implying that the nearby village of Suurupi was their ancestral home. Hans Surenpe married the daughter of a man named Palmedack, a member of the Great Guild in Tallinn, and was accepted into that same institution in 1501. Even as a wealthy merchant, Ants of Suurupi remained loyal to his nautical profession: he, along with Tallinn captain Olaff Sesenkari (< Seiskari), was listed in the Lübeck customs log in the years 1492–6. If the man's career was exceptional, then this was only in so far as he managed

429

SILVERWHITE

to reach the tip of the social pyramid while still in his home village: a feat that would've been easier elsewhere, especially amidst the conflicts of the 16th century. We would be amiss to stress ethnic discrimination, as there was no ethnic issue in pre-Reformation society. Tallinn's acrimony was directed not at Estonians, but at noblemen. According to law at that time, manor lords were (Estonian) peasants' legal beneficiaries. Tallinn feared that noblemen would claim a hereditary right to property that a deceased peasant had acquired in the city and would thus begin to have a say in municipal affairs. The danger was real. Skipper Michael Sturmann (Mihkel Tüür) was a former feudal peasant. He wisely did not build himself a house in his hometown, but in distant Travemünde, a port town that has since merged with Lübeck. As court records show, *huseken selbs up sine egene unkost buwen laten* – he had a house built at his own expense, as a feudal lord appealed to inherit the faraway property after the man's death. But what are more significant factors than those involving legal standards are Sturmann's primary sailing route and also his wealth, which was able to trump that of a manor lord. As we may recall, the list of Estonian boatman Hinrik Münrik's debtors was topped by a manor lord on Gotland.

Some events or catastrophes ignite like a spark, only to vanish into the darkness of oblivion together with the man involved. In 1462, Ubja Klaus took on a load of salt from Lübeck and sailed to Tallinn. According to documents, his name was Clawes Ubbias and he came from the village of Ubja, near Rakvere. – A ship docked in Sagadi in 1539, and (Estonians) Mattis Musta and Laurencz Witte disembarked into history. – (Estonian) Captain Kalli Matis was charged with fraud. – A ship sank near Bornholm and King John of Denmark did not take the necessary steps to salvage the flotsam. The name of the ship's captain, (Estonian) Ludeke Kaseke, flares and extinguishes in Lübeck's disgruntled correspondence with the monarch. – On a sultry July day in 1557, a captain came before the Tallinn city council and asked permission to transport hemp, flax, and grain to Gotland. He received it: *Tonnies Berch Tallepoiss, unse inwoner ... ein pasbort erloevet*, as the minutes of 12 July read. Let us pause here for a moment. Earlier, we stated that ethnicity was not an issue in the Middle Ages. Yet the period can't be labelled 'cosmopolitan', as that would still be the inverse of a mindset or position that did not exist at the time. Medieval society was class-based, above all. Rigid walls increased specialisation, accelerated the concentration of technical abilities in a single family or profession, and ensured the inheritance and accumulation of skills. Class-based societies were universally bilingual at a minimum. Latin in Bologna was

430

THE FIRST CAPTAINS

pronounced similarly and written exactly as it was in the Tallinn Cathedral School or the University of Tartu. The crusades turned language into a societal trait. Anglo-Saxon England became multilingual in the wake of the Norman Conquest and started an era of painful cultural integration. Just as in Estonia. Just as in Arab-conquered Khwarazm and Swedish-conquered Finland. Missionaries learned to speak Estonian in the early years of social stratification, but clergy abandoned the skill after German became the mark of high society. And vice versa: a German who fell to the lowermost social rung would adopt Estonian, which was a mark of that status. Medieval Tallinn offers ample examples of people who should have been German according to our assessment, but didn't speak a word of the language (Mühlen 1969: 649).

Spoken language did not determine ethnicity so much as it did status. No class was ethnically homogeneous, but class solidarity transcended ethnic differences. The latter only surfaced when they took the shape of an inter-class conflict, i.e. mainly in peasants' struggles against the nobility. Molière's bourgeois aristocrat is a later phenomenon that already portends the collapse of the walls and approaching revolution. The urban middle class didn't dream of becoming aristocrats, not only because of the unsurpassable class barrier, but above all because of its immense disdain for the nobility. A citizen's pride was no less than that of a nobleman, and the same could be said for a peasant. One theatrical illustration of the latter is a letter from one peasant of Meremõisa, Harju-Madise, to his manor lord: an invitation to a duel (Johansen 1951: 206). Let us return momentarily to Tonnies Berch Tallepoiss, who in later documents could have been recorded as 'Antonius Berg' (*Tallepoiss < talupoeg < peasant*) and thus would have been left out of our voyage, along with the silent majority. Ethnicity was recorded by fortune alone in the Middle Ages. And yet, as one historian notes: 'It's not surprising that Estonian captains could play such an important role in trade (even appearing in Danzig and Lübeck, apparently captaining German ships) if we recall their role as merchant sailors and Baltic "pirates" in the pre-conquest era' (Mühlen 1969: 649).

* * *

Often, spirits were heard telling of flaws in a ship that the men themselves hadn't noticed.

* * *

SILVERWHITE

Once, there were two ships anchored almost side by side on the Pärnu River. The first was crewed by Saaremaa men, the other by men from Pärnu. One night, the Pärnu man on watch heard two men talking near the anchor winch. He crept to the wall of the forecastle and saw two men in blue overalls sitting on the gunwales and chatting. The watch heard one of the men say: 'Our old guy (captain) is a good man – he keeps the ship in such fine order that it's a thrill to sail, and he feeds wonderfully to boot!' – 'Damn, we're so poorly off here that I don't want to be on the ship at all. In the last storm, I had to hold on to the main mast as tight as I could to keep from falling overboard. And the main mast's as rotten as soup! Mark my words: if he doesn't replace the mast, then I'll knock it over in the next storm!' The watch immediately realised they were Klabautermänner. Next morning, he told the captain what he'd heard. They checked and found that the main mast was indeed rotten. So, they did some repairs and put up a new one and set sail once again. (Pärnu)

FROM CAPTAIN TO ADMIRAL

Emir al-bahar is an Arab title that means 'commander of the seas'. Europeans simplified it to 'admiral'. This highest naval rank first appeared on the Mediterranean in 1181, and in Tallinn as *amirali* in 1451 (Hansen 1885). The first Estonians deemed worthy of the title would flourish soon after. Let us settle for a dotted line connecting the invisible roots of the primeval tree to its rustling crown, from Pujalke purchasing salt in La Rochelle to the first commanders of the seas. When did Tallinn's six-striped naval flag first cross the equator? When did the first Estonian vessel reach America? Which of our seawolves were Cape Horners? These questions will remain unanswered. They are not part of our Silverwhite waterway chronicle.

The ancient Silverwhite burst into life anew for nearly half a century. With a single quill stroke, Tsar Ivan IV connected the Volga to the Narva in 1569. It was not a pale shadow of the former waterway, but rather the distant dawn of a century of light. The Company of Merchant Adventurers of London was granted the right to trade in Narva, Kazan, and Astrakhan (Platonov 1922). The prehistoric waterway, nature's careless gift to those who could use and appreciate it, was reopened and, among much else, demonstrated the stately talent of such a cruel and controversial ruler. Cartographers sailed at the seafarers' heels. Reports of the remote Arctic

THE FIRST CAPTAINS

Sea, the tempting Northeast Passage, and the Orient, which had abruptly come closer to reality and legends alike, seeped through Kuressaare and Narva to the West. To this day, one branch of the Obi River is still called the Indian Channel. World exploration could begin again.

Again? Sinbad the Sailor and Pujalke never perished. They have work to do today and will have work to do tomorrow, for they embody a human quality that can only disappear along with humankind itself.

Times occasionally change faster than people, and the experience is a bitter one. The same was especially true for Tallinn's seafarers. In their glory years, the captains' flags flapped atop the masts of forty large ships – masts taller than nine-storey apartment blocks. By the mid-17th century, however, Tallinn's merchants were only able to acquire a single seagoing vessel after many a year, and Hans Ohm began building a 500-tonne pine ship in Kolga Bay, which would be captained by Simon Sitke's son Peeter (Soom 1971: 189).

The reasons for this decline extend beyond the bounds of the Silverwhite chronicle, and we will not pause upon them here. Or will we?

A new wind would need to billow new sails from the east, the direction of the Silverwhite waterway. And it did.

Meanwhile, Tallinn's seamen had been jealously watching the growth of Riga and St Peterburg. Both were connected to their surrounding lands by waterways. Fanciful plans were hatched to turn Lake Ülemiste into an interior-waterway port connected to Lake Peipus and Russia via the Pirita or Pedja rivers (Pullat 1966: 50).

Shipbuilding and sailing skills survived in rural harbours during these hard times. Already in the late 18th century, shipbuilders on Hiiumaa constructed seagoing vessels of over 100 tonnes, including two 250-tonne brigs. Does that make you smirk? It shouldn't. Shipbuilder Erik Malm's *Rurik*, on which a young Otto von Kotzebue would soon sail around the world, wasn't greater than 180 tonnes, and Vitus Bering's ships were even half that size. The fresh gust of a new era, one filled with the acrid smell of birch smoke, reached Tallinn's port in 1827 when its first steamship set sail (Schlegel 1819–34: IX, 54–5). Nine years later, a regular route to Helsinki was opened with the 45-horsepower steamships *Storfursten and Fürst Menschikow*, continued in 1966 by the 4,000-horsepower motorised *Tallinn*. The era of ironclads wasn't far: built in 1869, the *Dorpat* was launched and – egads! – floated (Estonian Maritime Museum Catalogue 1937, no. 408, 14).

SILVERWHITE

Tallinn was forced to swallow yet another bitter experience. Pärnu outpaced the city, which had to settle for second place and the title of 'provincial port'. But then the railroad reached Tallinn and the ancient Silverwhite waterway's gates opened at a dizzying speed: a mere decade later, Tallinn was competing with St Petersburg for first place. By the eve of the Russian Revolution, every eighth future sailor of the Russian Empire was trained in an Estonian maritime school at either Heinaste (Ainaži, founded 1864), Narva (1873), Paldiski (1875), Käsmu (1884), or Kuressaare (1891) (K.L. 1938; Past 1935: 172).

Seaways are our life's axis.

On 27 April 1803, the *Revalsche Wöchentliche Nachrichten* (Tallinn Weekly News) published a brief report: 'As I plan to embark on travels, I would ask all individuals who believe they have any demand to make of me to inform me at once. I live in the Tiedemann House on Karja St, no. 470. D. Carl Espenberg.' And so began the first trips around the world. The Estonian ship's doctor Karl von Espenberg accompanied Adam Johann von Krusenstern to Kamchatka, and along with them sailed the young Fabian Gottlieb von Bellingshausen and Otto von Kotzebue. Estonian sailors Siimon Tauts, Paul Jakobson, and Olev Rannakopli surveyed the outline of Antarctica along with Bellingshausen.

From Kotzebue's notes:

Today, January 22nd [1815], at three in the afternoon, I bid farewell to Tallinn, my city of birth, and began our march towards Turku with the crew. The government permitted me to recruit the very best men and I found more than enough necessary volunteers who, with genuine ardour, wish to attempt to accomplish all that is possible with me. Such fearless souls were welcome, of course, and filled me with joyful hope for the future [...]. When Tallinn was behind us, I felt that the first step had been taken in an august endeavour.

Kotzebue's crew re-enacted a traditional peasant wedding when crossing the equator.[1]

So far as we know, the premiere of *Ter Talkus*, a play written by the captain's father, August von Kotzebue, took place in Pärnu in 1816. Or could it have been one year earlier at the equator? Does the chronicle of the Silverwhite waterway ultimately intersect with Estonian theatre history?

New winds, new ships, new destinations.

Are they really new?

434

THE FIRST CAPTAINS

Roots intertwine in the dusky past, drinking the juices of life from common soil.

* * *

This is a long procession of persons and events, the beginning and end of which disappear from sight. Saaremaa ship funerals, Pujalke, Juminda Kaubipoike and admirals are mere links in a lengthy chain that will be continued tomorrow by those who are still dressed in school uniforms today.

> But if a link were lost,
> the thought would melt like frost.

The sea is too dynamic as an economic, political, psychological, and emotional factor to vanish quietly from the lives of this coastal people. There has not been the time. Perhaps the lie of 'oppressive, hopeless, blind slavery' has been the most dangerous of all used to try to extinguish the Estonian spark, for if we were to turn our backs on the past, then we would also be turning them on the future. History does not have merit in and of itself, and is anything but its most uninspired definition: a chain of events that once happened and then froze in place for all eternity. History is alive and transforms because it does not exist outside the person – it finds its point or pointlessness in the present, in our acts or inactions, in us ourselves, just as we in turn perpetually rework the blank fabric of time into history, adding our own colours and patterns which a century or five centuries from now will have blended beyond recognition with the historical marks of old Pujalke or of a young modern-day sailor.

Teen mina laeva lagleluista,	I make a ship like a black goose,
pardad teen mina pardiluista,	make gunwales like a duck,
aerud teen hanetiibadest …	make oars like a grey goose …
Siis lä'än laintel laksotama	Then I go rocking on the waves,
poolest merda pühkimaie,	sweeping half the sea,
siis lään merele mehele.	then I take the sea as my husband.
Meri meid söödab, meri meid joodab,	The sea gives us food, the sea gives us drink,
laine annab lapsukesi.	the wave gives little children.

XII

WHERE DOES POETRY END,
WHERE DOES HISTORY BEGIN?

WHERE DOES POETRY END, WHERE DOES HISTORY BEGIN?

Nõmme, Tallinn, 10 November 1975

Returning from my summer travels, I find the edited *Silverwhite* manuscript on my desk. The book is taking shape, and our paths are now parting. Poetry's ways, wind's ways, author's and book's ways. New characters were added and a few facts clarified meanwhile.

A gale struck Bremen in 1380, seizing a nearly finished cog from its slip and sinking it in the Weser River along with tools that had been left on board. After being raised, reconstructed, and installed in the German Maritime Museum in Bremerhaven in 1962, the vessel (23.5 metres long, 7.5 metres wide) confirms earlier hypotheses. Unlike the ancient Baltic ship, earlier (13th-century) cogs had the rudder not on the starboard, but on the port side in the Celtic nautical tradition, and the manner in which ship's nails were used implies the same direction of borrowing.

Researchers have published new opinions in favour of identifying Tacitus's Aesti with the ancient Prussians, and Heinz von zur Mühlen presented a considerable argument against *guild* being borrowed from the Baltic-Finnic *kild*. Nevertheless, he adds (Heinz Mühlen to the author, 15.9.1975):

Estonians gathering into associations similar to guilds in the pre-conquest era is a topic worth researching, though it may be difficult to go further than hypotheses. I'm most convinced by Miller's emphasis of the principle of mutual solidarity among guests of the Holy Body, including at sea. This allows us to view it as a voluntary association for self-defence primarily abroad. Such a union may only have pulled on a religious tunic later.

Tallinn's Old Town has been archaeologically researched less than any other capital on the Baltic Sea: about 0.3 per cent of the 35 hectares. The question anchoring every subsequent link still slips through our fingers: 'Did a city-like structure develop in the 10th, 11th, or 12th century?' (Jaan Tamm, 1974). Archaeology alone can provide an answer, not poetry.

Poetry, still? And what remains? It's difficult for one to weigh oneself in the palm of one's own hand.

I hope that the reader will leave with three conclusions remaining.

The meteoric origin of material found by Ivan Reinwald in the Kaali crater was only conclusively proven by laboratory analyses in London and Tartu (Leonard James Spencer, Andres Väärismaa) in 1938. Using Estonian lexicology, Andrus Saareste believed in 1935 that he had found traces of a cosmic catastrophe. It is a handsome example of the bridge connecting human and pure sciences, one much more striking than Percival Lowell discovering Pluto in an inkwell, though we were unable to even see that bridge itself! The walls dividing the sciences grow ever taller as our knowledge of them deepens. The scope of the Kaali event has now been precisely determined. It was the latest of its magnitude on the European continent. I believe that it was imprinted on the minds, languages, and worldviews of all peoples who lived along the Baltic Sea. This seems more probable than the opposite supposition: that such an extraordinary event could vanish into oblivion without leaving a single trace.

I believe that the role the sea played in Estonian cultural development is somewhat greater than what we have previously assumed. Mountains, tundras, and bare steppes echo in the poetry and handicraft of different nations without their members even realising it. A gaze from outside is necessary. Just glance at Asia, America, Australia: Estonia is a narrow strip of coastal land, almost a peninsula, two sides bounded by the sea.

I believe that cultural contacts are not a new phenomenon, but a new quality. They have always been functioning, as far as the eye can see. They may be scattered throughout history as sparsely as cosmic dust – time's flow gives them mass and weight. Against the backdrop of a common cultural history, we should observe permanent and temporal phenomena in national characteristics: ones that cannot be absolutised or stored in a jar, much less filtered pure of outside influences so as to arrive at some kind of genuine proto-culture. This nostalgically hued longing renders

WHERE DOES POETRY END, WHERE DOES HISTORY BEGIN?

sterile the fruit of imagination that has never truly existed. The passion of development and discovery lies in the observation and conscious advancement of cultural contacts.

* * *

'Dad, what's the point of history?'

* * *

While writing about winds and ancient folklore, I've sometimes, albeit rarely, also considered history. The past is not yet history, and fact is not yet truth. To be honest, they have quite little in common. History is a science, though it is hard to define. Unlike biology, it studies humans as a social phenomenon and does so using a scientific method that we call general sociology. The outcome is a statement: it might have been like that, and likely was. In a book of poetry, we're limited by a simpler question: could it have been that way? A few months before his heroic death, Marc Bloch, a famous Frenchman whose historical experience forced him to take up arms in a partisan struggle against fascism, wrote:

> It is sometimes said: 'History is the science of the past.' To me, this is badly put. For, to begin with, the very idea that the past as such can be the object of science is ridiculous. How, without preliminary distillation, can one make of phenomena, having no other common character than that of being not contemporary with us, the matter of rational knowledge? On the reverse side of the medal, can one imagine a complete science of the universe in its present state? (Bloch 1963)

Leaving definition aside: why, then, history? Figuratively, you could put it like this. We know from experience that the faster the engine, the stronger the brakes. The faster the development, the more crucial its tie to the past. If only yesterday, the most immediate past, were to determine the present, it would result in a society so labile that it would lose its skeleton and collapse. This has not happened; hence, a connection to the past is one trait of humans and human society. Our biological traits are hereditary; our social ones are (so far) not. We know of fate's cruel experiments on humans who have grown up outside a social environment, such as among wolves. Despite every effort, these individuals remained at an animalistic level even after returning to a social environment as adults. Thus, a living connection to the past is an inseparable characteristic of humans as social

SILVERWHITE

beings and has influenced humankind's development for millions of years. We may derive a simple truth: the faster the development, the more important history becomes.

To conclude a book of poetry with a poem: if you are not a tourist, if you are a traveller, then go and stroll past a 500-year-old castle of the Teutonic Order and pause before a field of grain that has been farmed perpetually for over two millennia. Take a nice photo of it. The history of that field is greater, as it is inhabited by both the past and the future. History begins where poetry ends. Sometimes, they brush one another:

> Who built seven-gated Thebes?
> The books keep the names of kings.
> Was it kings who hauled the chunks of rock?
> And Babylon, so many times destroyed –
> Who rebuilt it so many times?
> In what houses of gold-gleaming Lima did its builders live?
> Where did the masons go that evening
> When the Great Wall of China was finished? Great Rome
> Is full of triumphal arches. Who erected them? Over whom
> did the Caesars triumph? Had Byzantium,
> much praised in song, only palaces for its inhabitants? Even in
> fabled Atlantis, the night the ocean engulfed it, the drowning
> still called out for their slaves.
>
> The young Alexander conquered India.
> Was he alone?
> Caesar defeated the Gauls.
> Did he not even have a cook with him?
> Philip of Spain wept when his armada
> Went down. Was he the only to weep?
> Frederick II won the Seven Years War. Who
> Besides him won it?
>
> On every page, a victory.
> Who cooked the feast for the victors?
> Every ten years, a great man.
> Who paid the costs?
>
> So many reports.
> So many questions.
> (Bertolt Brecht, 'Questions from a Worker Who Reads')[1]

440

AFTERWORD TO *SILVERWHITER*

Nõmme, Tallinn, 31 December 1982

Silverwhiter (1984) was not a sequel to *Silverwhite* (1976) but its introduction and first chapter. When it turned out that Pytheas of Massalia's solar observations could be mathematically analysed and accurately positioned, I initially thought I would stick to the calculations' results. However, I discovered they could be used to elucidate the history of Baltic exploration along with Estonian folklore, language, customs, cultural contacts, and seafaring. I reckoned that some details could appropriately step out of unwritten literature and enter the precisely datable pages of written literature.

I'd like to add that Thule and Troy formed a pair, whose golden era passed in the nineteenth century. Addressing folklore as a historical source withdrew before critical methods. Heinrich Schliemann was an outlier and even a ridiculous figure until the city of Troy began to emerge under shovels. Alexander Ziegler had no such luck on the Shetland Islands. Nevertheless, he still joined Alexander von Humboldt, who equated Thule with Shetland. The body of literature is immense. 'Thule' is still a household name today.

Translation by Adam Cullen, 9 February 2024, Haapsalu, Estonia

pp. [1–202]

NOTES

I. SHIPS TAKE FLIGHT

1. Naissaar is a large island separating the Bay of Tallinn from the Gulf of Finland.
2. The Daughter Islands (Estonian *Tütarsaared*, Finnish *Tytärsaari*) lie between Estonia and Finland in the eastern Gulf of Finland.
3. Throughout this translation, I use the words Sámi and Sápmi in place of the dated Lapps and Lapland, respectively, and to avoid confusion with the section that explores the Estonian etymology of the related term.
4. *Odamus* (Finnish *Otava*) is the ancient Estonian asterism corresponding to the Big Dipper.
5. Revala (also Revälä, Revele, Reuælæ, Revalia, etc.) was an ancient county of north Estonia, whose stronghold was also known by many names before becoming present-day Tallinn.

II. BENEATH TWO SUNS

1. In Estonian folklore, a *tulihänd* is a magical being constructed of random everyday objects that flies as a spark or a flash to gather valuables for its owner.
2. An archaic unit measured by the distance between an outstretched thumb and index or middle finger.

V. RED SAILS

1. *Taar* was a traditional fermented flour- or bread-based drink similar to kvass.
2. The Seto are a Finnic people, similar to Estonians but linguistically and culturally distinct, who live in southeast Estonia.

VI. THE SILVERWHITE WAY

1. https://muslimheritage.com/al-muqaddasi-the-geographer-from-palestine/, accessed 6 January 2023.
2. https://www.degruyter.com/document/doi/10.18574/nyu/9781479826698.003.0011/html.

VII. 'THEIR GREATEST MIGHT IS IN THE SHIPS'

1. *The Ynglinga Saga*, http://mcllibrary.org/Heimskringla/ynglinga.html.

443

NOTES

XI. THE FIRST CAPTAINS

1. Jaan Undusk to the translator (2.3.2024): 'I reviewed the German- and Russian-language accounts of Kotzebue's first voyage around the world (1815–1818). I can't say which came first, as he wrote in both languages. However, the two provide some clarity. Firstly, the "peasants' wedding" wasn't performed when crossing the equator, but later, deep in the southern hemisphere, at the 35th meridian west and the 18th parallel south, i.e. in the Atlantic about 1,500 kilometres northeast of Rio de Janeiro ('near Rio de Janeiro' from an oceanic perspective). Secondly, it was a sketch that the sailors invented to make mundane ship life a little jollier. The event was accompanied by ample sauerkraut provided by an American company (a "fresh" luxury on the vessel!) and even more alcohol than usual: "A play was put on that evening. At midday, a notice was nailed to the main mast informing us that a 'peasants' wedding' (*Bauerhochzeit*) was to be performed. The sailors put the play together themselves and the audience was greatly satisfied; in the end, a ballet was performed, and the actors received the applause they deserved." Based on this, one can assume that the sailors were familiar with peasant life and able to parody it as well.'

XII. WHERE DOES POETRY END, WHERE DOES HISTORY BEGIN?

1. Adapted from a translation by A.Z. Foreman (https://poemsintranslation.blogspot.com/2021/10/bertolt-brecht-questions-from-worker.html) and an anonymous translation (https://www.marxists.org/archive/brecht/works/1935/questions.htm).

BIBLIOGRAPHY

Aaloe, A., A. Liiva and E. Ilves. 1963. 'Kaali meteoriidikraatrite vanusest', *Eesti Loodus* (Tartu, Tallinn), no. 5.

Aalto, P. and T. Pekkanen. 1975, 1980. *Latin Sources on North-Eastern Eurasia*, 2 vols. Wiesbaden.

Agroklimatitšeski atlas. 1972. Moscow.

Alamaa, E. and A. Kivi. 1966. *Tallinn: linna asustus- ja ehitusajaloolisi materjale*, 7 vols. Tallinn.

Aleksejev, N.N. 1972. *Istoričeskie svjazi mordovskogo fol'klora i fol'klora slavjanskih narodov: problemy izučenija finno-ugorskogo fol'klora.* Saransk.

al-Yaʿqubi. 2018. *The Works of Ibn Wāḍiḥ al-Yaʿqūbī*, edited by Matthew S. Gordon, Chase F. Robinson, Everett K. Rowson, and Michael Fishbein, 3 vols. Leiden.

Amelungen, F. and G. Wrangell. 1930. *Geschichte der Revaler Schwarzhäupter: ein Beitrag zur Geschichte des deutschen Kaufmanns im Osten.* Tallinn.

Anderson, W. 1923. *Nordasiatische Flutsagen.* Tartu.

Annist, A., J. Peegel, E. Laugaste, V. Pino, H. Laidvee and E. Piir. 1961–3. *Fr R. Kreutzwald: Kalevipoeg; tekstikriitiline väljaanne ühes kommentaaride ja muude lisadega*, 2 vols. Tallinn.

Antiquités russes. 1850–2. *Antiquités russes d'après les monuments historiques des Islandais et des anciens Scandinaves, editées par la Société Royale des Antiquaires du Nord*, 2 vols. Copenhagen.

Apollonius of Rhodes. 1912. *The Argonautica*, translated by R.C. Seaton. London.

Ariste, P. 1956. 'Läänemere keelte kujunemine ja vanem arenemisjärk', in *Eesti rahva etnilisest ajaloost.* Tallinn.

Ariste, P. 1959. 'Läänemerelaste vanast merevaigu nimetusest', *Etnograafia Muuseumi Aastaraamat* (Tallinn), vol. 16.

Ariste, P. 1969. *Vadja rahvakalender.* Tallinn.

Ariste, P. 1972. 'Sõna sealt, teine tealt', *Sõnasõel*, vol. 1. Tartu.

Atlas okeanov. 1977. 'Atlantičeskiĭ i Indiĭskiĭ okeany'. Moscow.

Bakhrushin, S.V. 1922. *Istoričeskiĭ očerk zaselenija Sibiri do poloviny 19 veka: očerki po kolonizacii Severa i Sibiri*, vol. 2. Petrograd

Bakka, E. 1971. 'Scandinavian Trade Relations with the Continent and the British Isles in Pre-Viking Times', *Early Medieval Studies* (Stockholm), vol. 3, no. 5; *Antikvariskt arkiv*, vol. 40, no. 5.

Biezais, H. 1967. 'Bischof Meinhard zwischen Visby und der Bevölkerung Livlands', in *Kirche und Gesellschaft im Ostseeraum und im Norden vor der Mitte des 13. Jahrhunderts*, edited by S. Ekdahl. Visby, 1969.

BIBLIOGRAPHY

Bloch, Marc. 1963. *The Historian's Craft*, translated by Peter Putnam. New York.

Blomqvist, R. 1963. 'Die Anfänge der Stadt Lund', in *Die Zeit der Stadtgründung im Ostseeraum*, edited by M. Stenberger. Visby, 1965.

Bodnarski, M.S. 1953. *Antičnaja geografija: kniga dlja čtenija*. Moscow.

Bonnell, E. 1862. *Russisch-Liwländische Chronographie von der Mitte des neunten Jahrhunderts bis zum Jahre 1410*. St Petersburg.

Braudel, F. 1981. *The Mediterranean and the Mediterranean World in the Age of Philip II*, 2 vols. London.

Brennsohn, I. 1922. *Die Aerzte Estlands vom Beginn der historischen Zeit bis zur Gegenwart: ein biografisches Lexikon nebst einer historischen Einleitung über das Medizinalwesen Estlands*. Riga.

Bruns, F. and H. Weczerka. 1962, 1967. *Hansische Handelsstrassen: Atlas, Textband*. (Quellen und Darstellungen zur hansischen Geschichte, vol. 13, parts 1–2). Cologne, Graz, and Weimar.

Busch, N. 1921. 'Zur Kosmographie des Aeticus Ister', I, *Mitteilungen aus der livländischen Geschichte* (Riga), vol. 21, no. 3.

Buxhöwden, P.W. von. 1838. *Beiträge zur Geschichte der Provinz Oesell*. Riga and Leipzig.

Cahen, M. 1921. 'Le mot "dieu" en vieux scandinave', I, *Collection linguistique: publiée par la Société de linguistique de Paris* (Paris), vol. 10.

Calissendorff, K. 1971. 'Place-Name Types Denoting Centres', *Early Medieval Studies* (Stockholm), vol. 3, no. 1; *Antikvariskt Arkiv*, vol. 40, no. 1.

Christensen, A.E. 1963. 'Über die Entwicklung der dänischen Städte von der Wikingerzeit bis zum 13. Jahrhundert', in *Die Zeit der Stadtgründung im Ostseeraum*, edited by M. Stenberger. Visby, 1965.

Churchill, W. 1957. *A History of the English-Speaking Peoples*. London.

Clark, J.G.D. 1953. *Doistoričeskaja Evropa* [translation of *Prehistoric Europe: The Economic Basis*]. Moscow.

Collinder, B. 1955. *Fenno-Ugric Vocabulary: An Etymological Dictionary of the Uralic Languages*. Stockholm.

Cook, James. 1803. *Jakob Cook's sämmtliche Reisen um die Welt*, 3 vols. Vienna.

Dalin, O. 1756. *Olof Dahlins Geschichte des Reiches Schweden*, part 1. Greifswald.

Daniel, H.A. 1859–63. *Handbuch der Geographie*, 3 vols. Frankfurt am Main and Stuttgart.

Delhaye, J. 1956. 'A propos de la Grande Ourse, *L'Astronomie*, February.

Diesner, H.J. 1978. *The Great Migration (4th to 7th Century)*. Leipzig.

Draviņš, K. 1960. Der Glauben an Zauberei und Drachen in Lettland im 17. Jahrhundert. *Årsbok 1955/56*. Lund.

Dreijer, M. 1960. *Häuptlinge, Kaufleute und Missionare im Norden vor tausend Jahren: ein Beitrag zur Beleuchtung der Umbildung der nordischen Gesellschaft während der Übergangszeit von Heidentum zum Christentum*. Mariehamn, Finland.

Ebel, W. 1963. 'Über skandinavisch-deutsche Stadtrechtsbeziehungen im Mittelalter', in *Die Zeit der Stadtgründung im Ostseeraum*, edited by M. Stenberger. Visby, 1965.

Eisen, M.J. 1894a. *Kolmas Rahwa-raamat: uus kogu wanu jutte*. Jurjev and Riga.

Eisen, M.J. 1894b. *Neljas Rahwa-raamat: uus kogu wanu jutte*. Jurjev and Riga.

Eisen, M.J. 1919. *Eesti mütoloogia*, vol. 1. Tallinn.

BIBLIOGRAPHY

Eisen, M.J. 1922. 'Esivanemate ohverdamised', *Eesti mütoloogia*, vol. 3.

Eisen, M.J. 1927. 'Eesti vana usk', *Eesti mütoloogia*, vol. 4.

Ekblom, R. 1925. 'Kolyván(n?): une contribution à l'histoire des noms de la capitale de l'Estonie', *Uppsala universitets årsskrift: Språkvetenskapliga sällskapets i Uppsala förhandlingar 1925–1927*. Uppsala.

Ekblom, R. 1931. 'Idrīsī und die Ortsnamen der Ostseeländer', *Namn och bygd: Tidskrift för nordisk ortnamnsforskning* (Uppsala), vol. 19.

Estonian Maritime Museum Catalogue. 1937. Tallinn.

Evans, James and J. Lennart Berggren. 2006. *Geminos's Introduction to the Phenomena: A Translation and Study of a Hellenistic Survey of Astronomy*. Princeton, NJ.

Feyerabend, K. 1801. *Kosmopolitische Wanderungen durch Preussen, Kurland, Liefland, Litthauen, Vollhynien, Podolien, Gallizien und Schlesien in den Jahren 1795–1798: in Briefen an einen Freund*, part 3. Danzig.

Finlay, Alison. 2004. *Fagrskinna: A Catalogue of the Kings of Norway*. Leiden.

Finsch, O. 1879. *Reise nach West-Sibirien im Jahre 1876*. Berlin.

Frähn, C.M. 1823. *Ibn-Foszlan's und anderer Araber Berichte über die Russen älterer Zeit: Text und Übersetzung mit kritisch-philologischen Anmerkungen, nebst drei Beilagen über sogenannte Russen-Stämme und Kiew, die Warenger und das Warenger-Meer, und das Land Wisu, ebenfalls nach arabischen Schriftstellern*. St Petersburg.

Frähn, C.M. 1848. 'Der orientalische Münzfund von Essemeggi in Estland: ein Nachtrag zu der topographischen Uebersicht den Ausgrabungen von alten Arabischen Gelde in Russland', *Bulletin de l'Académie impériale des Sciences de St.-Pétersbourg*, vol. 5, no. 8.

Frey, E. 1905. 'Der Grabstein des Heidenreich Sawijärwe', *Sitzungsberichte der gelehrten estnischen Gesellschaft*. Tartu, 1906.

Friedenthal, A. 1931. 'Die Goldschmiede Revals, *Quellen und Darstellungen zur hansischen Geschichte* (Lübeck), NS, vol. 8.

Gahlnbäck, J. 1926. 'Zinn bei den Esten und Finnen nach den Mythen des Kalewipoeg und des Kalewala und den noch erhaltenen Zinngegenständen späterer Zeit', *Sitzungsberichte der gelehrten estnischen Gesellschaft*. Tartu.

Gardberg, C.J. 1963. 'Über die älteste Geschichte der Stadt Åbo (Turku)', in *Die Zeit der Stadtgründung im Ostseeraum*, edited by M. Stenberger. Visby, 1965.

Genealogisches Handbuch der baltischen Ritterschaften, 'Oesel'. 1935–9. Tartu.

Gordeyev, N.P. 1973. 'Proishoždenie ètnonima "kolbjagi"', VI Vsesojuznoĭ Conference on the Study of the Scandinavian Countries and Finland, part 1. Tallinn.

Gordin, Isak. 1967. *Vanhoja Suomen karttoja / Old Maps of Finland*. Helsinki.

Greiffenhagen, O. 1927–8. 'Revals Wappen und Flaggen', *Beiträge zur Kunde Estlands* (Tallinn), vol. 13, nos. 1/2–3, 5.

Grekov, B.D. 1953. *Drevnjaja Rus': očerki po istorii SSSR*, vols. 9–15. Moscow.

Gustavson, H. 1969. *Meditsiinist vanas Tallinnas kuni 1816. a.* Tallinn.

Gustavson, H. 1972. *Tallinna vanadest apteekidest kuni 1917. a.* Tallinn.

Haack, 1968. *Haack Grosser Weltatlas. Kartenteil, Register*. Gotha and Leipzig.

Haavio, M. 1965. *Bjarmien vallan kukoistus ja tuho: historiaa ja runoutta*. Porvoo and Helsinki.

BIBLIOGRAPHY

Halikova, E. 1970. 'K voprosu o finskom komponente v formirovanii naselenija volžskoĭ Bulgarii', *Third International Congress of Finno-Ugric Scholars, Tallinn, 17–23 August 1970*, vol. 2. Tallinn.

Hämäläinen, A. 1938. 'Arapialaisten kirjailijain kuvaukset Itä-Euroopan kansoista, etenkin niiden poltto- hautauksista', *Kalevalaseuran vuosikirja* (Porvoo, Helsinki), vol. 18.

Hansen, G. von. 1885. *Die Kirchen und ehemaligen Klöster Revals*. Reval.

Heinsius, P. 1956. 'Das Schiff der Hansischen Frühzeit', *Quellen und Darstellungen zur hansischen Geschichte* (Weimar), NS, vol. 12.

Helmersen, L. von. 1876. 'Zur Streitfrage über das Wrangell-Land', *St. Petersburger Zeitung* (St Petersburg) no. 125, 12 (24); no. 126, 13 (26); no. 127, 15 (27). Reprinted as 'K voprosu ob otkrytii Vrangelevoĭ zemli', *Izvestija Imperatorskogo Russkogo Geografičeskogo Obšestva*, vol. 12, no. 6.

Hennig, R., trans. 1961. *Nevedomye zemli*, 4 vols. Moscow.

Henry of Livonia. 2004. *The Chronicle of Henry of Livonia*, translated by James A. Brundage. New York.

Herodotus. 1920. *Herodotus, with an English translation by A.D. Godley*. Cambridge, MA.

Herteig, A.E. 1963. 'Zur Frage der Stadtentwicklung: mit archäologischen Beispielen aus Norwegen', in *Die Zeit der Stadtgründung im Ostseeraum*, edited by M. Stenberger. Visby, 1965.

Hildebrand, H. 1887. *Livonica, vornämlich aus dem 13. Jahrhundert, im Vaticanischen Archiv*. Riga.

Hillkowitz, K. 1934, 1973. *Zur Kosmographie des Aethicus*, 2 vols. Bonn and Cologne; Frankfurt am Main.

Höhlbaum, C. 1874. 'Aus Revals Mittelalter: kulturhistorisches', *Beiträge zur Kunde Ehst-, Liv- und Kurlands* (Reval), vol. 2, no. 1.

Holmberg, U. 1914. *Tsheremissien uskonto*. Porvoo.

Holmqvist, W. 1970. 'Helgösymposium', *Early Medieval Studies* (Stockholm), vol. 1, no. 1; *Antikvariskt arkiv*, vol. 38, no. 1.

Holzmayer, J.B. 1867. *Das Kriegswesen der alten Oeseler: nebst einem Anhang und zwei Tafeln*. Arensburg.

Holzmayer, J.B. 1872. 'Osiliana', *Verhandlungen der gelehrten estnischen Gesellschaft* (Reval), vol. 7, no. 2.

Homich, L.V. 1966. *Nentsy: istoriko-etnografičeskie očerki*. Moscow and Leningrad

Hupel, A.W. 1774–82. *Topographische Nachrichten von Lief- und Ehstland*, 3 vols.. Riga.

Hupel, A.W. 1781. 'Kurlands alter Adel und dessen Landgüter oder Kurländische Adelsmatrikul und Landrolle: nebst andern kürzern Aufsätzen', *Der nordischen Miscellaneen* (Riga), 3 parts.

Hupel, A.W. 1795. *Idiotikon der deutschen Sprache in Lief- und Ehstland: nebst eingestreueten Winken für Liebhaber*. Riga.

Ibn Battuta. 1829. *The Travels of Ibn Batuta*, translated by Samuel Lee. London.

Ibn Fadlan. 2012. *Ibn Fadlān and the Land of Darkness: Arab Travellers in the Far North*, translated by Paul Lunde and Caroline Stone. London.

BIBLIOGRAPHY

Jaanits, L. 1975. 'Merevaigu esmasest levikust läänemere-soomlastel', *Tartu Riikliku Ülikooli toimetised* (Tartu), no. 344.

Jakobson, G., A. Kivi, H. Lond and A. Soik. 1967. *Tallinna vesi ja sajandid: 550 aastat Tallinna veevarustust.* Tallinn.

Jalasto, H. 1960. *Hiiumaa.* Tallinn.

Janin, V.L. and M.H. Aleškovski. 1971. 'Proishoždenie Novgoroda (k postanovke problemy)', in *Istorija SSSR*, vol. 2.

Jelnitski, L.A. 1962. *Drevneĭšie okeanskie plavanija.* Moscow.

Jensen, J. 1933. *Eesti ajaloo atlas.* Tartu.

Jevsejev, V. 1972. *Istoričeskie vzaimosvjazi fol'klora finno-ugorskih narodov: problemy izučenija finno-ugorskogo fol'klora.* Saransk.

Johansen, P. 1933. *Die Estlandliste des Liber Census Daniae ...*, 2 vols. Copenhagen and Reval.

Johansen, P. 1951. *Nordische Mission, Revals Gründung und die Schwedensiedlung in Estland.* Stockholm.

Johansen, P. 1963. 'Die Kaufmannskirche', in *Die Zeit der Stadtgründung im Ostseeraum*, edited by M. Stenberger. Visby, 1965.

Johansen, P. 1964. 'Saxo Grammaticus ja Itä-Baltia', reprinted in *Trükis ilmunud: Historiallinen aikakauskirja* (Helsinki), vol. 1, 1965.

Johansen, P. and H. von zur Mühlen. 1973. *Deutsch und undeutsch im mittelalterlichen und frühneuzeitlichen Reval.* Cologne and Vienna.

Jung, J. 1876. *Liiwlaste würst Kaupo, ja sõdimised tema päewil, kui ka Liiwi rahwast ja nende kadumisest siin maal.* Tartu.

Juva, M., V. Niitemaa and P. Tommila. 1968. *Suomen historian dokumentteja*, vol. 1. Helsinki.

K.L. [A. Gustavson, K. Larens, and E. Past]. 1938. *Ülevaade Eesti merekoolidest: 1864– 1935.* Tallinn.

Kahk, J. and A. Vassar, eds. 1960. *Eesti NSV ajaloo lugemik*, vol. 1: *Valitud dokumente ja materjale Eesti ajaloost kõige vanemast ajast kuni XIX sajandi keskpaigani.* Tallinn.

Kalima, J. 1919. *Die Ostseefinnischen Lehnwörter im Russischen.* Helsinki.

Kangro-Pool, R. 1946. *Eesti teater algaastail.* Tallinn.

Kangro-Pool, R. 1968. 'Tallinna all-linna tekkeskeemi analoogiatest', *Ehitus ja arhitektuur* (Tallinn), no. 4.

Kanivets, V.I. 1970. 'Drevnee svjatiliŝe v Bol'še-zemel'skoĭ Tundre', Third International Congress of Finno-Ugric Scholars, Tallinn, 17–23 August 1970, vol. 2. Tallinn.

Kemppinen, I. 1970. 'Beziehungen des Glaubens der karelischen Vorzeit zu den ural-altaischen, indo-europäischen und zu den alten Religionen des nahen Ostens', Third International Congress of Finno-Ugric Scholars, Tallinn, 17–23 August 1970, vol. 2. Tallinn.

Kinnier Wilson, J.V. 1979. *The Rebel Lands: An Investigation into the Origins of Early Mesopotamian Mythology.* Cambridge.

Kiparski, V. 1936. *Fremdes im Baltendeutsch.* Helsinki.

Kivi, A. 1972. *Tallinna tänavad.* Tallinn.

BIBLIOGRAPHY

Kivikoski, E. 1955. 'Hämeen rautakausi', in *Hämeen historia*, vol. 1: *Esihistoria ja keskiaika*. Hämeenlinna, Finland.

Kivikoski, E. 1970. 'Zu den axtförmigen Anhängern der jüngsten Eisenzeit', in *Studia archaeologica in memoriam Harri Moora*, edited by Marta Schmiedehelm, Lembit Jaanits and Jüri Selirand. Tallinn.

Klyuchevsky, V.O. 1956, 1957. *Sočinenija: v vos'mi tomah*, 2 vols. Moscow.

Kohl, J.G. 1841. *Die deutsch-russischen Ostseeprovinzen oder Natur- und Völkerleben in Kur-, Liv- und Esthland*, 2 vols. Dresden and Leipzig.

Körber, M. 1899. *Oesel einst und jetzt*, vol. 2: *Die Kirchspiele Mustel, Kielkond, Anseküll, Jamma,Wolde und Pyha*. Arensburg.

Kovalevsky, A.P. 1956. *Kniga Ahmeda Ibn-Fadlana o ego putešestvii na Volgu v 921–922 gg: stat'i, perevody i kommentarii*. Kharkiv.

Krachkovsky, I.J. 1957. *Izbrannye sočinenija*, vol. 4. Moscow.

Kreutzwald, F.R. 2011. *Kalevipoeg: The Estonian National Epic*, translated by Triinu Kartus. Tartu and Tallinn.

Kruse, F. 1842. *Bemerkungen über die Ostsee-Gouvernements in Beziehung auf J. G. Kohl's deutsch-russische Ostsee-provinzen ...* Leipzig.

Kruse, F. 1859. *Necrolivonica oder Geschichte und Alterthümer Liv-, Esth- und Curlands griechischen, römischen, byzantinischen, nortmannischen oder Waräger-russischen, fränkischen, angel-sächsischen, anglodänischen Ursprungs ...* Leipzig.

Kruus, H. 1924. *Eesti ajaloo lugemik*, vol. 1: *Valitud lugemispalad Eesti ajaloo alalt 1561. aastani*. Tartu.

Kruus, H., ed. 1936. *Eesti ajalugu*, vol. 1: *Esiajalugu ja muistne vabadusvõitlus*, by H. Moora, E. Laid, J. Mägiste, and H. Kruus. Tartu.

Kruusberg, A. 1920a. *Esiisade enneajalooline õigus*, vol. 1: *Perekond*. Tartu.

Kruusberg, A. 1920b. 'Esiisade enneajalooline õigus', *Eesti kirjandus* (Tartu), nos. 3–4.

Kumlien, K. 1970. 'Der Historiker und das Birka-problem', *Early Medieval Studies* (Stockholm), vol. 1, no. 2; *Antikvariskt arkiv*, vol. 38, no. 2.

Kurgo, R. 1965. *Rannalautritest ilmameredele: lehekülgi purjelaevandusest Pärnu kandi rannavetes*. Tallinn.

Kurrik, Juhan. 2013. *Ilomaile*. Tartu.

Kuusi, M. 1963. *Suomen kirjallisuus*, vol. 1: *Kirjoittamaton kirjallisuus*. Helsinki.

Lang, A.W. 1968. 'Seekarten der südlichen Nord- und Ostsee: ihre Entwicklung von den Anfängen bis zum Ende des 18. Jahrhunderts', *Ergänzungsheft zur Deutschen hydrographischen Zeitschrift* (Hamburg), series B, 10.

Laugaste, E. and E. Normann. 1959. 'Muistendid Kalevipojast', in *Monumenta Estoniae antiquae*, vol. 2: *Eesti muistendid: hiiu- ja vägilasmuistendid*. Tallinn.

Laugaste, E. 1963a. *Eesti rahvaluuleteaduse ajalugu*, vol. 1: *Valitud tekste ja pilte*. Tallinn.

Laugaste, E., E. Liiv, and E. Normann. 1963b. 'Muistendid Suurest Tõllust ja teistest', in *Monumenta Estoniae antiquae*, vol. 2: *Eesti muistendid: hiiu- ja vägilasmuistendid*. Tallinn.

Laugaste, E. 1970. 'Der Parallelismus in den älteren estnischen Volksliedern', Third International Congress of Finno-Ugric Scholars, Tallinn, 17–23 August 1970, vol. 2. Tallinn.

Lelewel, J. 1852. *Géographie du moyen âge*. Brussels.

BIBLIOGRAPHY

Lewis, A.R. 1958. *The Northern Seas: Shipping and Commerce in Northern Europe A.D. 300–1100.* Princeton, NJ.

Liiv, O. 1937. 'Dekabristide mäss ja Eesti', *Nädal pildis* (Tallinn), nos.. 1–4.

Lönnroth, E., comp. 1958: *Kalevala.* Tallinn, 1959.

Lönnroth, E. 1963. 'Probleme der Wikingerzeit', in *Die Zeit der Stadtgründung im Ostseeraum*, edited by M. Stenberger. Visby, 1965.

Loorits, O. 1931. 'Der norddeutsche Klabautermann im Ostbaltikum', *Sitzungsberichte der gelehrten estnischen Gesellscahft* (Tartu).

Loorits, O. 1939, 1941. *Endis-Eesti elu-olu*, 2 vols. Tallinn.

Loorits, O. 1949. *Grundzüge des estnischen Volksglaubens*, 3 vols. Lund.

Lõugas, V. 1970a. 'Über die Kulturbeziehungen der Bevölkerung des estnischen Gebiets in der frühen Metallzeit', Third International Congress of Finno-Ugric Scholars, Tallinn, 17–23 August 1970, vol. 2. Tallinn.

Lõugas, V. 1970b. 'Sõrve laevkalmed', in *Studia archaeologica in memoriam Harri Moora*, edited by Marta Schmiedehelm, Lembit Jaanits and Jüri Selirand. Tallinn.

Lõugas, V. 1972. 'Lääne-Eesti rahvastiku kultuurist rooma rauaajal', *Eesti NSV Teaduste Akadeemia toimetised. Ühiskonnateadused* (Tallinn), no. 2.

Lõugas, V. 1973. 'Helgö: põhjamaade vanim linn', *Horisont* (Tallinn), no. 1.

LUB, 1853. *Liv-, Esth- und Curländisches Urkundenbuch: nebst Regesten*, edited by F.G. v. Bunge. Reval, Riga, and Moscow.

Lucan. 1896. *The Pharsalia of Lucan*, translated by Sir Edward Ridley. London.

Luce, J.W.L. von. 1827. *Wahrheit und Muthmassung: Beytrag zur ältesten Geschichte der Insel Oesel.* Pärnu.

Lumiste, M. and R. Kangro-Pool. 1969. 'Sõdalane, sarv ja päikeseketas Saare-Lääne varases raidkunstis', *Kunst: kujutava ja tarbekunsti almanahh* (Tallinn), no. 1.

Luts, A. 1970. 'Soomlaste silgukaubandus Viru rannikul', *Etnograafia muuseumi aastaraamat* (Tallinn), vol. 24.

MacCana. P. 1973. *Celtic Mythology.* London and New York.

Mägiste, J. 1951. 'Tšuudien kansallisuusnimen alkuperän ongelma ja suomen suudin, vir. suue 'kiila', *Virittäjä* (Helsinki), vol. 1, no. 2.

Mägiste, J., trans. and ed. 1962. *Henriku Liivimaa Kroonika.* Stockholm.

Mägiste, J. 1970. *Vanhan kirjaviron kysymyksiä: tutkielmia viron kirjakielen varhaisvaiheista 1200-luvulta 1500-luvun lopulle.* Helsinki.

Mark, J. 1936. 'Soome-ugri rahvaste kaubandusest', V Soome-Ugri Cultural Congress. Tallinn.

Marshak, A. 1972. 'Cognitive Aspects of Upper Paleolithic Engraving', *Current Anthropology* (Chicago), vol. 13, nos. 3–4.

Masing, U. 1939. 'Taara päritolust', *Usuteaduslik ajakiri* (Tartu), no. 1.

Mathesius, P.N. 1734. *Dissertatio Geographica de Ostrobotnia.* Uppsala.

Matthia, G. 1723. *Novum Manuale Lexicon Latino-Germanicum et Germanico-Latinum.* Magdeburg.

Melnikova, E.A. 1973. 'Skandinavskie runičeskie nadpisi kak istočnik po istorii narodov Vostočnoĭ Evropy', VI Vsesojuznoĭ Conference on the Study of the Scandinavian Countries and Finland, part 1. Tallinn.

BIBLIOGRAPHY

Mette, H.J., ed. 1952. *Pytheas von Massalia*. Berlin.

Mikkola, J.J. 1938. 'Die älteren Berührungen zwischen Ostseefinnisch und Russisch', *Suomalais-Ugrilaisen Seuran toimituksia* (Helsinki), vol. 75.

Miller, K. 1895–6. *Mappae mundi: die ältesten Weltkarten*, 5 books. Stuttgart.

Miller, K. 1926, 1927. *Mappae Arabicae: arabische Welt- und Länderkarten des 9.-13. Jahrhunderts ...*, 2 vols. Stuttgart.

Miller, V. 1972. *Eestlane vanas Tallinnas; minevikust tulevikku; artikleid ja ettekandeid, 1940–1970*. Tallinn.

Molvõgin, A. 1970. 'Tamzeskiĭ klad monet načala 13 veka', in *Studia archaeologica in memoriam Harri Moora*, edited by Marta Schmiedehelm, Lembit Jaanits and Jüri Selirand. Tallinn.

Mongait, I.A.L. 1959. 'Abu Hamid al-Garnati i ego putešestvie v russkie zemli 1150–1153 gg', *Istorija SSSR*, vol. 1.

Moora, H. 1955. 'Varafeodaalsete suhete kujunemine: eesti poliitiline ühendus Muistse Vene riigiga', in *Eesti NSV ajalugu*, vol. 1: *Kõige vanemast ajast XIX sajandi 50-ndate aastateni*, edited by A. Vassar. Tallinn.

Moora, H. 1957. 'Eine steinzeitliche Schlangenfigur aus der Gegend von Narva', in *Studia neolithica in honorem Aarne Äyräpää*. Helsinki.

Moora, H. 1968. 'Über den ostbaltischen Händel im 12.-13. Jahrhundert', in *Liber Iosepho Kostrzewski [Józef Kostrzewski] octogenario a veneratoribus dicatus*, edited by Konrad Jażdżewski. Wrocław.

Moora, H. and H. Ligi. 1970. *Wirtschaft und Gesellschaftsordnung der Völker des Baltikums zu Anfang des 13. Jahrhunderts*. Tallinn.

Mühlen, H. von zur. 1969. 'Siedlungskontinuität und Rechtslage der Esten in Reval von der vordeutschen Zeit bis zum Spätmittelalter (mit 2 Karten)', *Zeitschrift für Ostforschung: Länder und Völker im östlichen Mitteleuropa* (Marburg), vol. 18, no. 4.

Müller, F.H. 1837. *Der ugrische Volksstamm, oder, Untersuchungen über die Ländergebiete am Ural und am Kaukasus in historischer, geographischer und ethnographischer Beziehung*. Berlin.

Must, G. 1951. 'Zur Herkunft des Stadtnamens Reval', *Finnisch-ugrische Forschungen: Zeitschrift für finnisch-ugrische Sprach- und Volkskunde* (Helsinki), vol. 30, no. 3.

Nansen, F. 1911. *Nebelheim: Entdeckung und Erforschung der nördlichen Länder und Meere*, vol. 2. Leipzig.

Németh, J. 1972. 'Zoltán Gombocz et la théorie des mots d'emprunts turks bulgares du hongrois', in *Mélanges offerts a Aurélien Sauvageot pour son soixante-quinzième anniversaire*, edited by J. Gergely et al. Budapest.

Nerman, B. 1958. *Grobin-Seeburg: Ausgrabungen und Funde*. Stockholm.

Neus, A.H. 1849. *Revals sämmtliche Namen, nebst vielen anderen, wissenschaftlich erklärt*. Reval.

Niitemaa, V. 1949. 'Die undeutsche Frage in der Politik der Livländischen Städte im Mittelalter', *Suomalaisen tiedeakatemian toimituksia* (Helsinki), series B, Humaniora, vol. 64.

Niitemaa, V. 1952. 'Der Binnenhandel in der Politik der Livländischen Städte im Mittelalter', *Suomalaisen tieteakatemian toimituksia* (Helsinki), series B, Humaniora, vol. 76, no. 2.

BIBLIOGRAPHY

Niitemaa, V. 1955. 'Das Strandrecht in Nordeuropa im Mittelalter', *Suomalaisen tiedeakatemian toimituksia* (Helsinki) series B, Humaniora, vol. 94.

Niitemaa, V. 1963. 'Die frühen Städte Finnlands', in *Die Zeit der Stadtgründung im Ostseeraum*, edited by M. Stenberger. Visby, 1965.

Niitemaa, V. 1972. 'Remeslennoe delo v srednie veka v raĭone Baltiĭskogo morja', Sovetsko-finskiĭ simpozium po istorii remesla i vozniknovenija manufakturnoĭ promyšlennosti, 11–15 December 1972. Leningrad.

Nissilä, V. 1972. 'Éléments varègues dans la toponymie de l'Est du Golfe de Finlande', in *Mélanges offerts a Aurélien Sauvageot pour son soixante-quinzième anniversaire*, edited by J. Gergely et al. Budapest.

Nordenskjöld, A.E. 1973 [1889]. *Facsimile Atlas of the Early History of Cartography with Reproductions of the Most Important Maps Printed in the XV and XVI Centuries*, translated from the Swedish original by Johan Adolf Ekelöf and Clements R. Markham, with a new introduction by J.B. Post. New York.

Ojansuu, H. 1920. 'Tallinnan kaupunkin vanhin virolainen nimi', *Uusi Suomi* (Helsinki), no. 22, 28 January 1920.

Olsen, O. 1963. 'Die Kaufschiffe der Wikingerzeit im Lichte des Schiffsfundes bei Skuldelev im Roskilde Fjord', in *Die Zeit der Stadtgründung im Ostseeraum*, edited by M. Stenberger. Visby, 1965.

Öpik, E. 1970. *Vadjalastest ja isuritest XVIII saj. lõpul: etnograafilisi ja lingvistilisi materjale F. Tumanski Peterburi kubermangu kirjelduses.* Tallinn.

Pall, V. 1969. *Põhja-Tartumaa kohanimed*, vol. 1. Tallinn.

Past, E. 1935. *Jooni eesti mereasjanduse minevikust.* Tallinn.

Peegel, J. 1970. 'Über die poetischen Synonyme in altestnischen alliterierenden Versen', Third International Congress of Finno-Ugric Scholars, Tallinn, 17–23 August 1970, vol. 2. Tallinn.

Pekkanen, T. 1968. 'The Ethnic Origin of the δουλοσπόροι', *Arctos: Acta Philologica Fennica* (Helsinki), Supplement 1.

Pekkanen, T. 1974. 'Adam of Bremen', *Arctos: Acta Philologica Fennica* (Helsinki), vol. 8.

Pekkanen, T., trans. and ed. 1976. *Tacitus: Germania.* Helsinki.

Pekkanen, T. 1980a. 'Suomi ja sen asukkaat latinan- ja kreikankielisessä kirjallisuudessa', *Trükis ilmunud: Suomalais-Ugrilaisen Seuran aikakauskirja* (Helsinki), 1984.

Pekkanen, T. 1980b. 'Exegetical Notes on the Latin Sources of Northern Europe', *Arctos: Acta Philologica Fennica* (Helsinki), vol. 14.

Pekkanen, T. 1981. 'Vanhin kirjallinen tieto suomalaisista', *Trükis ilmunud: Suomalais-Ugrilaisen Seuran aikakauskirja* (Helsinki), vol. 78.

Platonov, S.F. 1922. *Inozemcy na russkom Severe: očerki po istorii kolonizacii Severa i Sibiri*, no. 2. Petrograd.

Pliny the Elder. 1855. *The Natural History*, translated by John Bostock and H.T. Riley. London. www.perseus.tufts.edu.

Poetic Edda. 1936. *The Poetic Edda*, translated by H.A. Bellows. Princeton, NJ.

Pokorny, J. 1947. *Indogermanisches etymologisches Wörterbuch*, 3 vols. Bern and Munich.

Põldmaa, K. 1973. 'Rästik', *Edasi*, no. 199, 25 August 1973.

BIBLIOGRAPHY

Pomponius Mela. 1998. *Pomponius Mela's Description of theWorld*, translated by F.E. Romer. Ann Arbor, MI.

Popov, A.I. 1972. *Voprosy istoričeskogo izučenija fol'klora finno-ugorskih narodov: problemy izučenija finno-ugorskogo fol'klora.* Saransk.

Popov, A.I. 1973. *Nazvanija narodov SSSR: vvedenie v ètnonimiku.* Leningrad.

Popov, A.I. 1981. *Sledy vremen minuvših: iz istorii geografičeskih nazvaniǐ Leningradskoǐ, Pskovskoǐ i Novgorodskoǐ oblasteǐ.* Leningrad.

Povest. 1950. *Povest' vremennykh let* [Primary Chronicle], 2 vols. Moscow and Leningrad.

Primary Chronicle. 1953. *The Russian Primary Chronicle*, translated and edited by Samuel Hazzard Cross and Olgerd P. Sherbowitz-Wetzor. Cambridge, MA.

Prüller, P. 1968. 'Eesti rahvaastronoomia', in *Teaduse ajaloo lehekülgi Eestist*, vol. 1. Tallinn.

Pullat, R. 1966. *Tallinnast ja tallinlastest: nihkeid elanikkonna sotsiaalses koosseisus, 1871–1917.* Tallinn.

Rajandi, E. 1966. *Raamat nimedest.* Tallinn.

Ränk, G. 1979. *Sest ümmargusest maailmast: läbielatu, nähtu ja kuuldu põhjal kirja pannud.* Stockholm.

Rasmusson, N.L. 1963. 'Münz- und Geldgeschichte des Ostseeraumes vom Ende des 10. bis zum Anfang des 14. Jahrhunderts: ein numismatisch-geographischer Überblick', in *Die Zeit der Stadtgründung im Ostseeraum*, edited by M. Stenberger. Visby, 1965.

Roos, E. 1963a. '"Kalevipoja" Linda', *Keel ja kirjandus* (Tallinn), no. 8.

Roos, E. 1963b. 'Läti Henriku Lyndanise ja "Kalevipoja" Lindanisa', *Keel ja kirjandus* (Tallinn), no. 10.

Rosen, E. 1910. *Rückblicke auf die Pflege der Schauspielkunst in Reval: Festschrift zur Eröffnung des neuen Theaters in Reval im September 1910, herausgegeben vom Revaler Deutschen Theaterverein.* Melle (Hanover).

Rücker, C.G. 1854. *General-Karte der Russischen Ost-See-Provinze Liv-, Ehst- und Kurland, nach den vollständigsten astronomisch-trigonometrischen Ortsbestimmungen u. den speciellen Landesvermessungen auf Grundlage der Specialkarten v. C. Neumann, C.G. Rücker und J.H. Schmidt.* Reval.

Saareste, A. 1922. *Valitud eesti rahvalaulud: keelelise ja värsiõpetusliku sissejuhatuse ning sõnastikuga.* Tallinn.

Saareste, A. 1935. 'Tulihänna nimetustest: avec un résumé: des noms de tulihänd en estonien', *Eesti Keele Arhiivi toimetised* (Tartu), no. 1.

Saareste, A. 1951. 'Kalev, Kaleva sõna algupärast', *Virittäjä* (Helsinki), vol. 1, no. 2.

Saareste, A. 1958. *Eesti keele mõisteline sõnaraamat*, 4 vols. Uppsala, 1979.

Salištšev, K.A. 1948. *Osnovy kartovedenija: čast' istoričeskaja i kartografičeskie materialy.* Moscow.

Šaskolski, I.P. 1954. *Maršrut torgovogo puti iz Nevy v Baltiǐskoe more v IX–XIII vv. Geografičeskiǐ sbornik*, vol. 3: *Istorija geografičeskih znaniǐ i geografičeskih otkrytiǐ.* Moscow.

Šaskolski, I.P. 1971. 'Èstonija i drevnjaja Rus', in *Studia historica in honorem Hans Kruus.* Tallinn.

Saxo Grammaticus. 1979. *Saxo Grammaticus: The History of the Danes*, translated by Peter Fisher and edited by H.E. Davidson. Woodbridge, UK.

Scheffer, Johannes. 1673. *Argentoratensis Lapponia* ... Frankfurt am Main.

BIBLIOGRAPHY

Scheffer, Johann. 1674. *The History of Lapland* ... Oxford. https://old.no/samidrum/lapponia/chap-ix.html.

Scheffer, Johannes. 1675. *Lappland: das ist: neue und wahrhafftige Beschreibung von Lappland und dessen Einwohnern, worin viel bißhero unbekandte Sachen von der Lappen Ankunfft, Aberglauben, Zauberkünsten, Nahrung, Kleidern, Geschäfften, wie auch von den Thieren und Metallen so es in ihrem Lande giebet, erzählet, und mit unterschiedlichen Figuren fürgestellet worden.* Frankfurt am Main and Leipzig.

Schlegel, C.H.J. 1819–34. Reisen in mehrere russische Gouvernements in den Jahren 178*, 1801, 1807 und 1815, 10 vols. Meiningen.

Schoy, L. 1924. 'The Geography of the Moslems of the Middle Ages', *Geographical Review* (New York).

Schroeder, L. von. 1888. *Hochzeitsbräuche der Esten und einiger anderer finnisch-ugrischer Völkerschaften in Vergleichung mit denen der indogermanischen Völker: ein Beitrag zur Kenntniss der ältesten Beziehungen der finnisch-ugrischen und der indogermanischen Völkerfamilie.* Berlin.

Selirand, J. 1973. 'O severoevropeĭskih nakonečnikah kopiĭ iz damasskoĭ stali', VI Vsesojuznoĭ Conference on the Study of the Scandinavian Countries and Finland, part 1. Tallinn.

Selirand, J. and E. Tõnisson. 1963. *Läbi aastatuhandete: teaduse teedelt.* Tallinn.

Setälä, E.N. 1913. 'Bibliographisches Verzeichnis der in der Literatur behandelten älteren germanischen Bestandteile in den ostseefinnischen Sprachen', *Finnisch-ugrische Forschungen: Zeitschrift für finnisch-ugrische Sprach- und Volkskunde* (Helsingfors, Leipzig), vol. 13, nos. 1-3.

Siebmacher, J. 1895. *J. Siebmacher's grosses und allgemeines Wappenbuch in einer neuen, vollständig geordneten und reich vermehrten Auflage mit heraldischen und historisch-genealogischen Erläuterungen*, vol. 3. Nuremberg.

Sirelius, U.T. 1913. 'Primitive konstruktionsteile an prähistorischen schiffen', *Finnisch-ugrische Forschungen: Zeitschrift für finnisch-ugrische Sprach- und Volkskunde* (Helsingfors, Leipzig), vol. 13, nos. 1–3.

Sirelius, U.T. 1934. *Die Volkskultur Finnlands*, vol. 1: *Jagd und Fischerei in Finnland*. Berlin.

SKES. 1974–8. *Suomen kielen etymologinen sanakirja*, 6 vols. Published by Lexica Societatis Fenno-Ugricae.

SKVR. 1908–48. *Suomen kansan vanhat runot*. Published by Suomalaisen Kirjallisuuden Seuran [Finnish Literary Society], Helsinki.

Smirnov, I. 1970. 'O kul'turnyh svjazjah Pribaltiki so srednim Povolž'em i Priural'em v èpohu drevnosti i srednevekov'ja', in *Studia archaeologica in memoriam Harri Moora*, edited by Marta Schmiedehelm, Lembit Jaanits and Jüri Selirand. Tallinn.

Smirnov, I. 1972. *Predanija Evropeĭskogo Severa o čudi: problemy izučenija finno-ugorskogo fol'klora.* Saransk.

Sõgel, E. 1965. *Eesti kirjanduse ajalugu: viies köites*, vol. 1: *Esimestest algetest XIX sajandi 40-ndate aastateni*. Tallinn.

Solovjov, C.M. 1959. *Istorija Rossii s drevneĭših vremen: v pjatnadcati knigah*, book 1, vols. 1–2. Moscow.

Soom, A. 1940. 'Die Politik Schwedens bezüglich des russischen Transithandels über die estnischen Städte in den Jahren 1636–1656', *Õpetatud Eesti Seltsi* (Tartu), no. 32.

BIBLIOGRAPHY

Soom, A. 1971. *Die Zunfthandwerker in Reval im siebzehnten Jahrhundert*. Stockholm.

Spekke, A. 1957. *The Ancient Amber Routes and the Geographical Discovery of the Eastern Baltic*. Stockholm.

Spencer, L.J. 1938. 'The Kaalijärv Meteorite from the Estonian Craters', *Mineralogical Magazine* (London), vol. 25, no. 161.

Spreckelsen, A. 1907. 'Die Revalschen Freibeuter in den Jahren 1558–1561', *Beiträge zur Kunde Est-, Liv- und Kurlands* (Reval), vol. 6, no. 2.

Spreckelsen, A. 1927. 'Ein Steinzeitlicher Lagerplatz in der Sandwüste unter Nõmme bei Reval', *Beiträge zur Kunde Estlands* (Tallinn), vol. 13, no. 3.

Stahl, H. 1638. *Hand und Hauszbuches für die Pfarherren und Hausz-Väter Esthnischen Fürstenthumbs*, part 3 ... Revall.

Stahl, H. 1641. *Leyen Spiegel* ... Revall.

Stenberger, M. 1977. *Vorgeschichte Schwedens*. Berlin.

Stieda, W. 1910. 'Mündriche und Träger in Reval', *Beiträge zur Kunde Est-, Liv- und Kurlands* (Reval), vol. 7.

Sturlason [Sturluson], Snorre. 1907. *The Heimskringla: A History of the Norse Kings*, vol. 1, translated by Samuel Laing. London.

Suits, G. and M. Lepik. 1932. *Eesti kirjandusajalugu tekstides*, part 1. Tartu.

Taani. 1929. *Geodaetisk Instituts Kort over Danmark i 1:160000*. Copenhagen.

Tablitsõ prilivov. 1977. *Tablitsõ prilivov na 1979 god*, vol. 3: *Zarubežnye vody*. Moscow.

Tacitus. 1910. *Tacitus on Germany*, translated by Thomas Gordon. New York.

Tallgren-Tuulio, O.J. and A.M. Tallgren. 1930. 'Idrisi: la Finlande et les autres pays baltiques orientaux ...', *Studia Orientalia* (Helsingfors), vol. 3.

Tedre, Ü., ed. 1969–74. *Eesti rahvalaulud: antoloogia*, vols. I–IV. Tallinn.

Tedre, Ü. 1973. *Eesti pulmad: lühiülevaade muistsetest kosja- ja pulmakommetest*. Tallinn.

Thomsen, V. 1870. *Über den Einfluss der germanischen Sprachen auf die finnisch-lappischen: eine sprachgeschichtliche Untersuchung*. Halle.

Tihhanova, M. 1970. 'K voprosu o svjazjah Južnoï Skandinavii s Vostočnoï Evropoï v pervoï polovine I tysjačeletija n', in *Studia archaeologica in memoriam Harri Moora*, edited by Marta Schmiedehelm, Lembit Jaanits and Jüri Selirand. Tallinn.

Tiik, L. 1957. 'Keskaegsest mereliiklusest Balti merel ja Soome lahel', *Eesti Geograafia Seltsi aastaraamat* (Tallinn).

Toivonen, Y.H. 1924. 'Pygmäen und Zugvögel: alte kosmologische vorstellungen', *Finnisch-ugrische Forschungen: Zeitschrift für finnisch-ugrische Sprach- und Volkskunde* (Helsinki), vol. 24.

Tõnisson, E. 1970. 'Kumna hõbeaare', in *Studia archaeologica in memoriam Harri Moora*, edited by Marta Schmiedehelm, Lembit Jaanits and Jüri Selirand. Tallinn.

Tooley, R.V. and C. Bricker. 1976. *Landmarks of Mapmaking: An Illustrated Survey of Maps and Mapmakers*. New York.

Tretjakov, P.N. 1966. *Finno-ugry, balty i slavjane na Dnepre i Volge*. Moscow and Leningrad.

Trummal, V. 1970. 'O vozniknovenii poselenija gorodskogo tipa Tartu', Third International Congress of Finno-Ugric Scholars, Tallinn, 17–23 August 1970, vol. 2. Tallinn.

BIBLIOGRAPHY

Tšenekal, V.L. 1970. 'Starinnye solnečnye časy v Latvii: iz istorii estestvoznanija i tehniki Pribaltiki', 2 vols. Riga

Tusculum. 1963. *Tusculum-Lexikon: griechischer und lateinischer Autoren des Altertums und des Mittelalters*. Munich.

Tuulio, O.J. 1934. 'Le géographe arabe Idrīsī et la toponymie baltique de 1'Allemagne', *Suomalaisen tiedeakatemian toimituksia* (Helsinki), series B, vol. 30.

Udam, H. 1973. 'Al-Bīrūnī ja idamaine teadus: Usbekistani õpetlase tuhandenda sünni-aastapäeva puhul'. *Looming* (Tallinn), no. 11.

Uuspuu, V. 1938. 'Eesti nõiasõnade usulisest iseloomust', *Usuteadusline ajakiri* (Tartu), no. 1.

Vain, H. 1966. *Mööda Hiiumaad.* Tallinn.

Vanagas, A. 1970. 'K probleme finno-ugorskogo substrata v litovskoĭ toponimii', Third International Congress of Finno-Ugric Scholars, Tallinn, 17–23 August 1970, vol. 1. Tallinn..

Varep, E. 1957. *C.G. Rückeri Liivimaa spetsiaalkaardist 1839. aastal.* Tallinn.

Vasmer, M. 1953–8. *Russisches etymologisches Wörterbuch*, 3 vols. Heidelberg.

Vende, E. 1967. *Väärismetalltööd Eestis 15.–19. sajandini.* Tallinn.

Viires, A. 1967. 'Vaenukoged Läänemerel', *Horisont* (Tallinn), no. 7.

Vilkuna, K. 1963. 'Volkstümliche Arbeitsfeste in Finnland', *FF Communications* (Helsinki), vol. 80, no. 191.

Vilkuna, K. 1964. *Kihlakunta ja häävuode: tutkielmia suomalaisen yhteiskunnan järjestymisen vaiheilta.* Helsinki.

Vries, J. de. 1970. *Altgermanische Religionsgeschichte*, 2 vols. Berlin.

Wiedemann, F.J. 1893. *Ehstnisch-Deutsches Wörterbuch*, 2nd edn edited by Jacob Hurt. St Petersburg.

Winkler, R. 1900 [1896]. 'Ueber Kirchen und Capellen Ehstlands in Geschichte und Sage', *Beiträge zur Kunde Ehst-, Liv- und Kurlands* (Reval), vol. 5, no. 1.

Wuttke, H. 1853. *Die Kosmographie des Istrier Aithikos im Lateinischen Auszuge des Hieronymus aus einer Leipziger Handschrift zum erstenmale besonders herausgegeben von Heinrich Wuttke.* Leipzig.

Zobel, R. n.d. *Tallinna linnaehitusliku kujunemise põhijooned: ajalooline tuumik kuni 1219. aastani.* Tallinn.

Zobel, R. 1966. *Tallinna linnamüür.* Tallinn.

Zobel, R. 1970. 'Tallinna lõuna- ja idapiirdest 14. sajandi algul', *Ehitus ja arhitektuur* (Tallinn), no. 1.

Zutis, J. 1949. *Očerki po istoriografii Latvii: čast' pervaja; pribaltiĭsko-nemeckaja istoriografija.* Riga.

INDEX

Aaloe, Ago, 50–2, 53, 59

Absalon, Bishop of Roskilde, 297

Abu-Hamid al-Gharnati, 269

Adam of Bremen, 22, 33, 42, 43–4, 60, 64, 68, 102, 162, 167, 172, 175, 182, 195, 207, 208, 223, 261, 293, 308, 331, 347, 398

Aethelred, Anglo-Saxon king, 295–6

Agnes, Queen of Denmark, 421

Agricola, Julius, 144, 157–8

Agricola, Mikael, 78, 132, 374

Agrippa, Marcus Vipsanius, 176–7

Agur, Ustus, 51

Äiteki, guide, 118–19, 219

Akkatus, Helene, 323

Akkatus, Johannes, 323

Albert II, Archbishop of Riga, 421

Albert, Swedish king, 423

Alexander the Great, 23, 28–9, 101, 166, 269

Alfred the Great, Anglo-Saxon king, 156

Almıs, Volga Bulgarian ruler, 237, 243, 248, 289

Altema, Tytte, 429

Alvre, Paul, 179

Amundsen, Roald, 120

Anaximander, 22

Anderson, Magnus, 406

Anderson, Walter, 56

Andrew, Archbishop of Lund, 399, 401

Anglicus, Bartholomaeus, 161, 186

Annist, August, 268

Apollonius of Rhodes, 29, 93, 115, 145, 333

Ariste, Paul, 16, 58, 88–9, 101–2, 123, 156, 157, 179, 290, 333, 358–9, 360

Aristotle, 28

Askbrand, merchant, 285

Askold (Höskuldr), ruler of Kyiv, 285

Asso, Tallinn citizen, 388

Attata, Chukchi, 120

Augustus, Emperor, 138

Authun, merchant, 285

Avienius, Roman poet, 102

Bakhrushin, Sergei, 276

Bake, Olaf, sailor, 423

Barents, Willem, 25

al-Battani, Muhammad (Albategnius), 183, 234–5

Battuta, Abu Abdullah Muhammad ibn, 278

Beatus, Benedictine monk, 176–7

Beg, Ulugh, astronomer, 396

Bellingshausen, Fabian Gottlieb von, 434

Berg, Antonius, 431

Bering, Vitus, 347, 415, 433

Bergsson, Nikulás, Icelandic monk, 293

al-Biruni, Abu Raihan (Maître Aliboron), 183–5, 265–6, 269, 290, 322, 335

Bjørnbo, Axel Anthon, 175

Blaeu, Joan, 162

Bloch, Marc, 439

Blumfeldt, Evald, 154–5

Bogdanov, Prokop, Zyrian hunter, 15, 123

Bohr, Nils, 234

INDEX

Bonnell, Ernst, 298
Borich, merchant, 285
Bruni, merchant, 285
Budimirovich, Solovey, 324
Bul, Erick, sailor, 423–4
Bunge, Friedrich Georg von, 316
Büchner, Martin, 72–3

Caesar, Julius, 11, 100
Campanella, 103
Canutipoeg, Mart Johan, 214
Castrén, Matias Aleksanteri, 77
Charlemagne, 265–6, 357, 417
Chirikov, Aleksei, lieutenant, 415
Churchill, Winston, 33, 183
Clavus, Claudius, 336
Clementz, goldsmith, 352
Cnut the Great, 298, 309, 366
Columbus, Christopher, 227–8, 230,
    263, 398
Comnena, Anna, princess, 185, 235
Conemann, silversmith, 384
Cook, James, 214, 227–8
Cosmas Indicopleustes, 33, 38, 115

al-Dimasqi, 264
Diodorus Siculus, 33, 34, 38, 42, 140
Diogenes, Antonius, 102
Dir (Dyri, also Tiuri, Turi), ruler of
    Kyiv, 285
Dravin‚š, Karlis, 72
Dubler, César E., 270–1
Duc, Léouzon le, 418–19

Eelsalu, Heino, 244
Einstein, Albert, 228, 234
Eirik, Norwegian king, 293
Eisen, Matthias Johann, 69, 201, 240,
    317
Eistr, envoy, 285
Ekblom, Richard, 328–9
Erasmus of Rotterdam, 297
Eratosthenes, 22, 29, 40, 115, 143–5,
    177

Erik V Klipping, 367, 368
Erik the Red, 263
Eriksson, Harald, 271
Espenberg, Karl von, ship's doctor, 434
Este, Anders, shipbuilder, 414
Este, Andreas, shipbuilder, 414
Este, Mattis, shipbuilder, 414
Este, Nils, shipbuilder, 414
Eudoxus, astronomer, 107

Fadlan, Ahmed ibn, 237–45, 247–50,
    254, 257–64, 266–9, 271, 273,
    275, 282–3, 323, 331, 332
Faehlmann, Friedrich Robert, 216
al-Faqih, Ibn, 226
Feirgil (Aethicus Ister), 178–82, 375,
    377
Feyerabend, Karl, 61, 70
Ficke, merchant, 428–9
Fohrman, Jürgen, 404
Frazer, James George, 56
Frutan, merchant, 285
Frähn, Christian Martin, 280
Fulco (Folquinus), bishop, 366

Gama, Vasco da, 230
Geminus, astronomer, 38, 111, 113,
    115–16, 183
Gamal, merchant, 285
Gordeyev, N., 288
Grammaticus, Saxo, 87, 297, 307–8
Gregory IX, pope, 299
Grim, envoy, 285, 420
Grote, Clawes, 404
Gunnfast, merchant, 285
Gunnar, sailor, 375
Göseken, Heinrich, 359

Haavio, Martti, 213, 239
Halfdan, merchant, 285
Hallvarth, envoy, 285
al-Hamadani, Nadzhib, 236, 273
al-Hamawi, Yakut ibn Abdullah al-
    Rumi, 247, 262, 268–9

460

INDEX

Harek, sailor, 193, 194, 195

Harrison, John, watchmaker, 114

Hartwig, priest, 261

Harun al-Rashid, 170, 263, 269

Harva, Uno, 129

Hebbus, Danish magistrate, 261

Hein, Matz, 404

Heinsius, Paul, 286, 301–2, 406

Hecataeus of Miletus, 22, 23–5

Heming, merchant, 285

Hendrik of Uppsala, Bishop of Finland, 75

Hennig, Richard, 31–2

Henry of Livonia, 60, 62, 64, 69, 70–2, 73, 77, 80, 91, 161–2, 173, 186, 220, 256, 260–2, 298, 301, 303, 304–7, 321, 332, 333, 351, 353, 354, 362, 373, 376, 379, 381–2, 386, 389, 395–8, 400–1, 406, 413, 415–16, 425, 427

Herberstein, Sigismund von, 276

Hermann, Bishop of Saaremaa, 311

Herodotus, 22, 30, 95, 138, 166

Hesiod, 32

Hildelemb, Estonian feudal lord, 366, 380

Hiltinus, Johannes, Bishop of Estonia, 242, 366

Himilco, Carthaginian, 27

Hipparchus, astronomer, 31, 106–8, 109, 113–14, 116, 140, 143

Holm, Mathias, sailor, 423

Holmqvist, Wilhelm, 181

Holzmayer, Jean Baptiste, 353–4

Homer, 93, 130, 218

Honorius III, pope, 299

Hroald (I–II), merchant, 285

Humboldt, Alexander von, 441

Humboldt, Wilhelm, 330

Hupel, August Wilhelm, 236, 260, 384

al-Idrisi, Abu Abdullah Muhammad, 182–5, 325, 327–42, 345–50, 351, 363, 373, 374–5, 395, 397, 399

Ingivald, merchant, 285

Ingjald, merchant, 285

Igor (Ingvar), prince, 285

Ilves, Evald, 53, 59

Indreko, Richard, 156

Innocent III, pope, 397, 398–9

Isidorus, 34

Isgaut, envoy, 285

Itkonen, Erkki, 155–6

Ivan IV, 424, 432

Ivar, envoy, 285

Jaanits, Lembit, 88

Jacob, Petrus, Bishop of Roskilde, 399, 404

Jedvardsson, Erik, 75

Jenkinson, Anthony, 132

Johansen, Paul, 135, 369, 370, 371–2

John, Danish king, 430

Jordanes, Gothic historian, 156, 166

Jung, Jaan, 149

Kaalep, Ain

Kaharpe (Kaharpea), Jacob, 404

Kaikivalda (Kaugovalda *de Vinlandia*), Petrus, missionary, 173

Kallas, Oskar, 16

Matis, sailor, 404

Kalv, Peep, 217

Kanilu, Inuit, 219

Kanitzar, envoy, 285

Kappenberg, merchant, 428–9

Kari, envoy, 16, 285

Karjalainen, Kustaa, 129

Kaseke, Ludeke, captain, 430

Kassesepp, Jan, 429

Kassowe, salt transporter, 403

al-Kashgari, Mahmud, 265

Kaubapoeg (Kaubipoicke, Kaubipoike), Andreas (Andres) Jumetack, (Jummetack, Juminda), captain, 424

Kauke, Knuth, sailor, 423

Kettunen, Lauri, 290

461

INDEX

Khordadbeh, Abu-l-Kasim Ubaidullah ibn Abdallah ibn, 183
al-Khwarizmi, Muhammad ibn Musa, 235
Kinnier Wilson, James V., 55–7
Kiparsky, Paul, 156
Kipper, Aksel, 425
Kivikoski, Ella, 367
Klipping, Erik, 367–8
Kock, Tonnies, captain, 404
Koila, Pard (Albern de Kokaeli), 366
Kolskegg, sailor, 375
Kolyvanov, Samson, uss-man, 288
Kosterkin, Demnime, shaman, 121
Kosterkin, Dulsimaku, 124
Kotlyarevsky, Aleksander, 91
Kotzebue, August von, 434
Kotzebue, Otto von, 433–4
Kovalevsky, A., 322
Krasna, Max, 404
Krachkovsky, Ignaty, 328, 335
Kreutzwald, Friedrich Reinhold, 240–1, 262
Krusenstern, Adam Johann von, 240–1, 262, 268, 279, 281–2
Kränkel, Ernst, 275
Kurge, schipman, 403
Kussi, merchant, 285
Kuusi, Matti, 285
Künnapuu, Sulev, 316
Küti (von Kurküll), Ants (Hans), sailor, 404

Laid, Eerik, 131
Lazyamov, Aleksei, Khanty boatbuilder, 8
Leemet (Clemens Esto), 381
Leif, envoy, 263
Leif Erikson, 186
Lelewel, Joachim, 346
Lembitu, military leader, 80
Lewick, Tadeusz, 337
Lifflender, Andreas, sailor, 404
Lifflender, Jacob, sailor, 404

Liiva, Arvi, 53, 59
Lodewicus Rootslane (Rozleyne), Tallinn merchant, 404–5
Lodhe, Helmoldus de, vassal, 422
Loorits, Oskar, 216–17
Lotman, Juri, 66
Lowell, Percival, 438
Lucanus, Marcus Annaeus, 97
Luce, Johann Wilhelm Ludwig von, 303
Lõugas, Anu, 255
Lõugas, Vello, 20, 51, 73, 169
Lönnrot, Elias, 79, 141
Lönnroth, Erik, 196

MacCana, Proinsias, 187
Maekiuse (Mägiste), Bertald (Pärt), Maökula vassal, 389
Maekius (Mägiste), Johannes, 404
Malm, Erik, 433
Manni, merchant, 285
Maran, Rein, 74
Marcussen, Erich, 404
Margaret I, Danish queen, 291, 380
Markwart, Josef, 262
Marlinsky, Aleksander, 134
Masing, Otto Wilhelm, 60, 63, 415–16
Masing, Uku, 70
al-Masudi, Abu al-Hasan Ali, 228–30, 272, 395–6, 398
Mathesius, Peter Niklas, 288
Mathias, silver burner, 384
Matis, Kalli, sailor, 404, 430
al-Maqdisi, 230–1, 247, 322
Meil, Jurgen, 404
Meinhard, Bishop of the Livonians, 354, 362–3
Mekis, Otto, captain, 403–4
Melenthewe, silver burner, 384
Mellin, Ludvig August, 58
Meri, Kristjan, 74
Meri, Mart, 74
Mette, Hans Joachim, 165
Middendorff, Alexander Theodor von, 236

INDEX

Miller, Konrad, 175, 177, 346
Muhammad, Prophet, 225, 248
Moora, Aliise, 261
Moora, Harri, 67, 156, 353
More, Thomas, 103
al-Muqtadir, Ja'far, 237
Munthor, envoy, 285
Muresep, Jan Woress, 429
Must, Gustav, 376
Musta, Mattis, captain, 404, 430
Mägiste, Julius, 83–4
Mühlen, Heinz von zur, 135, 437
Münrik, Luder, 391
Münrik, Hinrik, 391, 430

Nansen, Fridtjof, 31, 33
Nerman, Birger, 149
Nestor, chronicler, 284
Neus, Alexander Heinrich, 376
Nicholaus, bishop, 367
Niitemaa, Vilho, 362
Nissilä, Viljo, 173, 290, 333
Nöldeke, Theodor, 327–8

Ohm, Hans, 433
Ohthere, Swedish king, 214, 219, 321, 332
Oleg (Helgi), prince, 284–5
Oleif, envoy, 285
Olleloic-Stroschnider (Õlelõikaja), Andreas, 404
Oloffson, Jacob, 404
Öpik, Ernst Julius, 260, 267
Orosius, Paulus, 156, 162, 172
Oseleer, 403
Osilia, Albern de, 366
Ovid, 94, 104

Pallas, Peter Simon, 267
Pall, Valdek, 334, 374
Palmedack, member of the Great Guild, 429
Paul the Deacon, 172–3
Peter I, 286, 288

Peeter, son of Simon Sitke, 433
Pekkanen, Tuomo, 115, 144, 166
Peredolsky, Vasiliy, 179
Pernstein, Friedrich von, Archbishop of Avignon, 428
Perttunen, Arhippa, 79, 81
Peterboi, Las, sailor, 423
Petersen, Hieronimo, 404
Petronius, 97
Philipp, Bishop of Ratzeburg, 307
Plato, 170
Plenisner, Friedrich, colonel, 347
Pliny the Elder, 23–6, 34–40, 42–3, 46, 102, 113, 137–42, 144, 145, 149–51, 154, 179
Podder, Jurgen, sailor, 404, 428
Polybius, 32–3, 143, 165
Pompey the Great, 100
Pomponius Mela, 22, 41–2, 162
Popov, Aleksander, 89
Posidonius, geographer, 36
Freystein (I–III), envoys, 285
Presser (Pres), Erick, sailor, 423
Procopius, Byzantine historian, 131, 171–3
Pryn, Ewertt, 404
Ptolemaios, Klaudios, 29
Pujalke (Puujalg), sailor, 403, 407, 428, 429, 432–3
Puseppe, Andrus, 429
Pytheas of Massalia, 26, 34, 107, 113, 441

al-Qazwini, Zakaria ibn Muhammed, 268–9, 275

Raam, Villem, 315
Rajandi, Edgar, 69, 367
Rannakopli, Olev, 434
Rebja, Andres, 377
Reinwald, Ivan, 58–9, 438
Reppele, Olaf, 377
Roger II, King of Sicily, 328
Roos, Eduard, 373

463

INDEX

Rousseau, Jean-Jacques, 322
Rüssow, Balthasar, 133, 407

Saareste, Andrus, 59, 65, 75, 328, 438
Sacke, Jürgen, 72
Sallam the Interpreter, 269
Savijärve (Sawijerwe), Bartholomaeus,
  Bishop of Tartu, 148
Savijärve, Heidenreich, 148
Scheherazade, 241–2
Schimmelpenninck, Simon (Symen)
Schliemann, Heinrich, 441
Schlözer, August Ludwig von, 324
Seneca, Lucius Annaeus, 97, 102
Sesenkar, Oleff (Olaff), captain, 403,
  429
Setälä, Emil Nestor, 83, 162
Sevenborgk, Clemendt, captain, 404
Shahriyar, Buzurg ibn, long-distance
  captain, 411
Shirzai, Saadi, 184
Sigbjorn, envoy, 285
Sigfrid, envoy, 285
Sigurd, uncle of Olaf Tryggvason, 296
Sinbad the Sailor, 433
Skalk, Swedish farming family on
  Hiiumaa, 310
Skötkonung, Olof, Swedish king, 298
Slothi, envoy, 285
Solinus, Gaius Julius, 162
Spence, Leonard James, 438
Spreckelsen, Arthur, 352
Ssuwe, Matz, sailor, 404
Staal, Lorentz, captain, 404, 428
Stahl, Heinrich, 374
Statius, Publius Papinius, 102
Stenberger, Mårten, 102
Stephen of Perm, 267, 275
Strabo, geographer, 9, 32–3, 38, 40
Strehlow, Theodor George Henry, 136
Struwe, Friedrich Georg Wilhelm von,
  398
Struwe, Otto Wilhelm von, 396
Sturluson, Snorri, 88–9

Sturmann, Michael, captain, 193, 195,
  292, 295, 304
Styr, merchant, 285
Suddenpe, Peter, 317
Sundy, captain, 403
Surenpe (Suurpea), Hans, captain and
  merchant, 403
Surenpc, Meynecke, 429
Surenpe, Peter, 429
Surepee (Surepe), captain, 403, 404
Sveinki, merchant, 285
Svein, merchant, 285
Sverre, Norwegian king, 405
Swadesh, Morris, 203–4
Syseke, Simen, captain, 404

Tacitus, Cornelius, 142
Taleyev, Aleksandr, 122
Tallepoiss (Tallepoisz), Tonnies Berch,
  430, 431
Tallgren, Aarne Michael, 328, 336
Tallgren-Tuulio, Oiva Johannes, 328
Tamm, Jaan, 438
Tatishchev, Vasily, 324
Tauts, Siimon, 434
Techler, Pauell, 391
Tedre, Ülo, 255
Theoderich, Bishop of Estonia, 60,
  261, 297, 307, 389, 399
Theodoric, King of the Ostrogoths,
  170
Thiodrek, saga author, 295
Thorbjorn, merchant, 285
Thorfrid, merchant, 285
Thorolf, Olaf Tryggvason's foster father,
  295
Thorstein, merchant, 285
Tiberius, Caesar Augustus, 137
Timaeus, 46
Tirr, merchant, 285
Togan, Zeki Validi, 263
Torricelli, Evangelista, 396
Tryggvason, Olaf, King of Norway,
  193, 295–300, 308

464

INDEX

Ubbias (Ubja, Ubies), Clawes (Klaus), salt transporter, 430
Uexküll, Johann von, 420
Unt, Jaan, 166

Valdemar II, 351, 366, 368, 380, 381, 389, 399–401, 417
Valdemar, King of Holmgard, 296, 351, 366, 368, 380, 381, 389, 399, 400–1, 417
Vallmysraea, Peter, 386
Varag, chief, 238
Vefast, envoy, 285
Velven, Andreas von, Teutonic Order representative, 409
Veniaminov, Ivan, missionary, 25
Verdi, Giuseppe, 160
Verginius, Adrian, 60
Vetseke (Vyachko), prince, 307
Vilde, Eduard, 323
Vililemb, Johannes, 380
Vililemb, lower vassal, 380
Vilkuna, Kustaa, 156, 348, 349
Vinci, Leonardo da, 167, 265
Vladimir, Prince of Novgorod, 298, 324
Vogel, Walter, 308
Voltaire, 231

Vos (Voss), Reinolt, captain, 404, 428
Voss, Tonnies, craftsman, 428
Vries, Jan de, 90, 114
Visleif, merchant, 285
Väärismaa, Andres, 59

Waytay, Peter, 386
Wegener, Alfred, 93
Wenceslaus, Slavic prince, 399
Wiedemann, Ferdinand Johann, 415
Wielo, Ado, 72
Wieselgren, Per, 376
William of Modena, 299, 316, 351
William the Conqueror, 19
Witte, Laurencz, captain, 404, 430
Wulfstan, 162

al-Yaʿqubi, Ahmed, 226, 227
Yaroslav, 298, 402
Yatving, envoy, 285
Yernykhov, Nikita, Khanty boatbuilder, 12
Yngvar, King of Uppsala, 292–3
ibn Yunus al Sadafi al-Misri, 231

Zedenpeyke, fisherman, 317
Ziegler, Alexander, 441
Zuddenpe, fisherman, 317